Springer Series in Geomechanics and Geoengineering

Series editors

About this Series

Geomechanics deals with the application of the principle of mechanics to geomaterials including experimental, analytical and numerical investigations into the mechanical, physical, hydraulic and thermal properties of geomaterials as multiphase media. Geoengineering covers a wide range of engineering disciplines related to geomaterials from traditional to emerging areas.

The objective of the book series is to publish monographs, handbooks, workshop proceedings and textbooks. The book series is intended to cover both the state-of-the-art and the recent developments in geomechanics and geoengineering. Besides researchers, the series provides valuable references for engineering practitioners and graduate students.

More information about this series at http://www.springer.com/series/8069

Kam-Tim Chau · Jidong Zhao
Editors

Bifurcation and Degradation of Geomaterials in the New Millennium

Proceedings of the 10th International Workshop on Bifurcation and Degradation in Geomaterials

Springer

Editors
Kam-Tim Chau
Department of Civil and Environmental Engineering
The Hong Kong Polytechnic University
Hong Kong
China

Jidong Zhao
Department of Civil and Environmental Engineering
The Hong Kong University of Science and Technology
Kowloon, Hong Kong
China

ISSN 1866-8755 ISSN 1866-8763 (electronic)
Springer Series in Geomechanics and Geoengineering
ISBN 978-3-319-13505-2 ISBN 978-3-319-13506-9 (eBook)
DOI 10.1007/978-3-319-13506-9

Library of Congress Control Number: 2014956208

Springer Cham Heidelberg New York Dordrecht London

Printed on acid-free paper

Springer International Publishing AG Switzerland is part of Springer Science+Business Media (www.springer.com)

Preface

Interests in localization and related instabilities in the field of geomechanics date back to the JMPS paper of Rudnicki and Rice (1975) "Conditions for the localization of deformation in pressure-sensitive dilatant materials". It models strain localization in rocks as material instability. Subsequently, strain localization in soils was considered as shear band by Vardoulakis et al. (1978) "Formation of shear bands in sand bodies as a bifurcation problem" in IJNAMG. Research interests in this area expanded considerably and resulted in the first international workshop on Localization of Soils organized in Karslrule, Germany, February 1988, and this international workshop became the first sequel to our current International Workshop on Bifurcation and Degradation in Geomaterials (IWBDG). This aroused so much enthusiasm and interest in the fundamental aspects of bifurcation theory to soils that the second workshop followed in Gdansk, Poland, September 1989. The topic was then extended to rock mechanics at the third international workshop in Aussois, France, September, 1993. In 1997, this international workshop series was expanded to include instabilities and degradations in geomaterials at the fourth workshop in Gifu, Japan, September 1997. Since then, the name of IWBDG was adopted and subsequent international workshops were held at Perth, Australia, November 1999 (fifth), at Minneapolis, USA, June 2002 (sixth), at Crete, Greece, June 2005 (seventh), at Lake Louise, Canada, May 2008 (eighth) and at Porquerolles, France, May 2011 (ninth). The tenth international workshop of this series continued this central theme of bifurcation and degradation of geomaterials, and was held in Hong Kong during May 28–30, 2014 (10th IWBDG) at the beautiful campus of the Hong Kong Polytechnic University.

The 10th IWBDG was attended by 66 participants representing 16 countries or regions, including Australia, Austria, Belgium, Canada, Chile, France, Greece, Hong Kong China, Iran, Japan, Mainland China, Norway, Poland, Sweden, UK and USA. A total of 55 presentations were delivered, covering three full days. Among them, 17 were registered as students. This proceedings published by Springer contains 54 peer reviewed full papers.

The workshop would not be possible without the help of qualified and diligent reviewers, and they include Mustafa Alsaleh, Ronaldo I. Borja, Jacques Desrues,

Zhiwei Gao, Ning Guo, Peijun Guo, Marte Gutierrez, Wenxiong Huang, Mingjing Jiang, Xia Li, Francois Nicot, Fusao Oka, Jacek Tejchman, Antoinette Tordesillas, Richard Wan, Gang Wang, Jeff Jianfeng Wang, Wei Wu, Zhenyu Yin, Jidong Zhao (in alphabetical order). Their helps are highly appreciated. The financial sponsors are Fong On Construction Limited (courtesy of Dr. James C.K. Lau, JP) and the Faculty of Construction and Environment, The Hong Kong Polytechnic University (through Conference Support Scheme). Non-financial sponsors include Geomechanics Committee, AMD of ASME, Elasticity Committee, EMI of ASCE, HKGES, Geotechnical Division of HKIE and TC103 Numerical Methods of ISSMGE. The clerical and logistics supports from the Department of Civil and Environmental Engineering are highly appreciated.

Kam-Tim Chau
Jidong Zhao

References

Rudnicki JW, Rice JR (1975) Conditions for the localization of deformation in pressure-sensitive-dilatant materials. J Mech Phys Solids 23(6):371–394

Vardoulakis I, Goldscheider M, Gudehus G (1978) Formation of shear bands in sand bodies as a bifurcation problem. Int J Numer Anal Meth Geomech 2(2):99–128

Contents

Simulations of Shear Zones and Cracks in Engineering Materials Using eXtended Finite Element Method

Jerzy Bobiński and Jacek Tejchman

Abstract Numerical simulations of cracks and shear zones in quasi-brittle materials are presented. Extended Finite Element Method is used to describe both cracks and shear zones. In a description of tensile cracks, a Rankine criterion is assumed. A discrete Mohr-Coulomb law is adopted for simulations of shear zones. Results of simple numerical tests: uniaxial tension, bending and biaxial compression are demonstrated.

1 Introduction

Localization of deformation is observed in many materials like concrete, glass, metals, polymers, soils and rocks. In soils, rocks and concrete this phenomenon is manifested by the presence of shear zones and cracks. The numerical modelling of strain localization within continuum mechanics requires the use of advanced constitutive models, based e.g. on an elasto-plasticity or hypoplasticity theory. All continuum constitutive laws have to include a characteristic length of micro-structure to obtain mesh-independent results of the width and spacing of localized zones. Shear or tensile zones can be also simulated more explicitly by using interface (cohesive) elements or the discrete element method (DEM).

Another method, which gains strong popularity in different areas to describe shear and tensile zones in quasi-brittle materials and soils as discontinuities in a displacement field is the eXtended Finite Element Method (XFEM). It is based on the Partition of Unity Method and it assumes an enrichment of displacements to capture jumps across localized zones. Extra degrees of freedom are added in regions where strain localization occurs. This approach allows for a placement of localized zones within finite elements.

J. Bobiński (✉) · J. Tejchman
Gdansk University of Technology, Narutowicza 11/12, 80-233 Gdansk, Poland
e-mail: bobin@pg.gda.pl

J. Tejchman
e-mail: tejchmk@pg.gda.pl

K.-T. Chau and J. Zhao (eds.), *Bifurcation and Degradation of Geomaterials in the New Millennium*, Springer Series in Geomechanics and Geoengineering,
DOI 10.1007/978-3-319-13506-9_1

The paper presents numerical simulations of the formation and growth of shear and tensile zones in soils with the aid of the eXtended Finite Element Method. To describe shear zones, the constitutive law was based on the Mohr-Coulomb model. To simulate cracks under tensile loading, the Rankine approach was used. Some tests like biaxial compression, uniaxial tension and three-point bending were simulated.

2 Extended Finite Element Method

2.1 General Description

The eXtended Finite Element Methods (XFEM) allows for simulating displacement jumps across finite elements (Belytschko and Black 1999). It can be used to simulate brittle materials (Moës and Belytschko 1999), cohesive cracks (Wells and Sluys 2001) or shear zones in soils (Song et al. 2006). The formulation used follows (with some slight modifications and improvements) the general idea presented by Wells and Sluys (2001). It is based on the so-called shifted-basis enrichment (Zi and Belytschko 2003) to describe a displacement field with discontinuous jumps. This modification has two advantages over the standard version: the total nodal displacements are equal to the standard displacements and the implementation of finite elements is simpler since two types of elements exist only.

In a non-cracked region, a linear elastic constitutive law between stresses and strains was always assumed. To create a new crack segment, a crack creation criterion has to be fulfilled at least in one point of the element at the front of the crack tip. Crack tips (end points) can be placed only at finite elements edges. In order to smoothen the stress field around the crack tip, the averaged stresses were used in determining the crack direction (Wells and Sluys 2001) defined as:

$$\sigma^* = \int_V \sigma\, w dV \text{ with } w(r) = \frac{1}{(2\pi)^{3/2} l_{av}^3} \exp\left(-\frac{r^2}{2 l_{av}^2}\right), \tag{1}$$

where the domain V is the semicircle at the front of the crack tip, w—the weight function, r—the distance between points and l_{av}—the averaging length, related to the size of finite elements.

2.2 Discrete Rankine and Mohr-Coulomb Model

A discrete Rankine model was defined to simulate tensile cracks. To activate a crack, the following condition was assumed:

$$\max\{\sigma_1, \sigma_2, \sigma_3\} > f_t, \tag{2}$$

where σ_1, σ_2 and σ_3 are the principal stresses and f_t is the tensile strength. The direction of the crack extension was assumed to be perpendicular to the direction of the maximum principal averaged stress (see Eq. 1). The following loading function within a discrete cohesive law was chosen:

$$f([[u_n]], \kappa) = [[u_n]] - \kappa \tag{3}$$

with the history parameter κ equal to the maximum value of the normal component $[[u_n]]$ of the displacement jump achieved during deformation. With $f > 0$, a loading case occurs, while $f < 0$ stands for a reloading or an unloading phase. The softening of the normal component of the traction vector was described using an exponential relationship:

$$t_n = f_t \exp\left(-\frac{f_t \kappa}{G_f}\right)\left(1 - \exp\left(-d_f \frac{f_t}{G_f}\kappa\right)\right), \tag{4}$$

where G_f is the fracture energy and d_f is the numerical drop factor (Cox 2009). In the tangential direction, a linear relationship between displacement jump and traction was defined by the stiffness T_s.

To simulate shear zone, a discrete version of the elasto-plastic Mohr-Coulomb law with the internal friction angle ϕ and dilatancy angle ψ was used. The activation function was also based on Mohr-Coulomb criterion:

$$\frac{1}{2}(\sigma_1 - \sigma_3) + \frac{1}{2}(\sigma_1 + \sigma_3)\sin\phi - c\cos\phi > 0. \tag{5}$$

Linear softening of material cohesion c was defined as:

$$c(\kappa) = \max\{c_{\max} - H\kappa, c_{res}\}, \tag{6}$$

where $c_{\max}$—the maximum cohesion, c_{res}—the residual cohesion and H—the softening modulus. The penalty stiffnesses: normal K_N and shear K_S were defined to calculate elastic displacement jumps. It allows to use standard plasticity algorithm. High value of K_N prevents over penetration of shear zone surfaces. The direction of the propagation θ was calculated based on a bifurcation analysis with respect to principal averaged stresses directions:

$$\tan^2\theta = \frac{2 + \sin\phi + \sin\psi}{2 - \sin\phi - \sin\psi} \tag{7}$$

3 Numerical Examples

First, a simple uniaxial tension test was simulated. The width of the specimen was 100 mm, height 150 mm and thickness 1 m (Fig. 1a). The starting point of the crack propagation was defined in the middle of the left edge. The modulus of elasticity was equal to $E = 30$ GPa, the Poisson's ratio was $\nu = 0.2$, the tensile strength was $f_t = 3$ MPa and the fracture energy $G_f = 100$ N/m with exponential softening (Eq. 4). The drop factor was chosen as $d_f = 10^4$. The stress averaging

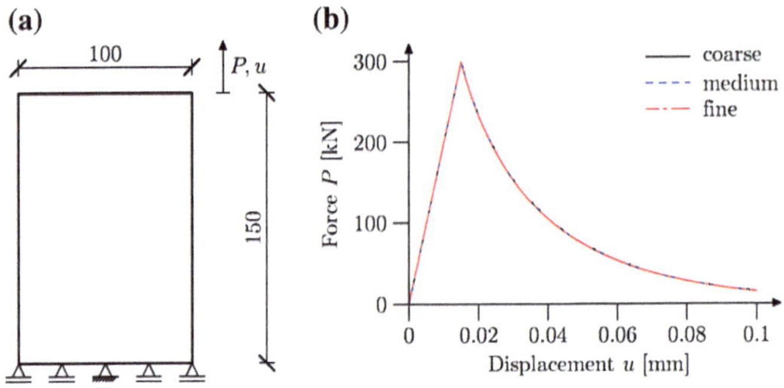

Fig. 1 XFEM results: for uniaxial tension **a** geometry and boundary conditions, **b** force-displacement diagrams

length was $l_{av} = 0$. The shear stiffness was $T_S = 10^{12}$ MPa/m. To examine the mesh insensitivity, simulations were performed with three different 3-node triangle FE-meshes: coarse (600 elements), medium (2,400 elements) and fine (5,400 elements). Almost the same force-displacement curves were obtained (Fig. 1b). A horizontal crack was properly reproduced.

Next, the simulations of a three-point bending test of notched concrete beams were carried out. The geometry was taken from experiments by Le Bellego et al. (2003) Three different beam sizes were numerically investigated: small ($h = 8$ cm), medium ($h = 16$ cm) and large ($h = 32$ cm). The span length of the beam was equal to $L = 3\,h$ (Fig. 2a). The loading was prescribed at the top edge at the mid-span via the vertical displacement. In the simulations, $E = 38.5$ GPa and $\nu = 0.2$ were taken with the tensile strength of $f_t = 3.2$ MPa. The exponential softening with the fracture energy $G_f = 80$ N/m was defined. The drop factor was chosen as $d_f = 10^4$. The stress averaging length was $l_{av} = 1$ cm. The material parameters were the same for all beams. Three different meshes with 3,068, 4,956 and 9,132 3-node constant strain triangles were defined for a small, medium and large beam, respectively. The crack starting points were located at the left side near the node at the line of the symmetry of the each beam.

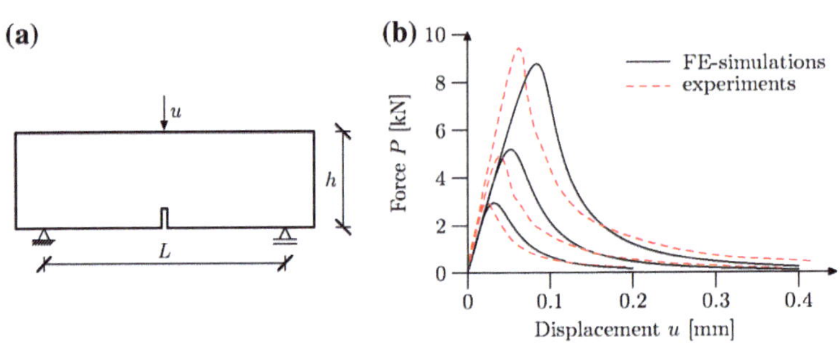

Fig. 2 XFEM results for three-point bending test: **a** geometry and boundary conditions, **b** force-displacement diagrams

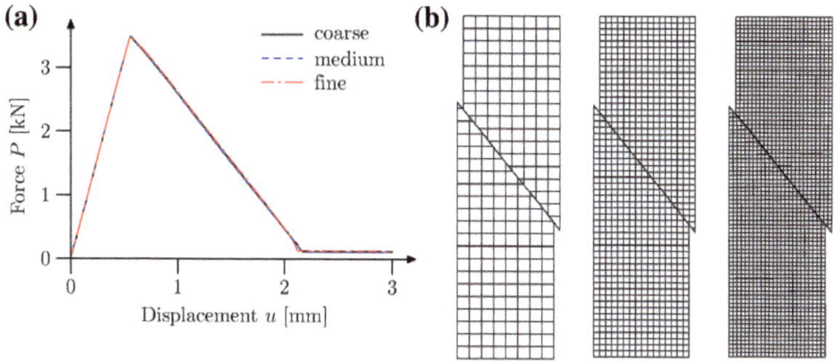

Fig. 3 XFEM result for biaxial compression: **a** force-displacement curve, **b** deformed meshes

Figure 2b shows the calculated force-displacement diagrams for notched concrete beams as compared with the experimental curves. Since it was not our intention to perfectly reproduce experiments, some differences can be seen, especially in a linear elastic regime. In turn, the softening behaviour of beams was properly captured. The maximum calculated force was equal to 2,945, 5,185 and 8,784 N for the small, medium and large beam, respectively. They were similar to the values obtained in experiments. Thus, a strong size effect in concrete beams (expressed by the increase of the load bearing capacity with decreasing size) was also properly reproduced.

Finally, biaxial compression was tested to simulate discrete shear zones. The specimen of 4 cm wide and 14 cm high was loaded via imposed vertical displacements on the top edge. The following material parameters were assumed: $E = 50$ MPa, $\nu = 0.3$, internal friction angle $\phi = 20°$, dilatancy angle $\psi = 0°$, maximum cohesion $c_{\max} = 30$ kPa, residual cohesion $c_{res} = 1$ kPa and softening modulus $H = 10$ MPa. The penalty stiffnesses were equal to $K_N = 5$ GPa and $K_S = 0.5$ GPa. The stress averaging length was $l_{av} = 0$. The starting point of the shear zone was located at the left edge (weak spot). Three meshes with 4-node quad elements were defined with 224, 896 and 2,016 elements for the coarse, medium and fine mesh, respectively. The obtained force-displacement curves are presented in Fig. 3a. Identical responses were achieved. The inclination of a shear zone was also properly reproduced (Fig. 3b).

4 Conclusions

The numerical results with the aid of the eXtended Finite Element Method have shown that this method is able to properly reproduce tensile cracks and shear zones as discontinuities in the displacement field under different loading conditions.

The present research activity is focused on combining XFEM with continuous cracks/shear zones descriptions using elasto-plastic and hypoplastic constitutive laws with non-local softening to describe the entire failure mechanism. More advanced boundary value problems with localization will be analysed.

Acknowledgments Scientific work has been carried out as a part of the Project: "Innovative resources and effective methods of safety improvement and durability of buildings and transport infrastructure in the sustainable development" financed by the European Union (POIG.01.01.02-10-106/09-01). The FE-calculations were performed at the Academic Computer Centre in Gdansk TASK.

References

Belytschko T, Black T (1999) Elastic crack growth in finite elements with minimal remeshing. Int J Numer Meth Eng 45(5):601–620

Cox JV (2009) An extended finite element method with analytical enrichment for cohesive crack modelling. Int J Numer Meth Eng 78(1):48–83

Le Bellego C, Dube JF, Pijaudier-Cabot G, Gerard B (2003) Calibration of nonlocal damage model from size effect tests. Eur J Mech A-Solid 22(1):33–46

Moës N, Belytschko T (1999) A finite element method for crack growth without remeshing. Int J Numer Meth Eng 46(1):131–150

Song JH, Areias PMA, Belytschko T (2006) A method for dynamic crack and shear band propagation with phantom nodes. Int J Numer Meth Eng 67(6):868–893

Wells GN, Sluys LJ (2001) A new method for modelling cohesive cracks using finite elements. Int J Numer Meth Eng 50(12):2667–2682

Zi G, Belytschko T (2003) New crack-tip elements for XFEM and applications to cohesive cracks. Int J Numer Meth Eng 57(15):2221–2240

Discrete Modelling of Micro-structural Phenomena in Granular Shear Zones

Michal Nitka, Jacek Tejchman and Jan Kozicki

Abstract The micro-structure evolution in shear zones in cohesionless sand for quasi-static problems was analyzed with a discrete element method (DEM). The passive sand failure for a very rough retaining wall undergoing horizontal translation towards the sand backfill was discussed. To simulate the behaviour of sand, the spherical discrete model was used with elements in the form of rigid spheres with contact moments.

1 Introduction

Earth pressure on retaining walls is one of the soil mechanics classical problems. In spite of intense theoretical and experimental research works over more than 200 years, there are still large discrepancies between experimental results and relevant theoretical solutions. The reason is the complexity of deformation field in granular bodies, especially near the wall, created by spontaneous emergence of shear localizations in a form of single or multiple narrow zones—the fundamental phenomenon characteristic for a granular material at shear deformation.

The patterning of shear zones is usually not taken into account in engineering calculations due to the lack of the basic knowledge on the phenomenon, which gives some practical importance to the research described in this paper. Its objective is to investigate, using the discrete element method DEM, the quasi-static evolution

M. Nitka (✉) · J. Tejchman · J. Kozicki
Faculty for Civil and Environmental Engineering, Gdańsk University of Technology, Narutowicza 11/12, 80233 Gdańsk-Wrzeszcz, Poland
e-mail: micnitka@pg.gda.pl

J. Tejchman
e-mail: tejchmk@pg.gda.pl

J. Kozicki
e-mail: jkozicki@pg.gda.pl

K.-T. Chau and J. Zhao (eds.), *Bifurcation and Degradation of Geomaterials in the New Millennium*, Springer Series in Geomechanics and Geoengineering,
DOI 10.1007/978-3-319-13506-9_2

of micro-structure within shear zones, created in initially medium dense sand under passive earth pressure conditions created by means of a rigid vertical wall moving towards the granular material. Several characteristic micro-structural events occurring in shear zones at the grain-level, such as: force chains, vortex structures, local void ratio fluctuations, strain non-uniformities were carefully studied.

2 DEM Results

To simulate the behaviour of cohesionless sand, a three-dimensional spherical discrete model YADE was developed at University of Grenoble (Kozicki and Donze 2008) by taking advantage of the so-called soft-particle approach (i.e. the model allows for particle deformation which is modelled as an overlap of particles). Spherical elements were used only. To approximately simulate a grain shape, additional moments were introduced into a discrete model, which were transferred through contacts and resisted particle rotations (Iwashita and Oda 1998). Particle breakage was not considered here, because of relatively low pressures adopted in simulations. The following five main local material parameters were needed for discrete simulations: E_c (modulus of elasticity of grain contact) υ_c (Poisson's ratio of grain contact), μ (the inter-particle friction angle), β (rolling stiffness coefficient) and η (moment limit coefficient) which were calibrated with corresponding triaxial laboratory test results with Karlsruhe sand (Wu 1992) ($E_c = 30$ GPa, $\upsilon_c = 0.3$, $\mu = 18^\circ$, $\beta = 0.7$ and $\eta = 0.4$). In addition, the particle radius R, particle density ρ and damping parameters α were required ($\rho = 2{,}550$ kg/m^3 and $\alpha = 0.08$).

The plane DEM calculations were performed with a sand body of a height of $H = 200$ mm and length of $L = 400$ mm. Along the depth, the granular specimen was composed of one grain layer. The height of the retaining wall was assumed to be $h = 200$ mm. The vertical retaining wall and the bottom of the granular specimen were assumed to be stiff and rough.

A typical particle configuration in the residual state at $u/h = 0.15$ (u-horizontal displacement of the wall) with the distribution of single sphere rotations ω is presented in Fig. 1a–c (red colour denotes the sphere rotation $\omega > +30^\circ$ and blue the sphere rotation $\omega < -30^\circ$, dark grey is related to the sphere rotation in the range $5^\circ \leq \omega \leq 30^\circ$ and light grey to the range $-30^\circ \leq \omega \leq -5^\circ$, the positive rotation means the clockwise rotation). All grains rotating within the range $-5^\circ \leq \omega \leq 5^\circ$ are medium grey. Accepting such colour convention makes shear zones clearly observable (only particles within shear zones significantly rotate). The thickness of the main curved shear zone is at the residual state: $t_s \approx 50$ mm ($10 \times d_{50}$) for $d_{50} = 5$ mm, $t_s \approx 33$ mm ($16 \times d_{50}$) for $d_{50} = 2$ mm and $t_s \approx 20$ mm ($20 \times d_{50}$) for $d_{50} = 1.0$ mm. The layout of shear zones depends on d_{50}. For $d_{50} = 1.0$ mm it is similar as in experiments with $d_{50} = 0.5$ mm (Niedostatkiewicz et al. 2011).

The distribution of void ratio across the main curved shear zone is strongly non-uniform and also has its maximum along the centre line (Fig. 1d). The void ratio strongly varies along a shear zone.

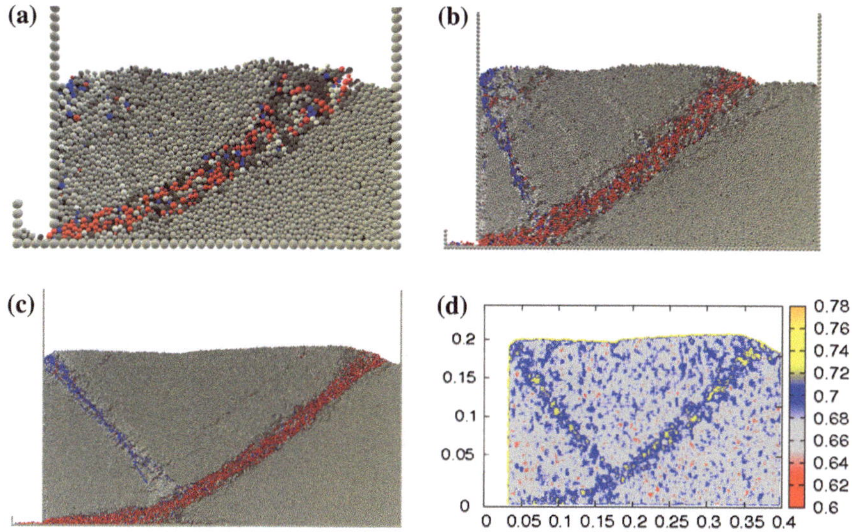

Fig. 1 Deformed granular body 0.2 × 0.4 m^2 with distribution of sphere rotation ('a'–'c') and void ratio ('d') for initially medium dense sand ($e_o = 0.62$) from DEM at residual state of $u/h = 0.15$: **a** $d_{50} = 5$ mm, **b** $d_{50} = 2$ mm, **c** and **d** $d_{50} = 1$ mm

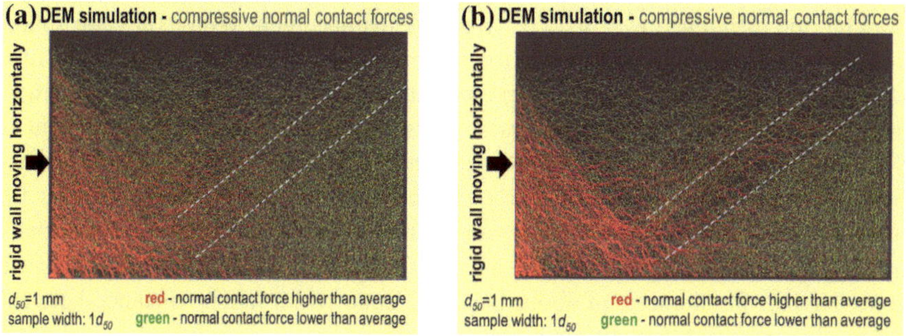

Fig. 2 Distribution of contact normal forces between spheres in entire granular specimen ($e_o = 0.62$, $d_{50} = 1$ mm) from DEM at: **a** $u/h = 0.02$ (without shear zone), and **b** $u/h = 0.15$ (full development of shear zone)

Figure 2 shows the two stages of the main shear zone evolution seen through the contact force network: (a) onset of shear localization and (b) shear zone fully developed. The location of the shear zone is indicated by the dashed lines. The 'force chains' bearing loads greater than average are marked red and those loaded below the average are marked green. The distribution of internal contact forces is non-uniform and continuously changes (Kozicki et al. 2013; Tordesillas et al. 2010). Force chains of heavily loaded grain contacts bear and transmit the compressive load on the entire granular system and are the predominant structure

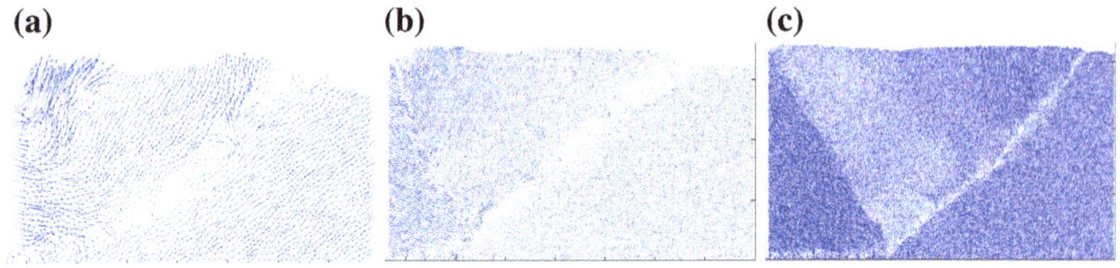

Fig. 3 DEM results ($e_o = 0.62$): formation of vortex structures in granular specimen at residual state of $u/h = 0.15$: **a** $d_{50} = 5$ mm, **b** $d_{50} = 2$ mm and **c** $d_{50} = 1.0$ mm

of internal forces at micro-scale. They continuously build up and collapse. The force chains are created mainly in the region between the wall, main and radial shear zone and along the main shear zone. The loads they transmit are the highest at the triangular region adjacent to the wall due to a great number of strong force chains (this region acts as a quasi-rigid body, Fig. 2). Inside the main curved shear zone at the residual state (Fig. 2b), the strongest force chains are approximately perpendicular to the shear zone line. The number of contacts decreases in a curved shear zone during wall translation due to sand dilatancy leading to a reduction of the number and stability of force chains.

Figure 3 presents spontaneous displacement changes within shear zones in the form of cells circulating as quasi-rigid bodies (so-called vortex structures) (Kozicki et al. 2013; Tordesillas et al. 2010). The plots in Fig. 3 were obtained by drawing the difference $\vec{V}_i - \vec{V}_{avr}$ between the displacement vector for each sphere and the average background translation corresponding to the homogeneous (affine) strain in the entire specimen ($\vec{V_i}$ represents the increment of sphere displacement during e.g. n = 1,000 iterations and $\vec{V}_{avr}$ is the average sphere displacement in the entire granular specimen for the same number of iterations). The vortex-like patterns are well recognized in the main curved shear zone, in particular, at the residual state for the highest mean grain diameter $d_{50} = 5$ mm. Several clockwise rotating vortices: 3 for $d_{50} = 5$ mm and 10 for $d_{50} = 1$ mm occur along the shear zone, having the diameter of about the shear zone width t_s. The distance between the vortices is variable (between t_s and $5 \times t_s$).

A link between force chains, vortex structures and void ratio changes in the region 70×100 mm^2 of the main curved shear zone at the residual state during the wall normalized displacement interval of $u/h = 0.01$ (from $u/h = 0.15$ up to $u/h = 0.16$) is demonstrated in Fig. 4 for the granular specimen built of $d_{50} = 5$ mm spheres. Two deformation stages are considered: (1) when the vortex exists at $u/h = 0.15$ (Fig. 4Aa) and (2) when the vortex does not exist at $u/h = 0.16$ (Fig. 4Ba). The force chain vanishes at $u/h = 0.15$ (Fig. 4Ba) and the new force chain is created at $u/h = 0.16$ (Fig. 4Bb). Small local dilatancy occurs close to a broken force chain (Fig. 4Ac–Ae) and small local contractancy takes place near a new force chain (Fig. 4Bc and Be). Thus, the occurrence and vanishing of vortex structures is related to the both force chain's creation and disappearance (Liu et al. 2012) and to the void ratio's decrease and increase

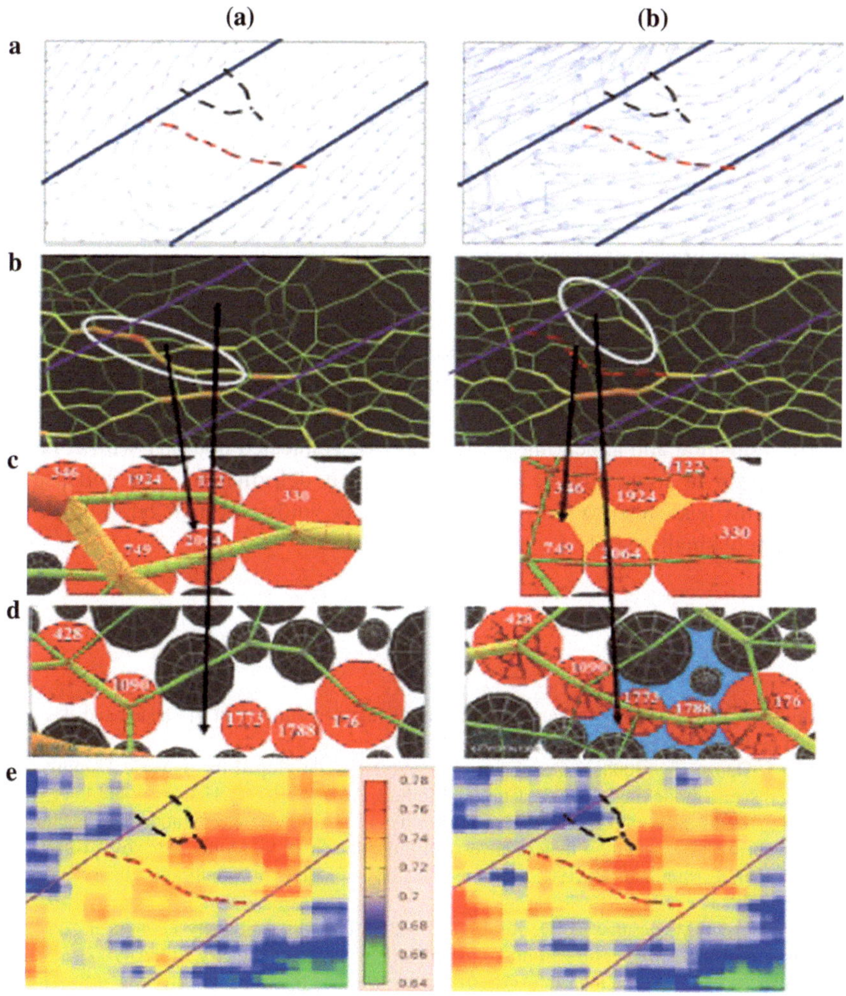

Fig. 4 DEM results ($e_o = 0.62$, $d_{50} = 5$ mm): evolution of micro-structures in main curved shear zone for normalized wall displacement at $u/h = 0.15$ (**a**) and $u/h = 0.16$ (**b**): (*solid lines* shear zone edges) (*a*) map of displacement fluctuations of Fig. 3a (*red dashed line* broken force chain, *black dashed line* new force chain), (*b*) geometry of force chains between spheres (*red dashed line* broken force chain), (*c*) and (*d*) zoom on geometry of force chains and spheres (*red spheres* build force chain, *yellow colour* between spheres denotes higher void ratio, *blue colour* between spheres denotes lower void ratio) and (*e*) map of void ratio

(Iwashita and Oda 1998). However, we have also observed that the occurrence and vanishing of vortices may correspond to the force chain softening and hardening only. The number of vortices increases with decreasing mean grain diameter and the distance between them increases with increasing mean grain diameter.

3 Conclusions

The discrete element method realistically simulates the experimental complex pattern of shear zones in the interior of initially medium dense sand. The results depend on the mean grain diameter.

The distribution of internal compressive contact forces is non-uniform due to a build-up and collapse of force chains. The number of contact forces continuously decreases in a granular specimen due to material dilatancy.

Vortex structures and local void ratio fluctuations spontaneously appear within the displacement field of shear zones and seem to have a periodically organized structure. The vortex diameter corresponds to the shear zone thickness. The vortices move as rigid bodies with small displacement fluctuations and insignificant rotations of single spheres.

The vanishing and appearing vortices may be connected not only with collapse and build-up of force chains, but also with their deformation. The collapse of force chains leads also to the formation of larger voids and their build-up to the formation of smaller voids.

References

Iwashita K, Oda M (1998) Rolling resistance at contacts in simulation of shear band development by DEM. ASCE J Eng Mech 124(3):285–292
Kozicki J, Donze FV (2008) A new open-source software developed for numerical simulations using discrete modelling methods. Comput Methods Appl Mech Eng 197:4429–4443
Kozicki J, Niedostatkiewicz M, Tejchman J, Mühlhaus HB (2013) Discrete modelling results of a direct shear test for granular materials versus FE results. Granular Matter 15(5):607–627
Liu X, Papon A, Mühlhaus HB (2012) Numerical study of structural evolution in shear band. Phil Mag 92(28–30):3501–3519
Niedostatkiewicz M, Leśniewska D, Tejchman J (2011) Experimental analysis of shear zone patterns in sand for earth pressure problems using particle image velocimetry. Strain 47(s2):218–231
Tordesillas A, Walker DM, Qun Lin Q (2010) Force cycles and force chains. Phys Rev E 81:011302
Wu W (1992) Hypoplastizität als mathematisches Modell zum mechanischen Verhalten granularer Stoffe. Ph.D. thesis, University of Karlsruhe, Karlsruhe

Modelling Hydromechanical Dilation Geomaterial—Cavitation and Localization

Y. Sieffert, O. Buzzi, F. Collin and R. Chambon

Abstract This paper presents an extension of the local second gradient model to multiphasic materials (solid particles, air, water) and including the cavitation phenomenon. This new development was made in order to model the response of saturated dilatant materials under deviatoric stress and undrained conditions and possibly, in future, the behaviour of unsaturated soils.

1 Introduction

A characteristic of geomaterials is to develop inelastic volume change. Clays, sands, rocks and concrete are dilatant materials, i.e. the porosity increases in the plastic regime.

In the case of saturated state, the pores are saturated with fluid. With a material permeability dependence, the pore volume increases more rapidly than the fluid can flow inside. Then the fluid is in tension which leads to a decrease of the pore pressure until negative pore pressure could be achieved.

This well-known problem in numerical modelling leads to a dilatant hardening behaviour because the decrease of pore water pressure is coupled with an increase of effective stresses (compression). This is particularly problematic with constitutive

Y. Sieffert (✉) · R. Chambon
Laboratoire 3S-R, Grenoble-INP, CNRS, Grenoble Université Joseph-Fourier, B.P. 53X, 38041 Grenoble Cedex, France
e-mail: yannick.sieffert@3sr-grenoble.fr

O. Buzzi
Priority Research Centre for Geotechnical and Materials Modelling, The University of Newcastle, Callaghan, NSW, Australia

F. Collin
Department of ArGEnCo Institut de Mécanique et Génie Civil, Université de Liège, 1 Chemin Des Chevreuils, 4000 Liège, Belgium

K.-T. Chau and J. Zhao (eds.), *Bifurcation and Degradation of Geomaterials in the New Millennium*, Springer Series in Geomechanics and Geoengineering,
DOI 10.1007/978-3-319-13506-9_3

equations modelling the degradation of the strength of materials as pressure of pore water may have an influence on shear band formation (Loret and Prevost 1991).

2 Presentation of the Model

Before starting, it is important to specify the main limitations of this work. The first restriction of this study is that we deal only with quasi-static problems in unsaturated conditions, under Richard's assumptions (vapour water pressure is constant). Furthermore for the sake of simplicity, isothermal condition, incompressible solid grains, incompressible vapour water, no flow for gas (no entry of air) are assumed. However phase changes between fluid and vapour water are considered. In all the computation large strains effects are taken into account.

As in Collin et al. (2006), the pore fluid and water vapour are assumed no influence at the microstructure level, micro kinematic gradient is not generated by pore pressure and vapour water variations.

The unknowns of the second gradient mechanical and the flow problems are respectively the (macro) displacement u_i, the micro kinematic v_{ij} and the pore water pressure p^w (possibly negative in unsaturated case).

In order to get second gradient models, we add the assumption that the micro kinematic gradient v_{ij} is equal to the macro displacement gradient F_{ij}. This implies similar relations for virtual quantities.

$$v_{ij}^* = \frac{\partial u_i^*}{\partial xj} = F_{ij}^*$$

In the framework of Schrefler's stress, the effective stress is:

$$\sigma_{ij}^t = \sigma_{ij}^{\prime t} + S_r^{w,t} p^{w,t} \delta_{ij} + \left(1 - S_r^{w,t}\right) p^v \delta_{ij}$$

σ_{ij}^t is the total stress, $\sigma_{ij}^{\prime t}$ is the effective stress, $p^{w,t}$ is the fluid pressure, p^v is the vapour pressure, $S_r^{w,t}$ is the water relative saturation and δ_{ij} is Kronecker's delta.

$$S_r^{w,t} + S_r^{v,t} = 1$$

$S_r^{v,t}$ is the vapour relative saturation.

The mass density of the mixture is:

$$\varrho^{mix,t} = \varrho^s\left(1 - \phi^t\right) + S_r^{w,t} \varrho^{w,t} \phi^t + \left(1 - S_r^{w,t}\right) \varrho^v \phi^t$$

ϱ^S is the solid grain density (assumed to be incompressible, i.e. $\varrho^S = cte$), $\varrho^{w,t}$ is the fluid density, ϕ^t is the porosity defined as $\phi^t = \Omega^{p,t} / \Omega^t$ where Ω^t is the current volume of a given mass of skeleton and $\Omega^{p,t}$ the corresponding porous volume.

In weak form (virtual work principle), the momentum balance for the mixture can thus be written as:

$$\int_{\Omega^t}\left(\sigma_{ij}^t\frac{\partial u_i^*}{\partial x_j^t}+\sum_{ijk}^t\frac{\partial^2 u_i^*}{\partial x_j^t\partial x_k^t}\right)d\Omega^t=\int_{\Omega^t}\varrho^{mix,t}g_i u_i^* d\Omega^t+\int_{\Gamma_\Omega^t}(\bar{t}_i u_i^*+\bar{T}_i Du_i^*)d\Gamma^t$$

where u_i^* is any kinematically admissible virtual displacement field, σ_{ij}^t is the Cauchy stress (total stress), $\sum_{ijk}^t$ is the double stress dual of the virtual second micro kinematic gradient, x_i are the current coordinates, g_i is the gravity acceleration, $\bar{t}_i$ is the external (classical) forces per unit area and $\bar{T}_i$ an additional external (double) force per unit area, both applied on a part Γ_Ω^t of the boundary of $\Omega^t \cdot D$ denotes the normal derivative of any quantity q.

In order to use C^0 functions for the displacement field (i.e. only first derivatives of the displacement), the following equation with λ_{ij} Lagrange multipliers is used.

$$\int_{\Omega^t}\left(\sigma_{ij}^t\frac{\partial u_i^*}{\partial x_j^t}+\sum_{ijk}^t\frac{\partial v_{ij}^*}{\partial x_k^t}\right)d\Omega^t-\int_{\Omega^t}\lambda_{ij}^t\left(\frac{\partial u_i^*}{\partial x_j^t}-v_{ij}^*\right)d\Omega^t$$
$$=\int_{\Omega^t}\varrho^{mix,t}g_i u_i^* d\Omega^t+\int_{\Gamma_\Omega^t}(\bar{t}_i u_i^*+\bar{T}Du_i^*)d\Gamma^t,$$
$$\int_{\Omega^t}\lambda_{ij}^*\left(\frac{\partial u_i^t}{\partial x_j^t}-v_{ij}^t\right)d\Omega^t=0$$

In volume Ω^t, the liquid fluid mass is equal to $M_{\Omega^t}^{w,t}=S_r^{w,t}\varrho^{w,t}\phi^t\Omega^t$ and the vapour fluid mass is equal to $M_{\Omega^t}^{w,t}=S_r^{v,t}\varrho^v\phi^t\Omega^t$.

In weak form, the mass balance equation for the fluid and water vapour can thus be written as:

$$\int_{\Omega^t}\left((\dot{M}^{w,t}+\dot{M}^{v,t})p^{w^*}-\mathrm{m}_i^{w,t}\frac{\partial p^*}{\partial x_i^t}\right)d\Omega^t$$
$$=\int_{\Omega^t}Q^{w,t}p^{w^*}d\Omega^t-\int_{\Gamma_q^t}\bar{q}^{w,t}p^{w^*}d\Gamma^t,$$
$$\dot{M}^{w,t}=\varrho^{w,t}\left(\frac{\dot{p}^{w,t}}{k^w}S_r^{w,t}\phi^t+\dot{S}_r^{w,t}\phi^t-S_r^{w,t}\frac{\dot{\Omega}^t}{\Omega^t}\right),$$
$$\dot{M}^{v,t}=\varrho^v\left(-(1-S_r^{w,t})\frac{\dot{\Omega}^t}{\Omega^t}-\dot{S}_r^{w,t}\phi^t\right),$$
$$m_i^{w,t}=-\varrho^{w,t}k\frac{k_r^{w,\tau_1}}{\mu^W}\left(\frac{\partial p^{w,t}}{\partial x_i^t}+\varrho^{w,t}g_i\right)$$

With $\dot{M}^{w,t}$ is the time derivative of the fluid phase, $m_i^{w,t}$ is the mass flow of water, k^w is the fluid bulk modulus, k is the intrinsic permeability, $k_r^{w,t}$ is the water relative permeability, μ^w is the fluid viscosity, $Q^{w,t}$ is a sink term and Γ_q^t is the part of the boundary where the input fluid mass per unit area $\overline{q}^{w,t}$ is prescribed.

2.1 Mechanical Constitutive Law

In order to reproduce the progressive decrease of the material strength, the mechanic constitutive law used in this study is an elastoplastic strain-softening Druker-Prager model with yield criterion given by the follow equation (Barnichon 1998):

$$f = II_{\widehat{\sigma}} - m\mathrm{I}_{\sigma} + k = 0$$

where m and k is defined by: $m = \frac{2\sin\phi_c}{\sqrt{3}(3-\sin\phi_c)}, k = \frac{6c\cos\phi_c}{\sqrt{3}(3-\sin\phi_c)}$. ϕ_c is the friction angle, c is the cohesion, $I_\sigma = \sigma_{ii}$ is the first invariant and $II_{\widehat{\sigma}} = \sqrt{\frac{1}{2}\widehat{\sigma}_{ij}\widehat{\sigma}_{ij}}$ is the second deviatoric invariant.

A general non-associated plasticity framework is considered to define the rate of plastic flow as perpendicular to the plastic potential g : $\dot{\varepsilon}_{ij}^p = \dot{\lambda}\frac{\partial g}{\partial\sigma_{ij}}$ and $g = II_{\widehat{\sigma}} - m'\mathrm{I}_\sigma = C_1$ with $m' = \frac{2\sin\psi}{\sqrt{3}(3-\sin\psi)}$ with, ψ is the dilatancy angle and c_1 is a constant. The softening process during plastic flow is introduced via an hyperbolic variation of the cohesion between initial c_0 and final c_f values as a function of the Von Mises equivalent plastic strain $\varepsilon_{eq}^p : c = c_0 + \frac{(c_f - c_0)\varepsilon_{eq}^p}{\beta_c + \varepsilon_{eq}^p}$.

2.2 Model Parameters

The Drucker-Prager model presents the advantage to use simple formulation and does not require enough parameters. All parameter values are presented in Tables 1 and 2.

Table 1 Mechanic parameters

Geomechanical characteristics	
Young's elastic modulus (MPa)	300
Poisson's ratio (–)	0.125
Initial cohesion (kPa) c_0	300
Final cohesion (kPa) c_f	100
Softening parameter (–) β_c	0.01
Friction angle (°) ϕ_c	18
Dilation angle (°) ψ	10
Solid specific mass (kg/m^3) ϱ^s	2,026
Second gradient parameter (N) D	150

Table 2 Hydraulic and air parameters

Hydraulic characteristics	
Initial porosity (–) ϕ	0.39
Intrinsic permeability (m/s) k	10^{-7}
Initial relative water permeability (–)	1
Water specific mass (kg/m^3) ϱ^w	1,000
Water dynamic viscosity (Pa·s) μ^w	10^{-3}
Water compressibility coefficient (MPa^{-1}) $1/k^w$	5.10^{-4}
Air characteristics	
Gas pressure (kPa) p^v	-100

2.3 Cavitation Model

In the model, before cavitation, the stress state governs the pore pressure and the specimen is saturated. After cavitation, a phase change takes place and the pore pressure is related to the relative degree of saturation as per the cavitation equation below: $p^{w,t} = p_c - C_2 + C_2 \cdot \exp^{-\left(\frac{1-\left(s_r^{w,t}/100\right)}{0.02}\right)}$ for $p^{w,t} < p_c$ with $C_2 = 40$ kPa in this paper.

3 Results

Visualization of the localization shear bands is performed by observing the loading index of a Gauss point for a given time step. When a Gauss point undergoes a plastic loading, a small square is plotted. No marker appears if the element undergoes elastic loading or unloading (Fig. 1).

The numerical results are displayed in the form of load-displacement curve and pore pressure inside the shear band curve in Fig. 2. First, the biaxial test leads to a homogenous solution, i.e. the plasticity behavior is identical in the entire sample (Fig. 1a). As all elements of the model enter into plasticity, global dilation takes place and pore pressure starts dropping steadily. The reduction in pore pressure turns into an increase of effective stress. The homogeneous solution is stable until the pore pressure achieves the cavitation pressure. Then, the water starts to change in vapour phase (Figs. 1b and 2). The effective stresses can decrease freely and localized bands are obtained. In accord with the experimental results given by Mokni and Desrues (1998), when the plasticity is first obtained with an homogeneous behavior, the cavitation triggers strain localization and then cavitation occurs before localization.

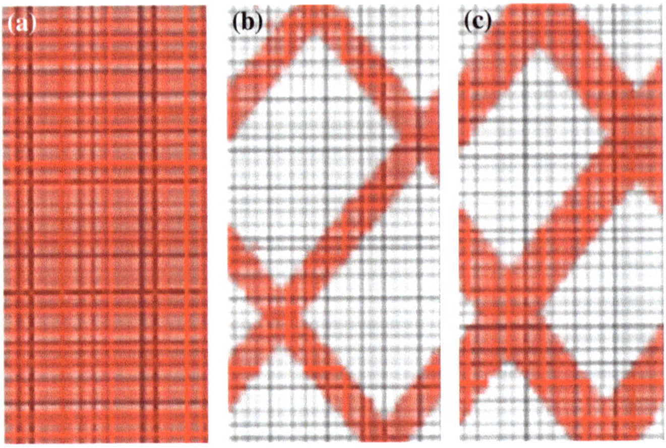

Fig. 1 Visualization of the shear bands using the loading index at the gauss point **a** axial strain = 0.14 %, **b** axial strain = 1.8 %, and **c** axial strain = 2.5 %

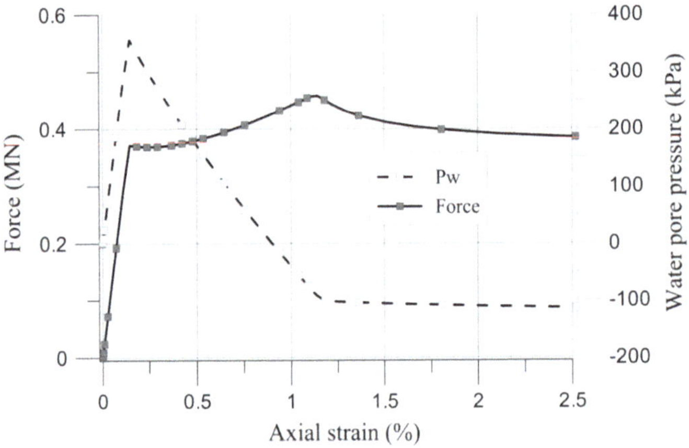

Fig. 2 Global curves of the loading force and water pressure versus the specimen shortening

4 Conclusion

This paper presents numerical investigations of strain localization for a dilatant material. To ensure objectivity of the formulation, i.e. no mesh dependency, a second gradient hydromechanical model is used. This paper gives a numerical implementation of the cavitation phenomena to resolve the challenge of the unrealistic negative pore pressure obtained typically with a dilatant porous material which could lead at a global hardening response. Based on the numerical simulation, the cavitation is capable of restoring the shear bands formation.

References

Barnichon J (1998) Finite element modelling in structural and petroleum geology. Ph.D. thesis, University of Liège, Belgium

Collin F, Chambon R, Charlier R (2006) A finite element method for poro mechanical modelling of geotechnical problems using local second gradient models. Int J Numer Methods Eng 65:1749–1772

Loret B, Prevost JH (1991) Dynamics strain localisation in fluid-saturated porous media. J Eng Mech 11:907–922

Mokni M, Desrues J (1998) Strain localization measurements in undrained plane-strain biaxial tests on Hoston RF sand. Mech Cohesive-Frictional Mater Struct 4:419–441

Fabric Evolution and Its Effect on Strain Localization in Sand

Zhiwei Gao and Jidong Zhao

Abstract Fabric anisotropy affects importantly the overall behaviour of sand including its strength and deformation characteristics. While both experimental and numerical evidence indicates that soil fabric evolves steadily with the applied stress/strain, how evolving fabric influences the initiation and development of shear band in sand remains an intriguing question to be fully addressed. In this paper, we present a numerical study on strain localization in sand, highlighting the special role played by soil fabric and its evolution. In particular, a critical state sand plasticity model accounting for the effect of fabric and its evolution is used in the finite element analysis of plane strain compression tests. It is found that the initiation of shear band is controlled by the initial fabric, while the development of shear band is governed by two competing physical mechanisms, namely, the structural constraint and the evolution of fabric. The evolution of fabric generally makes the sand response more coaxial with the applied load, while the structural constraint induced by the sample ends leads to more inhomogeneous deformation within the sand sample when the initial fabric is non-coaxial with the applied stress. In the case of smooth boundary condition, structural constraint dominates over the fabric evolution and leads to the formation of a single shear band. When the boundary condition is rough, the structural constraint may play a comparable role with fabric evolution, which leads to symmetric cross-shape shear bands. If the fabric is prohibited from evolving in the latter case, a cross-shape shear band pattern is found with the one initiated first by the structural constraint dominant over the second one.

Z. Gao (✉)
School of Engineering, University of Glasgow, Rankine Building, Oakfield Avenue, Glasgow G12 8LT, UK
e-mail: Zhiwei.Gao@glasgow.ac.uk

J. Zhao
Department of Civil and Environmental Engineering, Hong Kong University of Science and Technology, Clearwater Bay, Kowloon, Hong Kong

K.-T. Chau and J. Zhao (eds.), *Bifurcation and Degradation of Geomaterials in the New Millennium*, Springer Series in Geomechanics and Geoengineering,
DOI 10.1007/978-3-319-13506-9_4

1 Introduction

Strain localization is frequently observed in sand and is considered an important precursor of the failure of soil and relevant geostructures. Numerous investigations have been carried out on strain localization in sand but relatively less attention has been paid to the correlation of strain localization with the presence of an evolving fabric (e.g., Borja et al. 2013). Fabric anisotropy has been widely regarded to affect the key behaviour of sand including dilatancy, liquefaction and critical state. Based on plain strain compression tests, Tatsuoka et al. (1990) found that the shear band development in sand was dependent on the initial bedding plane orientation, or the fabric, of the sample. Meanwhile, micromechanical studies indicate that fabric evolves steadily with deformation and the evolving fabric exhibits unique characteristics when a sand sample reaches liquefaction, critical state and other states (Guo and Zhao 2013; Zhao and Guo 2013a, b). As an important type of instabilities, strain localization has been commonly observed in sand. The correlation between fabric and fabric evolution is an interesting topic but remains less explored. Inherent fabric anisotropy has recently been considered in a hypoplastic model by Tejchman et al. (2007) in the simulation of shear band development in sand but the interplay between fabric evolution and the development of shear band has not been properly considered. In this study, we employ an anisotropic sand model developed recently by the authors (Gao et al. 2014) and finite element method to investigate strain localization in sand under plane strain compression wherein the special role of fabric and its evolution is highlighted.

2 Finite Element Analysis of Strain Localization in Plane Stain Compression

The constitutive model used in this study was developed by Gao et al. (2014). It has been implemented in the finite element package ABAQUS through the user-material interface. The test data presented by Tatsuoka et al. (1990) will be employed to benchmark the model simulations. The model parameters have been calibrated based on the plane strain test results on Toyoura sand. The initial degree of anisotropy F_0 is set to be 0.45. The model simulations for single element tests can be found in Gao et al. (2014) and Gao and Zhao (2013).

In the finite element analysis, the sample setup is the same as that in Tatsuoka et al. (1990) (Fig. 1). Uniform 4-noded plane strain elements of 2.5 mm × 2.5 mm in size are used and constant confining pressure is applied in the horizontal direction of the sample. Vertical displacement Δh is applied to the top end of the sample by increment to ensure quasi-static loading. Both smooth (the top and bottom ends are free to move horizontally) and rough (the horizontal displacements of the top and bottom ends are restricted) boundary conditions with and without fabric evolution are considered and the initial void ratio distribution is assumed to be uniform throughout the entire sample.

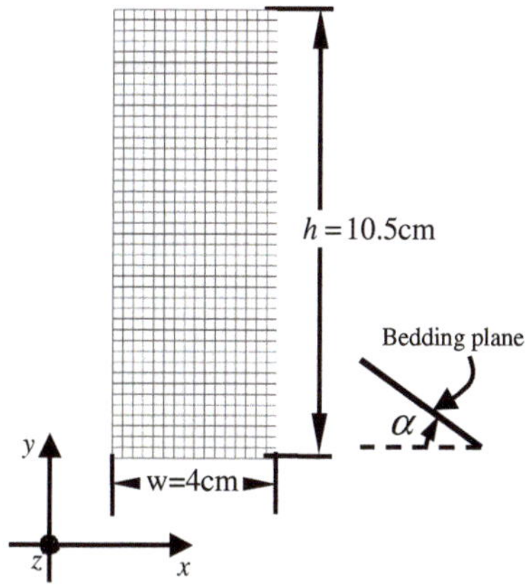

Fig. 1 Sample setup, element size, orientation of bedding plane and the reference coordinate system

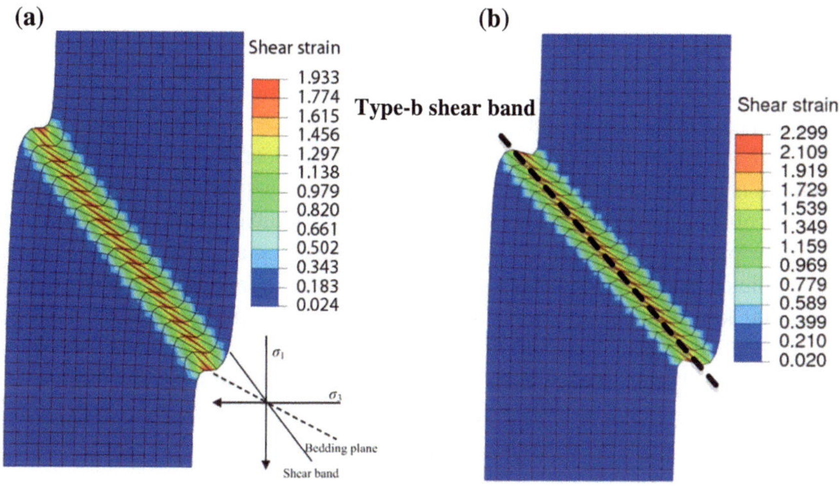

Fig. 2 Predicted strain localization for the cases with smooth boundary condition and **a** evolving fabric; **b** constant fabric at vertical strain $\varepsilon_h = \Delta h / h = 12\,\%$

Figure 2 shows the predicted shear band pattern in smooth boundary condition cases with initial void ratio $e_0 = 0.7$, confining pressure $\sigma_c = 400$ kPa and bedding plane orientation $\alpha = 45°$. It can be seen that the simulated shear band and bedding plane lie on the same side (left side in the upper part of the sample for the present study) of the major principal stress direction (Fig. 2a) and Tatsuoka et al. (1990) terms similar shear band pattern they observed in laboratory tests as Type-b shear band. Indeed, Type-b shear band is the most commonly observed pattern in laboratory tests with smooth boundary (Tatsuoka et al. 1990). When the fabric is assumed constant (the fabric is prohibited from evolving with deformation), the

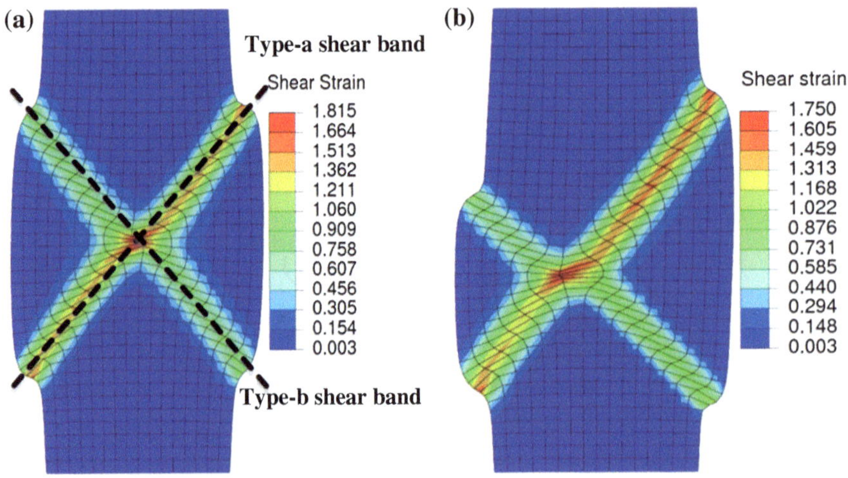

Fig. 3 Predicted strain localization for the cases with rough boundary condition and **a** evolving fabric; **b** constant fabric at vertical strain $\varepsilon_h = \Delta h / h = 15.5\,\%$

strain concentration in the shear band (Fig. 2b) will be more intense than the fabric evolution case (Fig. 2a). In addition, it is observed that the predicted vertical strain level corresponding to shear band initiation is independent of initial bedding plane orientation if the fabric is assumed constant, which appears to be inconsistent with the observations in Tatsuoka et al. (1990). This indicates that fabric evolution should be properly accounted for in the strain localization analysis. Figure 3 shows the predicted shear band pattern in rough boundary condition cases ($e_0 = 0.7$, $\sigma_c = 400$ kPa and $\alpha = 45°$). Two symmetric and asymmetric shear bands are observed for the cases with and without fabric evolution, respectively. The shear strain level in Type-a shear bands [according to the definition by Tatsuoka et al. (1990)] is higher than that in the Type-b shear bands in both cases.

3 Mechanisms Governing the Shear Band Patterns

Our study shows that the development of a shear band is essentially governed by two competing mechanisms—the *fabric evolution* which reduces the non-coaxial strain increment when the fabric and applied stress are initially non-coaxial and alleviates the strain localization and the *structural constraint* imposed by the boundary conditions which promotes the shear band development. The structural constraint can be better described by the reaction forces at the top and bottom sample ends as shown in Fig. 4.

In the smooth boundary condition cases, the initial bedding plane constitutes a natural weakened plane along which the sample can develop shear strain concentration and the vertical reaction forces on the two ends drive the upper and lower

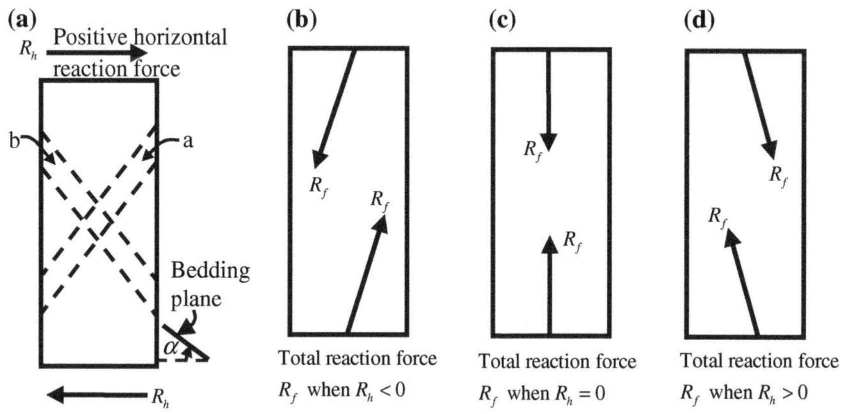

Fig. 4 Definition of **a** the positive direction of the horizontal reaction force R_h, **c–d** three cases of the total reaction force R_f imposed on the sample by the boundary

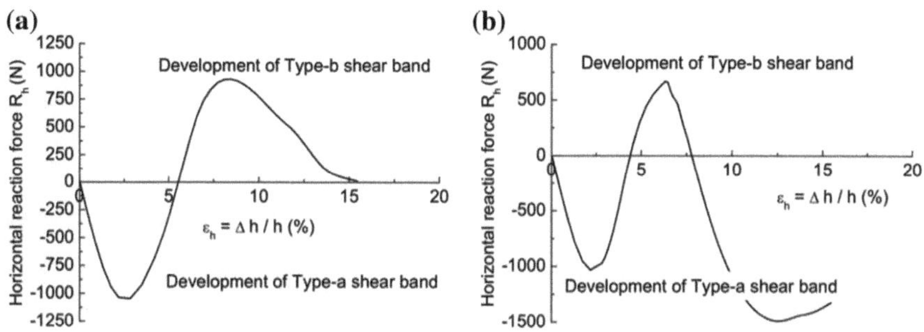

Fig. 5 Evolution of the horizontal reaction force at the top end of the sand sample with $\alpha = 45°$ and rough boundary condition ($e_0 = 0.7$, $\sigma_3 = 400$ kPa): **a** considering fabric evolution; **b** considering constant fabric

parts of the sample to the right and left respectively. This causes the occurrence of a single Type-b shear band. Since the strain localization initiates at relatively low strain level, the effect of fabric evolution is not large enough to prevent the shear band development in this case.

In the rough boundary condition cases, negative R_h develops to prevent the horizontal displacement of the top and bottom ends due to the non-coaxial strain increment at the initial loading stage (Fig. 5), which leads to a Type-a shear band initially. As the sample is weakened along direction-a due to stain localization (Fig. 5a), the magnitude of R_h decreases and becomes positive subsequently. This causes the initiation of Type-b shear band. Meanwhile, the evolution of fabric can eventually lead to rather symmetric geometry of the two bands as well as sample shape (Fig. 4a). If the fabric is fixed without evolution, the final sample shape and shear bands are asymmetric as R_h will become negative again later on (Fig. 5b).

4 Conclusions

The effect of fabric evolution on strain localization in plane strain compression has been studied numerically. It is found that the evolution of fabric generally tends to make the sand response more coaxial with the applied stress and alleviate the strain localization, while the structural constraint induced by the sample ends leads to more inhomogeneous deformation within the sand sample when the initial fabric is non-coaxial with the applied stress. In smooth boundary condition cases, structural constraint dominates over the fabric evolution and leads to the formation of a single shear band. When the boundary condition is rough, the structural constraint may play a comparable role with fabric evolution, which leads to symmetric cross-shape shear bands. If the soil fabric is not allowed to evolve, asymmetric cross-shape shear bands develop.

Acknowledgements The study was financially supported by RGC/GRF 622910 and DAG08/09.EG04.

References

Borja RI, Song XY, Rechenmacher AL, Abedi S, Wu W (2013) Shear band in sand with spatially varying density. J Mech Phys Solids 61:219–234
Gao ZW, Zhao JD (2013) Strain localization and fabric evolution in sand. Int J Solid Struct 50:3634–3648
Gao ZW, Zhao JD, Li XS, Dafalias YF (2014) A critical state sand plasticity model accounting for fabric evolution. Int J Numer Anal Meth Geomech 38:370–390
Guo N, Zhao JD (2013) The signature of shear induced anisotropy in granular media. Comput Geotech 47:1–15
Tatsuoka F, Nakamura S, Huang CC, Tani K (1990) Strength anisotropy and shear band direction in plane strain tests of sand. Soils Found 30(1):35–54
Tejchman J, Bauer E, Wu W (2007) Effect of fabric anisotropy on shear localization in sand during plane strain compression. Acta Mech 189:23–51
Zhao JD, Guo N (2013a) Unique critical state characteristics in granular media considering fabric anisotropy. Géotechnique 63(8):695–704
Zhao JD, Guo N (2013b) A new definition on critical state of granular media accounting for fabric anisotropy. In: Powders and grains 2013: AIP conference proceedings, vol 1542, pp 229–232. doi: 10.1063/1.4811909

Micro-mechanical Observations of Strain Localization in Granular Soils During Simple Shear Loading

Marte Gutierrez and Abdalsalam Muftah

Abstract This paper examines the micro-mechanical origins of strain localization in granular materials during simple shear loading using the two-dimensional Discrete Element Method (DEM) as implemented in the Particle Flow Code (PFC). A key feature of the DEM model used in the study is the implementation of a rolling resistance component in the contact behavior between particles. Analysis of the results of the DEM study was performed in terms of the evolution of force chains, buckling of columns of particles, particle rolling, and formation of voids in the shear band. The results show that at bifurcation fewer force chains carry the loads inside the shear band, and as the frictional resistance between particles is exceeded, the force chains collapse. This collapse is seen in the buckling of columns of particles in narrow zones in the shear band. High degrees of particle rotations and large voids are produced in the shear band as a result of particle column buckling.

1 Introduction

Strain localization is one of the most important phenomena occurring in granular materials. Strain localization occurs due to instability and bifurcation in the stress-strain response of materials during loading. A material which undergoes homogeneous deformation reaches a bifurcation point at which the material experiences instability and deformation starts to become non-homogenous. Non-homogeneous

---

M. Gutierrez (✉)
Khalifa University, Abu Dhabi, UAE
e-mail: marte.gutierrez@kustar.ac.ae; mgutierr@mines.edu

M. Gutierrez
Colorado School of Mines, Golden, CO 80401, USA

A. Muftah
Sirte University, Sirte, Libya

K.-T. Chau and J. Zhao (eds.), *Bifurcation and Degradation of Geomaterials in the New Millennium*, Springer Series in Geomechanics and Geoengineering,
DOI 10.1007/978-3-319-13506-9_5

deformation results from localized deformation, and since the predominant yielding mechanism for soils is shearing, the localized deformation zone is called a shear band. For strain hardening materials, strain localization and instability can occur before the peak shear strength corresponding to homogeneous response is reached. Localization is then followed by strain softening characterized by reduction in load carrying capacity during continued deformation.

One of the most common types of loading of soils is simple shear. Failure in soils occurs in narrow shear zones that follow simple shear loading, and the deformation inside the shear band itself follows simple shear loading (Gutierrez 2011). Understanding of behavior of soils in simple shear, particularly strain localization, is complicated by the inherent difficulties in simple shear testing. Direct simple shear testing of soils faces the difficulties of applying uniform loading conditions and of measuring the lateral stress during shearing.

A powerful technique to study the micro-mechanical behavior of granular materials during loading is numerical modeling using the Discrete Element Method (DEM). In DEM, granular materials are treated as ensembles of particles which interact at contact points (Cundall and Strack 1979). A major improvement in DEM is the modeling of particle rolling resistance. In the original DEM, particle normal contact or separation, and sliding are considered to play the dominant roles in the microscopic deformation of granular materials. Particles are assumed to be perfectly circular and to rotate freely and offer no resistance to rotation. In reality, real particles are non-circular and have rough surface textures (Oda 1982), and particles cannot roll freely when in contact with other particles. Rolling resistance can be defined as a force couple that is transferred between the particles through the contacts, and this force couple inhibits particle rotation. Oda et al. (1982) and Mohammed and Gutierrez (2010) have shown the important influence of rolling resistance in shear band formation during biaxial loading of granular soils. The main objective of this paper is to present the results of a study of strain localization during simple shear loading in granular soils using DEM which accounts for rolling resistance.

2 Methodology

The study of strain localization in granular soils in simple shear is performed using the two-dimensional DEM code PFC (Particle Flow Code) developed by Itasca (2008). PFC models granular materials using distinct rigid spherical or disk-shaped particles that interact only at contacts or interfaces between the particles. In DEM, mechanical behavior is expressed in terms of the movements of each particle and the internal forces generated at each contact point. The DEM simulation is performed under two-dimensional simple shear loading conditions with a model that is rectangular, 1.2 cm wide and 0.6 cm high, and consists of an assembly of poly-dispersed disk-like particles (see Fig. 1). Three ranges of particle radii were used: 0.3–0.4 mm, 0.3–0.6 mm and 0.3–0.8 mm. For these particle radii, the size

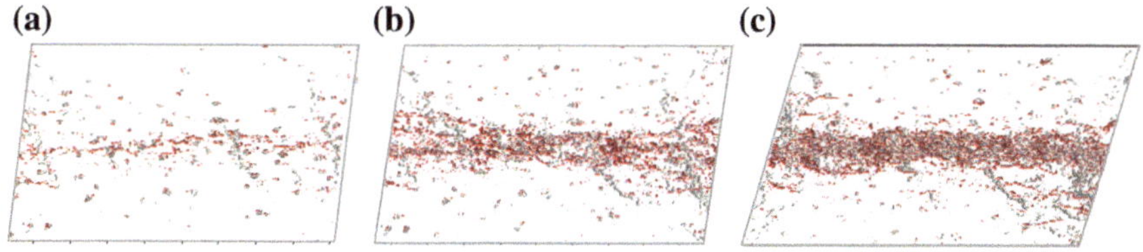

Fig. 1 Set-up of the simple shear DEM model for the simulation of the stress-strain and strain localization response of granular materials. **a** γ_{xy}=0.10 **b** γ_{xy}=0.13 **c** γ_{xy}=0.15

of the model is deemed large enough to simulate a representative element volume of granular material, but not too large to require extensive calculation times. The sample is contained by four frictionless walls. The top and bottom boundaries are confined by rigid platens and by the vertical normal stress σ_y. The left and right lateral walls are confined by semi-rigid walls and by the minor principal stress σ_x. The sample is first loaded isotropically during consolidation, then sheared by rotating the lateral walls. Shearing is performed under constant angular velocity and constant vertical and horizontal normal stresses. The constant normal stresses applied to the top and lateral boundaries are applied by a numerical servo-control.

To generate the model, a random particle generation procedure called the "expansion method" (Itasca 2008) is adapted to achieve a desired sample particle size distribution and porosity. The model parameters values used in the study consist of density ρ, normal and shear stiffness K_n and K_s, and friction coefficient μ, with values similar to those used by Mohammed and Gutierrez (2010). The simple shear simulations were run for range of vertical normal stress with σ_y varying from 0.1 to 2 MPa. It is known in the literature that a DEM model with more than 5,000 particles provides a representative element volume for the modeling of the stress-strain response of granular materials. In this study, the number of randomly generated particles ranged from 6,556 to 16,190. Circular particles are used but, as discussed above, rolling resistance is employed to partly account for the effects of non-circular and irregular particle shapes (Mohammed and Gutierrez 2010). The rolling resistance parameter α is affected by the particle shape, normal stiffness and normal force at a point of contact and typically ranges from 0 to 1 (Mohammed and Gutierrez 2010). For $\alpha = 0$, no rolling resistance (i.e., free rolling) exists. Based on previous studies, particle rotation is effectively prevented (i.e., no rolling) when $\alpha = 1$, which is equivalent to prescribing a high rolling friction angle of 45°. Parameter α is given a value of 0.5 in this study.

3 Results and Discussions

Figure 1 shows the evolution of shear strain within the sample during shearing for three levels of simple shear deformation. Shear strain levels were calculated using the "mesh-free" method developed by Wang et al. (2007). As can be seen,

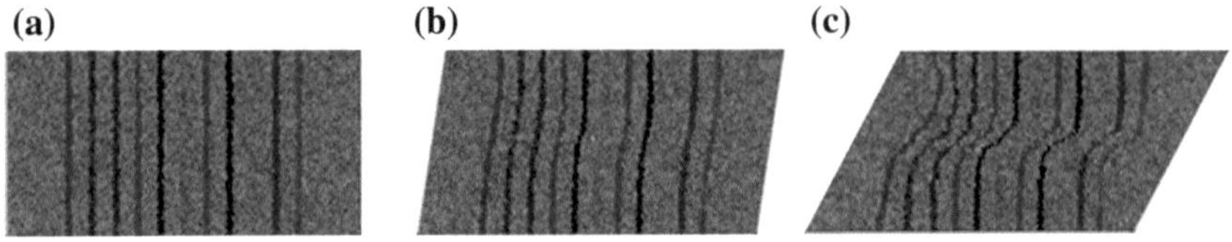

Fig. 2 Buckling of column of particles at different levels of shear strain. **a** γ_{xy}=0.0. **b** γ_{xy}=0.12. **c** γ_{xy}=0.50

strain localization started to become evident at a shear strain level of $\gamma_{xy} = 0.13$, and the shear band is fully developed and spans the full width of the sample at $\gamma_{xy} = 0.15$. The shear band is nearly horizontal although some waviness in the shape and wider bands near the boundaries can be observed. Figure 2 shows the development of buckled columns of particles as indicated by the differently colored column of particles. The colored columns used to track the movements of particles are slightly bent and curved at the same locations along the column. The columns buckled despite the fact that the lateral boundaries are constrained to deformed along straight lines. Rolling resistance endows the columns with bending stiffness which increases the overall shear strength of the granular material. However, the particle rolling resistance enhances instability due to the tendency of the columns to buckle during loading. In the case of low or no rolling resistance, particles are free to move and are unable to form columns with sufficient rigidity to buckle locally. As can be seen, the locations of the buckling correspond to where the shear bands are formed.

The development of force chains is a common feature in the stress-strain response of granular materials. Force chains are quasilinear arrangements of group of particles by which compressive loads are transmitted. Their average orientations are more or less sub-parallel to the major principal stress, and they form solid column-like structures which provide resistance to shear and serve as the load-bearing network responsible for the strength and stability of granular materials. The reconfiguration of the network of force chains during simple shear loading of the modeled granular material is shown in Fig. 3. Before failure, the network of force chains is dense and evenly distributed throughout the volume of the granular material. Once the peak shear stress is reached, the density of the number of force chains is reduced inside the shear band, and there are fewer particle-to-particle contacts available with in the shear band to transmit the loads. As the number of particle-to-particle contacts is decreased, the frictional resistance between particles may be exceeded causing the force chains to collapse. As a result, the load carrying capacity of the granular material is reduced and strain softening occurs.

It is seen in Fig. 4 that at the end of consolidation, particles have not rotated over the granular assembly as shown by the horizontal radial lines drawn from the center of each particle. After shearing to $\gamma_{xy} = 0.50$ when a distinct shear band has formed, particle rotations occur extensively within the shear band, as can be seen by wide orientations of the radial lines. These radial lines were initially noted as horizontal before shearing. In comparison, no significant rotations are observed for

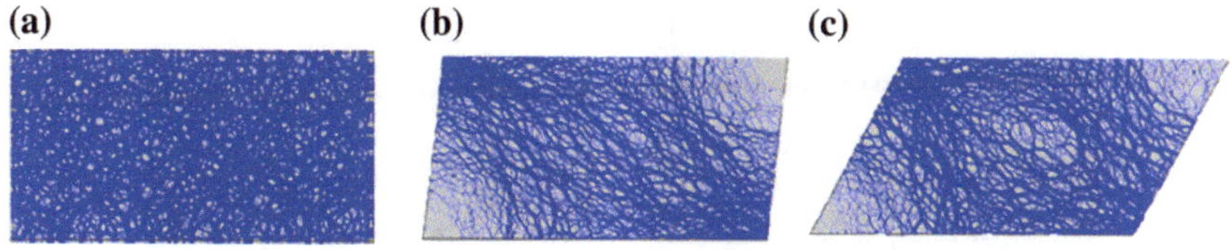

Fig. 3 Development of force chains at different levels of shear strain. **a** γ_{xy}=0.0. **b** γ_{xy}=0.12. **c** γ_{xy}=0.50

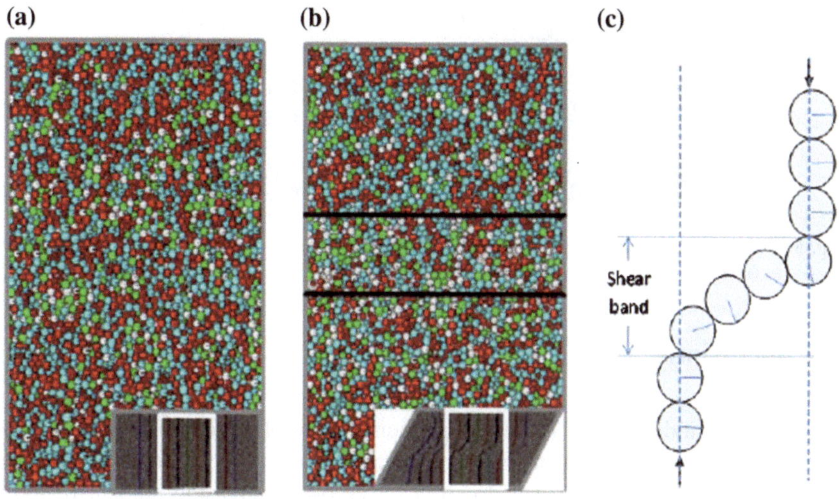

Fig. 4 Particle rotations at **a** $\gamma_{xy} = 0$ and **b** $\gamma_{xy} = 0.50$, and **c** magnified illustration of the particle rotations inside the shear band

the particles outside the shear band. As a result, the buckling of the columns and the shear band formation are accompanied by large difference in the magnitude of particle rotations inside and outside the shear band.

Another important observation is the formation of large voids within the shear band. Initially, before shearing, only small voids exist randomly all over the sample (Fig. 5a). After shearing, it can be seen that shear band formation, column buckling, reduced number of force chains, and particle rotations within the shear band, are associated with the formation of large voids within the shear band (Fig. 5b), while almost no voids are visible outside the shear band. The occurrence of areas of large voids in the case where particles have rolling resistance is attributed to the fact that particles in the presence of rotational moments tend to push each other apart, causing the formation of a "strong discontinuity." The rotational moment and particle rotations cause the granular assembly to expand and increase the overall volume locally in the shear band. In comparison, in the case of free rotation, particles are free to move locally and accommodate each other resulting in a more uniform distribution of porosity.

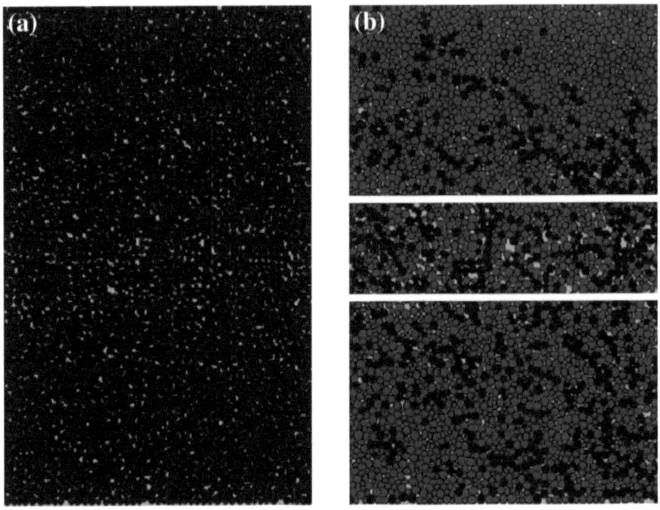

Fig. 5 Formation of voids inside the band which are not observed outside the band before shearing $\gamma_{xy} = 0.0$ (**a**), and after shearing $\gamma_{xy} = 0.50$ (**b**)

4 Conclusions

Based DEM modeling, the following phenomena are observed during shear band formation during simple shear loading of granular soils: (1) Columns of particles buckle in a narrow zone, (2) Load is transferred by a decreasing number of force chains as strains localize, (3) Particles rotate inside the shear band but not outside the band, and (4) Large voids are formed inside the band but not outside the band.

References

Cundall PA, Strack OL (1979) A discrete numerical model for granular assemblies. Geotechnique 29:47–65

Gutierrez M (2011) Effects of constitutive parameters on strain localization in sands. Int J Num Analy Meth Geomech 31:161–178

Itasca Consulting Group (2008) Particle flow code in 2D (PFC-2D): user's manual, version 4.0, Minneapolis, Minnesota

Mohamed A, Gutierrez M (2010) Comprehensive study of the effects of rolling resistance on the stress-strain and strain localization behavior of granular materials. Gran Matls 12:527–541

Oda M, Konishi J, Nemat-Nasser S (1982) Experimental micromechanical evolution of strength of granular materials: effects of particle rolling. Mech Mater 1:269–283

Wang J, Dove JE, Gutierrez M (2007) Discrete-continuum analysis of shear banding in the direct shear test. Geotechnique 57:513–526

Grainsize Evolution in Open Systems

Benjy Marks and Itai Einav

Abstract Granular flows are often characterised by spatial and temporal variations in their grainsize distribution. These variations are generally measured by geologists and geotechnical engineers after a flow has occurred, and two limiting states are commonly found; either a power law or log-normal grainsize distribution. Here, we use a lattice model to study how the grainsize distribution evolves in granular systems subject to grain crushing, segregation and mixing simultaneously. We show how the grainsize distribution evolves towards either of these grainsize distributions depending on the mechanisms involved in the flow.

1 Introduction

There exist a large number of physical processes in which granular materials advect in space, and simultaneously undergo changes to their grainsize distribution—whether spatial rearrangement, or by changing the physical size of constituent particles. These include geophysical flows, such as avalanches, rock falls, landslides (both submarine and subaerial), mud flows, pyroclastic flows, earthquake faulting and debris flows. Such flows are also present in many industrial processes aiming at size reduction (comminution), mixing, or de-mixing of powders, grains and ores.

Spatially, grains can either mix (Utter and Behringer 2004) or de-mix (for example size segregation is a relatively well studied de-mixing phenomenon) (Savage and Lun 1988). Spatial changes in the grainsize distribution over time have received significant attention from the granular materials community recently, generally for

B. Marks (✉)
Advanced Materials and Complex Systems, University of Oslo, 0371 Oslo, Norway
e-mail: benjy.marks@fys.uio.no

I. Einav
School of Civil Engineering, The University of Sydney, Sydney 2006, Australia
e-mail: itai.einav@sydney.edu.au

K.-T. Chau and J. Zhao (eds.), *Bifurcation and Degradation of Geomaterials in the New Millennium*, Springer Series in Geomechanics and Geoengineering,
DOI 10.1007/978-3-319-13506-9_6

the particular case of bi-mixtures (Savage and Lun 1988; Dolgunin and Ukolov 1995; Gray and Thornton 2005; Gray and Chugunov 2006).

Comminution has also been a hot topic of study, but in general this study has been limited to closed systems where particles are prohibited from advecting in space (Steacy and Sammis 1991; McDowell et al. 1996). This is very rarely the case in experiment, industry, or in the field. When particles crush, the void spaces near the crushed sites rearrange. This can cause compaction in some areas, which must inevitably cause dilation in others.

In a general sense, there exists a fundamental problem with comminution modelling in that for any crushing event, we expect a change in the local arrangement of neighbours surrounding the crushing site. Conversely, local crushing events are often attributed to different modes of fracture—which are dependent on the loading state of individual particles, and strongly dependent on the local arrangement of particles. If particle arrangement is of paramount importance, then for any continuum approach, there is a fundamental issue which needs to be addressed: How does one characterise the local arrangement of particles, well below the resolution of the continuum? In this work, previous comminution models (Steacy and Sammis 1991; McDowell et al. 1996) are extended to allow open systems to be studied, in conjunction with simple models of mixing and segregation (Dolgunin and Ukolov 1995; Gray and Thornton 2005; Gray and Chugunov 2006; Marks and Einav 2011; Marks et al. 2012).

2 Continuous Open Systems

We will describe here the evolution of a system of grains with no interstitial fluid, intrinsically involving grainsize distribution at every point in space. For any given volume V in space $\underline{r}= \{x, y, z\}$ and time t, there exists a subset which is filled with grains V_s. We can then describe the volume fraction Φ that is filled by a given grainsize range $(s_a, s_b]$ as $\Phi(s_a < s \leq s_b, \underline{r}, t) = V(s_a < s \leq s_b, \underline{r}, t)/V_s(\underline{r}, t)$, where $V(s_a < s \leq s_b, \underline{r}, t)$ is the volume occupied by the given grainsize range. Furthermore, we can define a probability density function $\Phi(s, r, t)$ such that its' integral over the grainsize coordinate is the volume fraction, $\Phi(s_a < s \leq s_b, \underline{r}, t) = \int_{s_a}^{s_b} \phi(s, \underline{r}, t)ds$. With these definitions, it is possible to define conservation of mass of such a system (Ramkrishna 2000), given that it is homogeneous (constant density ρ), as

$$\frac{\partial \phi}{\partial t} + \nabla \cdot (\phi \underline{u}) = h_b + h_d,$$

where $u(s, r, t)= \{u, v, w\}$ is the velocity of the material, $h_b(s, \underline{r}, t)$ is the birth rate, and $h_d(s, \underline{r}, t)$ is the death rate. These rates could, for instance, represent the birth of new particles of a given grainsize due to the death (fracture) of a larger particle. In this case, for some breakage rate $b\ (s, \underline{r}, t)$, the death rate could be expressed simply as

$$h_d = b\phi$$

and the corresponding birth rate would involve a summation over the deaths of all particles larger than a given size, as

$$h_b = \int_s^{\infty} P(s|s')b(s', \underline{r}, t)\phi(s', \underline{r}, t)ds',$$

where $P(s|s')$ is the probability that a volume of particles of grainsize s will fragment into grainsize s'. In a similar manner to conservation of mass, it is possible to define conservation of momentum for the same system (Dolgunin and Ukolov 1995; Gray and Thornton 2005; Gray and Chugunov 2006; Marks et al. 2012) such that

$$\rho \frac{D}{Dt}(\phi \underline{u}) = -\phi f \nabla \circ \sigma + \rho\phi \underline{g} - \frac{\rho\phi c}{\dot{\gamma}}\hat{\underline{u}} - d\nabla\phi,$$

where $\sigma(\underline{r}; t)$ is the size independent stress field, $f(s, \Phi$ is some grainsize dependent scaling of the stress, g is the acceleration due to gravity,) $\dot{\gamma}$ is the shear strain rate, c a coefficient of inter-particle friction, $\hat{u}(s, \underline{r}, t) = \underline{u}(s, \underline{r}, t) - \int \underline{u}\phi ds$ is the segregation velocity and d the diffusivity. This can be integrated over the full grainsize direction to describe the motion of the bulk (barycentric) flow, as $\rho \frac{D\overline{u}}{Dt} = -\nabla \cdot \sigma + \rho \underline{g}$, where we have required that $\int \phi f ds = 1$ so that conservation of bulk momentum satisfies the standard conservation law. Assuming that there is no bulk flow normal to the slope and that the segregative flow is slow, conservation of segregative momentum can be used to find the segregation velocity, $\hat{w}$, normal to the flow direction as

$$\hat{w} = |\dot{\gamma}|\frac{g\cos\theta}{c}(\frac{s}{\bar{s}} - 1) - \frac{d|\dot{\gamma}|}{c\phi}\nabla\phi,$$

where we have assumed that $f(s, \phi) = s/\bar{s}$, *and* $\bar{s} = \int \phi s ds$. We have considered here three mechanisms for grainsize change—mixing, segregation and crushing, as continuum phenomena. We will now present their analogous formulations as discrete phenomena in a simple lattice model.

3 Discrete Open Systems

3.1 Mixing

For the case of mixing of grains, there exists a strong, well established connection between the uncorrelated motion of individual constituents in a mixture, and large scale mixing of the system. This has been modelled extensively in a large variety of ways, and here we wish to describe a simple, two dimensional lattice to replicate this behaviour. Consider a system made of *NxM* cells, with position

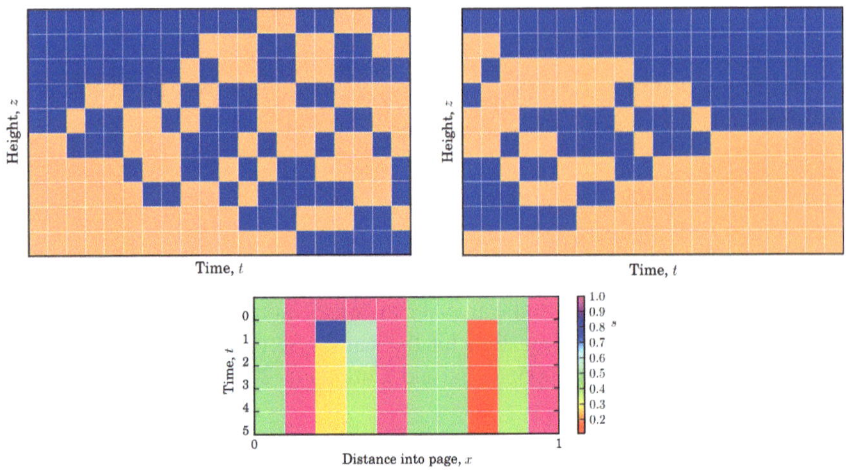

Fig. 1 *Top left* The mixing lattice model. Initially two phases are separated spatially, but over time they mix together. *Top right* The segregation lattice model. Initially two phases are well mixed (*blue* is large and *yellow* is small). Over time the two species de-mix such that the larger particles are spatially above the smaller ones. *Bottom* The crushing lattice model. Initially a bidisperse mixture is composed of two phases which crush over time such that local neighbours are of different size

$i, j = \{1...N, 1...M\}$ of equal volume, each filled by particles of a single grainsize $s_{i,j}$. The i direction represents a spatial coordinate, such as the perpendicular distance from the base of a flow towards the free surface. The j direction represents an internal coordinate which can be averaged over to describe the grainsize distribution. For this system, we can evolve in discrete time steps Δt. At each time, for each location $\{i, j\}$, a coin is flipped. Depending on the result, the grainsize is swapped with the cell either above or below it; Flip a coin, if heads: $s_{i,j} \Leftrightarrow s_{i-1,j}$, if tails: $s_{i,j} \Leftrightarrow s_{i+1,j}$. An example of this lattice model is shown in Fig. 1. By choosing an appropriate rate of swapping, this model is equivalent to Fickean diffusion (Chopard and Droz 1991; Utter and Behringer 2004), described by

$$\frac{\partial \phi}{\partial t} = \frac{d|\dot{\gamma}|}{c}\frac{\partial^2 \phi}{\partial z^2}.$$

3.2 Segregation

Segregation also describes advection, but now at a rate that is grainsize dependent (Savage and Lun 1988). We can capture this in a lattice model by swapping in a direction that depends on the local mean grainsize. Our rule of segregation is then:

$$\text{If}\left(s_{i,j} < \overline{s_{i,j}}\right) \wedge \left(s_{i,j} < s_{i-1,j}\right): \quad s_{i,j} \Leftrightarrow s_{i-1,j}.$$

$$\text{If}\left(s_{i,j} < \overline{s_{i,j}}\right) \wedge \left(s_{i,j} > s_{i+1,j}\right): \quad s_{i,j} \Leftrightarrow s_{i+1,j}.$$

Figure 1 shows an example of segregation, where an initially well mixed bidisperse system separates so that all of the large particles are arranged above to smaller ones. By carefully choosing the grainsize dependence of the rate of swapping, we can use this lattice model as a discrete representation for the following continuum equation,

$$\frac{\partial \phi}{\partial t} = \frac{g \cos \phi}{c} \frac{\partial}{\partial z} (\phi |\dot{\gamma}| (\frac{s}{\bar{s}} - 1).$$

3.3 *Crushing*

Many cellular automata have been motivated to describe crushing in granular materials, but generally only for closed systems (Steacy and Sammis 1991; McDowell et al. 1996). They typically rely on two competing mechanisms to facilitate realistic behaviour. Firstly, smaller particles are more difficult to crush—because of their inherent smaller volume, they contain less large cracks, increasing their crushing stress (Weibull 1951). Secondly, particles which are surrounded by neighbours of a similar size have in general a lower coordination number, and are therefore more likely to crush due to the anisotropy of their loading condition. These two mechanisms are included in the following simple rule for crushing

$$\text{If} \left| s_{i,j} - \overline{s_{i,j}} \right| \leq \beta s_{i,j} : \quad s_{i,j}(t + \Delta t) = X s_{i,j}(t),$$

where β is a non-dimensional coefficient controlling the crushability, and X represents an i.i.d. random variable drawn uniformly from the interval (0,1]. Using this rule we can obtain an example such as that pictured in Fig. 1. Other simple numerical models have shown that the grainsize distribution evolves towards a power-law from such rules (Steacy and Sammis 1991; McDowell et al. 1996).

Each of these mechanisms alters the grainsize distribution either spatially or locally. In the future, the choice of a lattice model as the basis for the model will allow us to simply combine these rules, and observe possible complex interactions which may arise.

4 Conclusion

A continuum description for the evolution of the grainsize distribution in open systems has been developed for the three mechanisms of mixing, segregation and crushing. Analogous lattice models have also been described, and examples of their operation have been shown. These complimentary views of open systems give insight into the evolution of the complex phenomena present in many industrial and geophysical granular flows.

References

Chopard B, Droz M (1991) Cellular automata model for the diffusion equation. J Stat Phys 64(3–4):859

Dolgunin V, Ukolov A (1995) Segregation modeling of particle rapid gravity flow. Powder Technol 83(2):95

Gray JMNT, Chugunov VA (2006) Particle-size segregation and diffusive remixing in shallow granular flows. J Fluid Mech 569:365

Gray JMNT, Thornton AR (2005) A theory for particle size segregation in shallow granular free-surface flows. Proc Roy Soc A 461(2057):1447–1473

Marks B, Einav I (2011) A cellular automaton for segregation during granular avalanches. Gran Matter 13(3):211–214

Marks B, Rognon P, Einav I (2012) Grainsize dynamics of polydisperse granular segregation down inclined planes. J Fluid Mech 690:499

McDowell G, Bolton M, Robertson D (1996) The fractal crushing of granular materials. J Mech Phys Solids 44(12):2079

Ramkrishna D (2000) Population balances: Theory and applications to particulate systems in engineering (Academic Press San Diego)

Savage S, Lun C (1988) Particle size segregation in inclined chute flow of dry cohesionless granular solids. J Fluid Mech 189:311

Steacy S, Sammis C (1991) An automaton for fractal patterns of fragmentation. Nature 353(6341):250

Utter B, Behringer RP (2004) Self-diffusion in dense granular shear flows. Phys Rev E 69(3):031308

Weibull W (1951) A statistical distribution function of wide applicability. J Appl Mech 18(3):293

Numerical Modeling of Failure Process in Cohesive Geomaterials

C. Yao, Q.H. Jiang and J.F. Shao

Abstract This paper is devoted to numerical modeling of failure process in anisotropic rocks. An extended Rigid Block Spring Method (RBSM) is proposed to describe the onset and propagation of cracks. The continuous rock material is replaced by an equivalent discrete assembly of rigid blocks using the Voronoi scheme. The material strength and deformation are described by an appropriate constitutive law and failure criterion of interfaces between blocks. The proposed method is applied to an anisotropic rock containing parallel bedding planes. The influence of the initial anisotropy on the failure process of material is investigated.

1 Introduction

The failure process of cohesive geomaterials such as rocks and concrete is mainly driven by the initiation, propagation and coalescence of microcracks. The failure mode is affected by the initial heterogeneity and microstructure of materials such as mineral inclusions, bedding planes etc. Further, the spatial distribution of microcracks is generally anisotropic. The classic macroscopic models are not able to take into account all these complex physical phenomena. The micromechanical models, formulated with homogenization techniques, tried to describe the consequences of microstructural evolution on the macroscopic behavior of heterogeneous materials. However, these models are generally obtained by some specific

C. Yao · J.F. Shao (✉)
LML, UMR8107 CNRS, University of Lille I, Villeneuve d'Ascq, France
e-mail: jian-fu.shao@polytech-lille.fr

C. Yao · Q.H. Jiang
School of Civil Engineering and Architecture, Nanchang University, Nanchang, China

J.F. Shao
College of Civil Engineering, Hohai University, Nanjing, China

K.-T. Chau and J. Zhao (eds.), *Bifurcation and Degradation of Geomaterials in the New Millennium*, Springer Series in Geomechanics and Geoengineering,
DOI 10.1007/978-3-319-13506-9_7

homogenization schemes which are related to strong simplifications in terms of material morphology. Mathematical difficulties exist when complex microstructures have to be considered. In order to evaluate the macroscopic behavior of geomaterials with complex microstructures, various numerical methods have been developed such as fast Fourier transformation (FFT), discrete element method (DEM), etc. The main idea is to perform the so-called complete field simulation of a representative volume element (RVE) of the material.

In the present work, we present the improvement and application of a rigid body spring method (RBSM), initially proposed by Kawai (1987). The cohesive geomaterial is considered as a composite material constituted by a number of polygonal discrete blocks which are cemented together through the interfaces. The spatial distribution of blocks can be either random (Voronoi diagram) or ordered according the microstructural analysis of materials (Fortune 1987; Nagai et al. 2004). Under the application of mechanical loads, some cemented interfaces are broken, inducing normal opening and tangential sliding. Compared with the initial method, a more realistic interface model is proposed in the improved RBSM presented in this work, allowing the description of progressive debonding of interfaces.

2 Fundamental Theory of the Modified Rigid Block Spring Method

Suppose an assembly of rigid blocks cemented by interfaces. Normal and tangential springs are uniformly distributed along the interfaces. There are three degrees of freedom for the centroid of each block, i.e. displacement in x direction u_{cx}, displacement in y direction u_{cy}, and a rotation θ_c. Assuming that only a small rotation can take place, the displacement of an arbitrary point on block $\{u\}$ can be expressed as:

$$\{u\} = [N]\{u_c\}, \{u\}^T = \left\{u_x, u_y\right\}, \{u_c\}^T = \left\{u_{cx}, u_{cy}\right\} \tag{1}$$

The interpolation function is given by:

$$[N] = \left\{ \begin{matrix} 1 \; 0 \; -(y-y_c) \\ 0 \; 1 \; (x-x_c) \end{matrix} \right\} \tag{2}$$

(x, y) are the global coordinates of one point on the block boundary while (x_c, y_c) are those of the block centroid. The relative displacements of one point on block interface are determined from those of the two neighboring blocks:

$$\{\Delta u\} = [B]([N]_1\{u_c\}_1 - [N]_2\{u_c\}_2) = [B][N]_{12}\{u_c\}_{12} \tag{3}$$

The vector $\{\Delta u\}^T = \{\Delta u_n, \Delta u_s\}$ denotes the relative displacements respectively in the normal and tangential direction. The matrix $[B]$ is defined by:

$$[B] = \begin{bmatrix} l & m \\ -m & l \end{bmatrix} \tag{4}$$

(l, m) are the components of the unit normal vector of the interface. The local stress is related to the relative displacement by the constitutive relation:

$$\{\sigma\} = \begin{Bmatrix} \sigma_n \\ \tau_s \end{Bmatrix} = [D]\{\Delta u\}, \quad [D] = \begin{bmatrix} k_n & 0 \\ 0 & k_s \end{bmatrix} \tag{5}$$

The parameters k_n and k_s are respectively the stiffness of normal and tangential springs. Using the Virtual Work Theorem and neglecting the body force, the following integral form stands:

$$-\sum_e \int_{l_0^e} \{\delta(\Delta u)\}^T \{\sigma\} dl = \sum_e \int_{l_\sigma^e} \{\delta u\}^T \{p\} dl \tag{6}$$

The variables l_0^e and l_σ^e respectively denote the interfaces and stress boundaries. $\{p\}$ is the applied loads. After mathematical operations, one obtains the global system of equilibrium equations:

$$[K]\{U\} = \{Q\} \tag{7}$$

$[K]$ is the global stiffness matrix, $\{U\}$ the global displacement vector and $\{Q\}$ the global force vector.

3 Constitutive Relation of Interface

A linear elastic law is employed to describe the mechanical behavior of an intact interface. Under compression (negative), the normal stress σ_n is given by:

$$\sigma_n = k_n \Delta u_n \tag{8}$$

The parameter k_n is the normal stiffness coefficient. In tension, the normal stress is still computed with the same normal stiffness but limited to the tensile strength T, such that:

$$\sigma_{nmax} = T \tag{9}$$

After the tensile strength is reached, the failure of interface occurs and the normal stress is set to zero. On the other hand, the shear stress is computed by:

$$\sigma_s = \{\sigma_s\}_{updated} + k_s \Delta u_s \tag{10}$$

$\{\sigma_s\}_{updated}$ is the updated shear stress at the end of previous loop. k_s is the tangential stiffness. Note that the stiffness coefficients k_n and k_s can be related to the macroscopic elastic properties (Yao 2013).

To describe the failure process of interface, a modified Mohr-Coulomb criterion is adopted (Fig. 1). The mechanical strength of interface is defined by the tensile strength T, the cohesion C, the frictional coefficient φ_1. Further, it is generally observed that the shear strength tends towards a stationary value for high normal stress due to the diminution of frictional effect. Therefore, we have introduced a

Fig. 1 Failure criterion for the modified RBSM

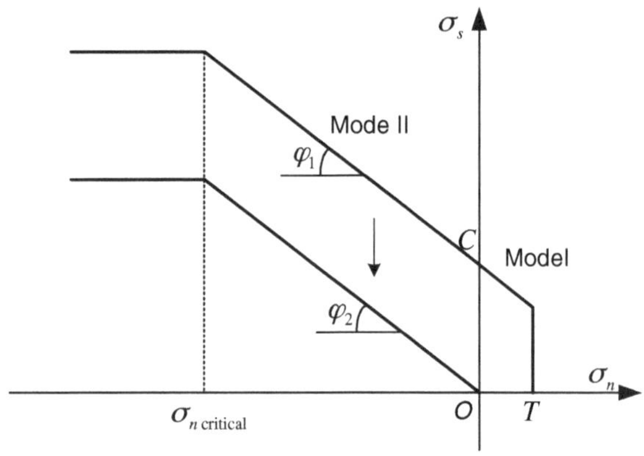

critical normal stress $\sigma_{n-\mathrm{critical}}$. The failure process of interface is described by the destruction of material cohesion and the possible diminution of frictional coefficient. Therefore, the residual shear strength of failed interface is described by the residual friction coefficient φ_2.

4 Effects of Bedding Planes on Macroscopic Failure Strength

We apply now the modified RBSM to modeling the mechanical behavior of a transversely isotropic rock. The structural anisotropy is due to the presence of parallel bedding planes. Due to the paper length limitation, only the macroscopic failure strength is discussed here. Consider first a uniaxial compression on a representative volume element (RVE) of rock. The objective here is to study the variation of the uniaxial compression strength (UCS) with the loading orientation. For this purpose, a series of parallel weak interfaces are introduced in the RVE to represent the bedding planes of rock. The loading orientation angle α is defined as the angle between the bedding planes and the loading axis. Various values of α are considered. The RVE size is 100 mm $\times$ 50 mm, the spacing of bedding planes is 12.5 mm and the width is 0.8 mm.

The effect of cohesion of bedding planes is first studied. For convenience, we have considered different ratios between the cohesion of bedding planes and that of intact interfaces (C_f/C). In Fig. 2, we show the variation of UCS with loading orientation. It is observed that the anisotropy of UCS is strongly influenced by the ratio C_f/C. When the value of C_f/C is smaller (very weak bedding planes), the anisotropy of UCS is stronger. But for all the values of C_f/C considered, the minimum of UCS is found for $\alpha = 30°$.

Consider now the influence of frictional angle. For the sake of simplicity, the initial friction coefficient φ_1 is equal to the residual one φ_2 for both intact

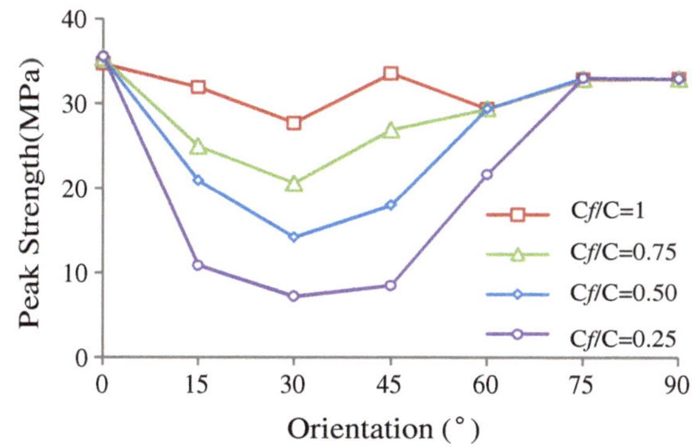

Fig. 2 Influences of cohesion of bedding planes on UCS anisotropy

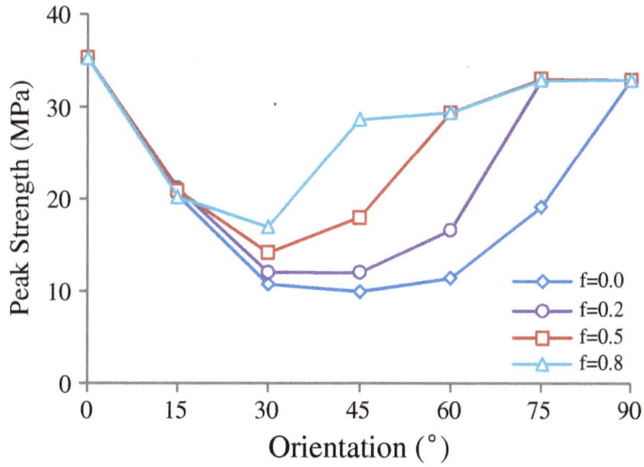

Fig. 3 Influences of frictional angle of bedding planes on UCS ($f = \tan \varphi_1 = \tan \varphi_2$)

interfaces and bedding planes. In Fig. 3, the variation of UCS with loading orientation is presented. Again, it is clear that the anisotropy of UCS depends on the frictional coefficient of bedding planes. The anisotropy of UCS is more significant with smooth bedding planes than for rough ones.

In Fig. 4, we show the variation of UCS with loading orientation for three different densities (spacing) of bedding planes, respectively 25, 12.5 and 6.25 mm. It is observed that the variation of UCS nearly not affected by the bedding planes spacing.

In order to verify the capability of the modified RBSM to reproduce the mechanical strength of anisotropic rocks, the numerical results are compared with typical experimental data. In Fig. 5, we show the comparisons of peak deviatoric stress under different confining pressures and with different loading orientations for a hard claystone (Niandou et al. 1997). It is observed that there is a good agreement between numerical results and experimental data.

In Fig. 6, we present the deformed configurations of RVE in triaxial compression tests with two different confining pressures. One can see that under low confining pressure, there is the onset and growth of many microcracks and a splitting

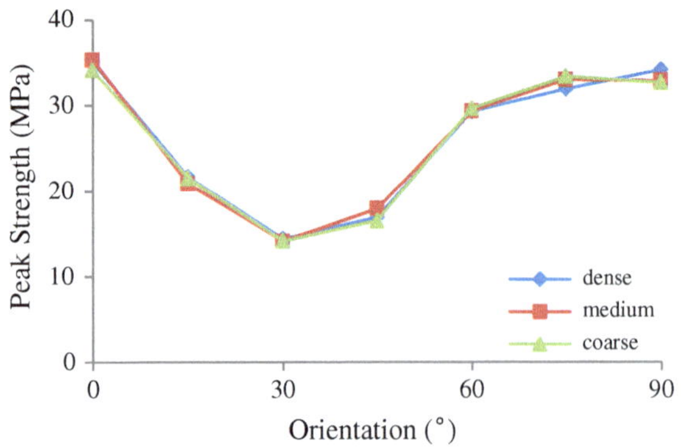

Fig. 4 Influence of bedding planes spacing on UCS anisotropy

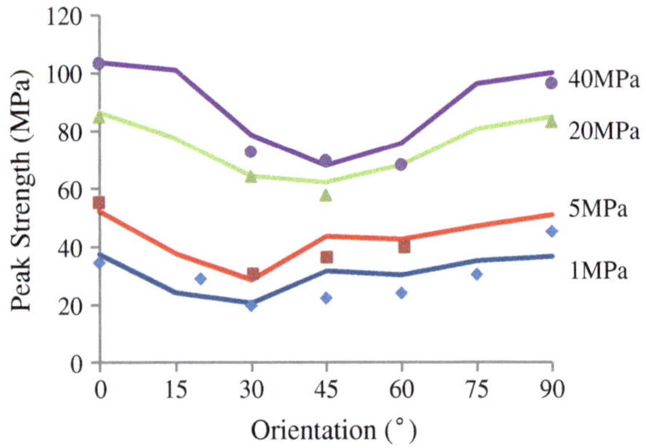

Fig. 5 Comparison of peak deviatoric stress between numerical results (*continuous lines*) and experiments under different confining pressures

Fig. 6 Deformed Specimen for $\alpha = 60°$ after peak strength under confining pressures 1 and 20 MPa (displacement in y direction ×10)

failure mode is obtained. However, when the confining pressure is higher, the failure is entirely controlled by the slipping along bedding planes.

5 Conclusion

A modified RBSM is proposed to describe the deformation and failure process of rock materials, based on a Voronoi diagram. The macroscopic behavior is controlled by the local deformation and failure criterion of interfaces. The proposed method is able to describe influences of microstructure on macroscopic responses. The propose model is applied to a transversely isotropic claystone. The numerical results have shown that the macroscopic strength is strongly affected by the bedding planes. The numerical results are in agreement with typical experimental data.

References

Fortune S (1987) A sweepline algorithm for Voronoi diagrams. Algorithmica 2:153–174
Kawai T (1987) New element models in discrete structural analysis. J Soc Naval Architects Japan 141:187–193
Nagai K, Sato Y, Ueda T (2004) Mesoscopic simulation of failure of mortar and concrete by 2D RBSM. J Adv Concr Technol 2(3):359–374
Niandou H, Shao JF, Henry JP, Fourmaintraux D (1997) Laboratory investigation of the mechanical behavior of tournemire shale. Int J Rock Mechanics Min Sci 34:3–16
Yao C (2013) Numerical modeling of hydromechanical coupling in fractured rock mass. Ph.D. thesis, University of Lille, France

Hydromechanical Modelling of an Initial Boundary Value Problem: Studies of Non-uniqueness with a Second Gradient Continuum

F. Marinelli, Y. Sieffert and R. Chambon

Abstract A non-uniqueness study for a hydromechanical boundary value problem is performed. A fully saturated porous medium is considered using an elasto-plastic constitutive equations to describe the mechanical behavior of the skeleton. A real hydromechanical experiment which consists in a hollow cylinder test on a Boom Clay sample is modelled. It is shown that the time step discretisation of the numerical problem has an effect on the initialisation of the Newton-Raphson algorithm on a given time step. Different solutions for the same initial boundary value problem can consequently be found.

1 Introduction

In the simulation of initial boundary value problems using constitutive equations for geomaterial behavior, it is well known that some difficulties can arise, particularly if degradation of the materials occurs. These problems have been studied for the case of single phase materials and some theoretical results have been established. Within the small strain assumption and for a rate problem, the uniqueness can be proved using the so-called Hill exclusion functional (Hill 1978). From a numerical standpoint, in case of non-uniqueness, the solution can be influenced by different numerical inputs such as the spatial discretization of the problem, the size of the time step, the mesh spacing, some tolerance values and so on (Chambon et al. 2001).

In this study, a real hydromechanical experiment is modelled using a finite element code Lagamine (University of Liège). A local hydromechanical second gradient model (see Collin et al. 2006) is considered to obtain regularised solutions of this initial boundary value problem. No other regularization is necessary in terms of the fluid phase or the solid-fluid coupling.

F. Marinelli (✉) · Y. Sieffert · R. Chambon
Laboratoire 3S-R, UJF, G-INP, CNRS, Grenoble Université Joseph Fourier,
B.P. 53, 38041 Grenoble Cedex, France
e-mail: ferdinando.marinelli@3sr-grenoble.fr

K.-T. Chau and J. Zhao (eds.), *Bifurcation and Degradation of Geomaterials in the New Millennium*, Springer Series in Geomechanics and Geoengineering,
DOI 10.1007/978-3-319-13506-9_8

We show that, changing the time discretisation of the numerical problem, different solutions can be found. In function of the shear bands path during the time, a classification of all computed solutions is proposed.

2 Local Hydromechanical Second Gradient Model

It is well known that classical continuum medium cannot be considered to model the localization of plastic strains. An internal length has to be introduced into the model in order to obtain mesh independent solutions of an initial boundary value problem. In this respect, the problem can be considered regularised. The regularization (that means the introduction of an internal length into the model) cannot restore the uniqueness of the solution as it is proved in Chambon et al. (1998) for a local second gradient continuum model.

In the framework of microstructure continuum (Germain 1973), a local hydromechanical second gradient model can be considered (Collin et al. 2006). The problem is solved by enforcing (in a weak form) the balance equations for the mixture Eq. (1), and the mass conservation for the fluid Eq. (2). Both equations are written in the current solid configuration denoted Ω^t (updated Lagrangian configuration).

$$\int_{\Omega^t} \left(\sigma_{ij}^t \frac{\partial u_i^*}{\partial x_j^t} + \overset{t}{\sum_{ijk}} \frac{\partial^2 u_i^*}{\partial x_j^t \partial x_k^t} \right) d\Omega^t = \int_{\Omega^t} \rho^{mix,t} g_i u_i^* d\Omega^t + \int_{\Gamma_\sigma^t} (\bar{t}_i u_i^*) d\Gamma^t, \quad (1)$$

$$\int_{\Omega^t} \left(\dot{M}^t p^* - m_i^t \frac{\partial p^*}{\partial x_i^t} \right) d\Omega^t = \int_{\Omega^t} Q^t p^* d\Omega^t - \int_{\Gamma_q^t} \bar{q}^t p^* d\Gamma^t. \quad (2)$$

In Eq. (1), Terzaghi's principle of effective stress is assumed to compute the Cauchy stress of the mixture σ_{ij}^t. The mass density of the mixture at the time t $\rho^{mix,t}$ is defined as follows:

$$\rho^{mix,t} = \rho^s \left(1 - \phi^t\right) + \rho^{w,t} \phi^t, \quad (3)$$

where ρ^s is the solid phase density (assumed to be incompressible, i.e., ρ^s is a constant); $\rho^{w,t}$ is the fluid density and ϕ^t is the porosity defined as $\phi^t = \Omega^{w,t} / \Omega^t$. Ω^t is the current volume of a given mass of the solid skeleton and $\Omega^{w,t}$ the corresponding volume of fluid. The fluid is assumed to be compressible. All variables appearing in Eqs. (1) and (2) are explained as follows:

- u_i^* is any kinematically admissible virtual displacement field.
- $\overset{t}{\sum_{ijk}}$ is the double stress dual of the virtual second micro kinematic gradient.
- x_i is the current coordinate.

- g_i is the gravity acceleration.
- $\bar{t}_i$ is the external (classical) force per unit area.
- m_i^t is the mass flow of water, which is governed by Darcy's law.
- $\dot{M}^t$ is the time derivative of the fluid mass of the skeleton.
- p^* is any kinematically admissible virtual pore pressure.
- Q^t is a sink term.
- Γ_q^t is the part of the boundary where the input fluid mass per unit area $\bar{q}^t$ is prescribed.

To solve the Eqs. (1) and (2) at time $\tau = t + \Delta t$ with a Newton-Rapshon algorithm a consistent linearisation of the system has to be performed. All the developments of the linearization for a local hydromechanical second gradient model are detailed in Collin et al. (2006).

The classical part of the constitutive equations, giving the effective stress of the skeleton, is described by a non-associated elasto-plastic model based on the Drucker-Prager yield field (Plasol model, Barnichon 1998). Cohesion c and friction angle for a compressive triaxial path φ, represent the hardening/softening variables for these elasto-plastic constitutive equations. The evolution of c and φ depends on the Von Mises equivalent plastic strain ε_{eq}^p, which represents the only internal variable of the model and describes degradation due to the shear. Dilatancy is considered as a constant during plastic flow. A hyperbolic variation of the friction angle and cohesion is proposed as hardening law. The elastic behavior is modelled by isotropic linear elasticity. These constitutive equations are computed with a return mapping algorithm (Ortiz and Simo 1986).

The other part (giving the double stress $\sum_{ijk}^t$) is a particular case of a more general isotropic linear relation derived by Mindlin (1964) that involves five independent parameters, in the interests of simplicity only one parameter (namely D, see Bésuelle et al. 2006) is used.

3 Non-uniqueness Studies

In order to perform our non-uniqueness studies, we use realistic material properties related to the behavior of Boom Clay (Horseman et al. 1987). In our study we choose to model a real hydromechanical hollow cylinder experiment performed in Lausanne (Labiouse et al. 2014). This experiment is divided in two phases, that are described in Fig. 1. A non-uniqueness study is carried out for Plasol constitutive equations by considering phase A of boundary conditions that represents a hydromechanical unloading. In phase B, the evolution of the shear bands with time is modelled. All computations presented are performed under plane strain hypothesis to model the behavior of a horizontal section of the hollow cylinder.

From a numerical point of view, non-uniqueness can be studied by changing the initialization of the Newton-Raphson algorithm for a given time step. In our case, the initialization of the numerical algorithm is obtained from the nodal velocity field v_j^{n-1} (that one of the previous step) at the first iteration of the current step.

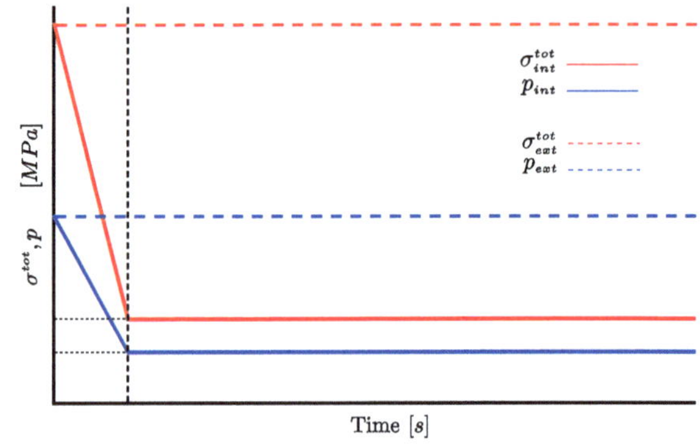

Fig. 1 The initial boundary conditions on the inner and external side of the hollow cylinder: a continuous line represents boundary conditions on the inner hollow cylinder, a dot line on the external hollow cylinder

$$x_j^{n^{i=1}} = x_j^{n-1} + v_j^{n-1}\Delta t, \tag{4}$$

where Δt represents the step size, x_j^{n-1} is the nodal coordinate at the previous step and $x_j^{n^{i=1}}$ is the nodal coordinate for the first iteration of the current step. Different sequences of time steps can cause a change in the initialisation of the numerical problem for each time step. In this way, non-uniqueness can be studied for the same initial boundary value problem.

In order to visualize the non-uniqueness of this initial boundary value problem, the evolution of the shear strains are shown in terms of the loading index. This quantity is defined as follows: if a Gauss point undergoes plastic loading, a small blue square is plotted, otherwise (in the case of elastic unloading or reloading of the Gauss point), a small yellow square is plotted.

Basically, all the localized deformation patterns present two kinds of shear band configurations: the first is characterized by some crossing shear bands (the so-called conjugated shear bands), while the second is characterized by spiral shear bands (the so-called periodic shear bands).

Among the great number of computations performed five solutions, which are representative of the non-uniqueness problem, are shown in Fig. 2. All the solutions are presented in Fig. 3 and can be classified qualitatively according to shear bands evolution as follows:

- Computations with radial symmetry ISO-0.
- Computations with shear bands that start from a thin plastified region around the inner hollow cylinder, which then grow towards the external part of the sample (solutions ISO-1 and ISO-4).
- Computations with shear bands that appear by an elastic unloading in a thick plastified zone of the hollow cylinder at the end of phase A (solutions ISO-2 and ISO-3).

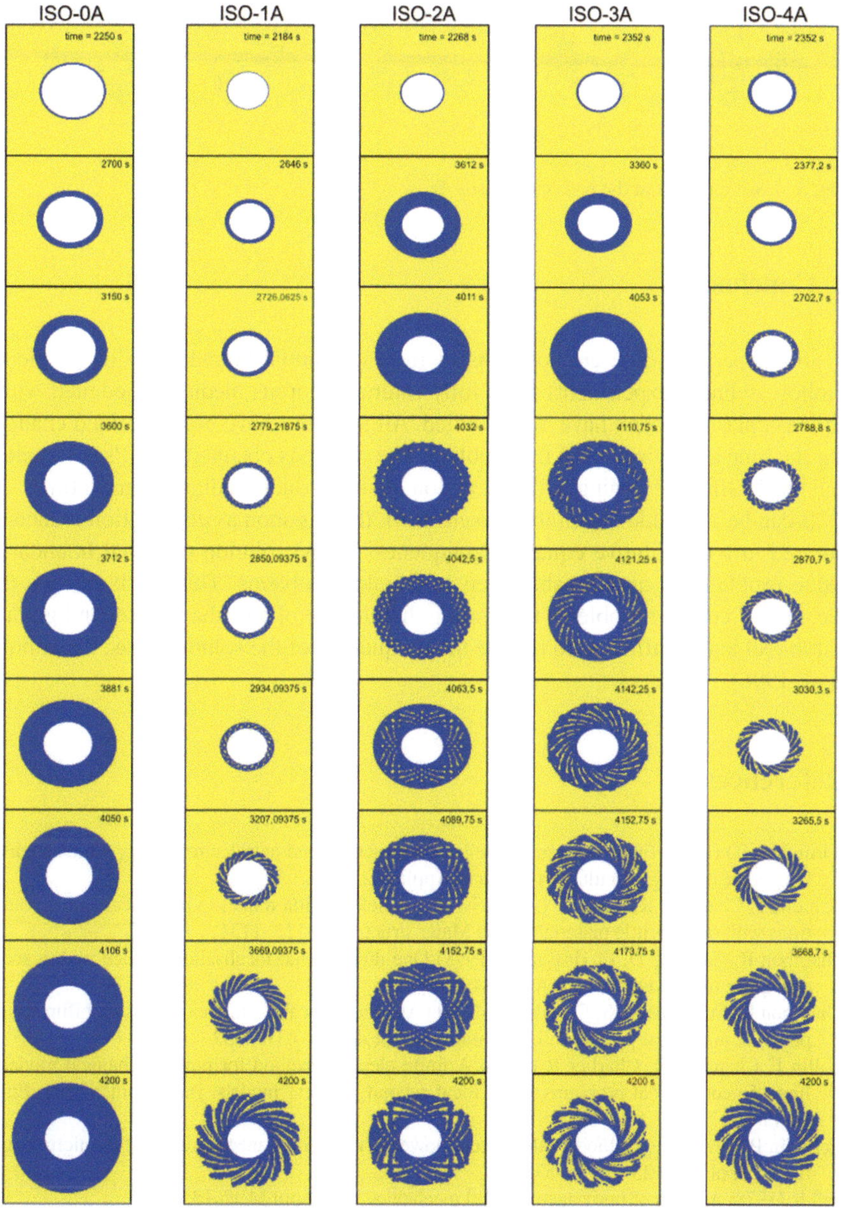

Fig. 2 Loading index of five solutions during phase A of boundary conditions

The development of the shear bands during the phase B of the boundary conditions is shown in Fig. 3.

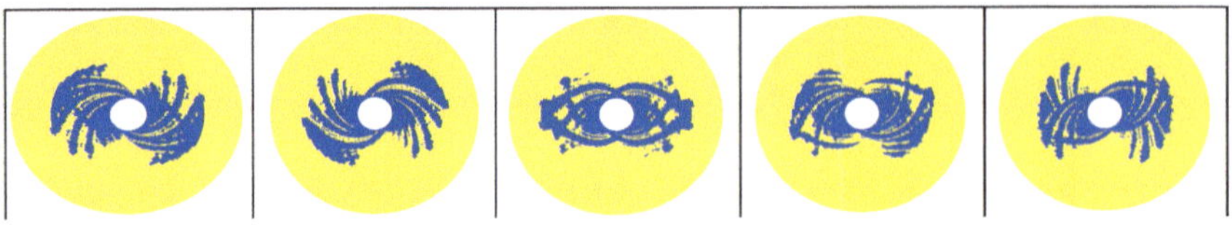

Fig. 3 Loading index at the end of the phase B

4 Conclusions

In this paper, the different solutions of the same initial boundary value problem (hollow cylinder experiment) for a fully saturated porous medium, modelled with an elastoplastic model, have been studied. All solutions have been obtained changing the time discretization of the problem. The solutions obtained with Plasol model can be classified according to the evolution of shear bands configurations in time.

It can be concluded from the present work that, as soon as degradation is incorporated in a constitutive equation, uniqueness of the solution of initial boundary value problems is questionable even in coupled problems. This work extends to the case of coupled problems the finding that the use of an enhanced second gradient model is not sufficient to restore the uniqueness of the solutions (see Chambon et al. 1998).

References

Barnichon JD (1998) Finite element modelling in structural and petroleum geology. Ph.D. thesis, Université de Liège Faculté des Sciences Appliquées

Bésuelle P, Chambon R, Collin F (2006) Switching deformation modes in post-localization solutions with quasi-brittle material. Mech Mater struct 1(7):1115–1134

Chambon R, Caillerie D, El Hassan N (1998) One-dimensional localisation studied with a second grade model. Eur J Mech A Solids 4:637–656

Chambon R, Crochepeyre S, Charilier R (2001) An algorithm and a method to search bifurcation points in non-linear problems. Int J Numer Meth Eng 51(3):315–332

Collin F, Chambon R, Charlier R (2006) A finite element method for poro mechanical modelling of geotechnical problems using local second gradient models. Int J Numer Meth Eng 65:1749–1772

Germain P (1973) The method of the virtual power in continuum mechanics. Part 2: microstructure. J Appl Math 25(3):556–575

Hill R (1978) Aspects of invariance in solid mechanics. Adv Appl Mech 18:1–75

Horseman ST, Winter MG, Entwistle DC (1987) Geotechnical characterization of boom clay in relation to the disposal of radioactive waste (page 87). Technical report, Commission of the European Communities, EUR 10987

Labiouse V, Sauthier C, You S (2014) Hollow cylinder simulation experiments of galleries in boom clay formation. Rock Mech Rock Eng 47:43–45

Mindlin RD (1964) Micro-structure in linear elasticity. Arch Ration Mech Anal 16(1):51–78

Ortiz M, Simo JC (1986) An analysis of a new class of integration algorithms for elastoplastic constitutive relations. Int J Numer Meth Eng 23(3):353–366

A FE^2 Model for Hydro-mechanical Coupling

B. van den Eijnden, F. Collin, P. Bésuelle and R. Chambon

Abstract A new approach is investigated for the modelling of the hydro-mechanical behaviour of Callovo-Oxfordian claystone, a potential host rock for radioactive waste repositories in France. This approach is a double-scale finite element method with a micro and a macro scale. At the micro level a representative elementary volume (REV) is used to model the material behaviour. The global response of this REV serves as an implicit constitutive law for the macro scale. On the macro scale, a poro-mechanical continuum is defined with fully coupled hydro-mechanical behaviour; the microscale contains a model that takes into account the material micro structure and fluid/solid interaction to provide the material responses and associated stiffness matrices. Computational homogenization is used to retrieve these stiffness matrices from the micro level. This double scale approach is applied in the simulation of a biaxial deformation test and the response at the macro level is related to the micro-mechanical behaviour.

1 Introduction

The principle of deep geological repositories for the disposal of high-level and intermediate-level long-lived radioactive wastes relies among others on the low permeability of the host rock. As the permeability is influenced by mechanical alteration of the material, the coupling between hydraulic and mechanical

B. van den Eijnden
Andra, 1/7 Rue Jean Monnet, Chatenay-Malabry, France

F. Collin
ArGEnCo, Université de Liège, Liège, Belgium

P. Bésuelle (✉) · R. Chambon
Laboratoire 3SR, CNRS/UJF-Grenoble 1/Grenoble INP, Grenoble, France
e-mail: pierre.besuelle@3sr-grenoble.fr

K.-T. Chau and J. Zhao (eds.), *Bifurcation and Degradation of Geomaterials in the New Millennium*, Springer Series in Geomechanics and Geoengineering,
DOI 10.1007/978-3-319-13506-9_9

behaviour of the host rock is an important factor in the study of radioactive waste disposal. This behaviour on the engineering scale is controlled by micromechanical material behaviour. In this context a fully-coupled double-scale hydromechanical model is developed in which a macro scale is defined at engineering level and a micro scale at the material level. The micro scale computations provide the local material behaviour for the macro scale. This requires the microscale computations to capture the micromechanical phenomena of the material by means of a representative elementary volume (REV). Homogenization of the REV response to enforced deformation provides the constitutive relations for the macro scale computation.

Different homogenization techniques exist to retrieve the global response from the REV. In this work, computational homogenization by static condensation (Kouznetsova et al. 2001) is applied on periodic REVs taking into account the hydro-mechanical couplings. Finite element methods are used to solve both the microscale and the macroscale problem using the so-called finite element squared (FE^2) method.

This paper gives a summary of the double scale coupled model. The scale transition by computational homogenization for hydromechanical coupling is discussed after which an example is given of the application in the modelling of a biaxial compression test with localization. The results of this experiment form the basis of the concluding remarks.

2 Macro Scale Balance Equations

On the macro level, a poromechanical continuum under quasi-static assumption is defined. To obtain objective (mesh independent) results for localization problems the macroscale continuum is enriched using a local second gradient model (Matsushima et al. 2002; Mindlin 1964). This allows decomposing the constitutive relation in a classical first gradient part (derived from the micro scale), and a second gradient part that links a double stress Σ_{ijk} to the second gradient of displacement. The balance equations for this continuum in domain Ω can be written using the principle of virtual work by any two kinematically admissible virtual fields of displacement u_i^* and fluid pressure p^* (Collin et al. 2006) as

$$\int_{\Omega} \left(\sigma_{ij} \frac{\partial u_i^*}{\partial x_j} + \Sigma_{ijk} \frac{\partial^2 u_i^*}{\partial x_j \partial x_k} \right) d\Omega - \int_{\Gamma} \left(\bar{t}_i u_i^* + \bar{T}_i \frac{\partial u_i^*}{\partial x_j} n_j \right) d\Gamma = 0 \tag{1}$$

$$\int_{\Omega} \left(m_j \frac{\partial \bar{p}^*}{\partial x_j} - \dot{M} p^* \right) dV - \int_{\Gamma} \bar{q} p^* dS = 0 \tag{2}$$

with σ_{ij} the Cauchy total stress, Σ_{ijk} the double stress, $\bar{t}_i$ the external traction, $\overline{T}_i$ the external double traction, $\dot{M}$ the time derivative of the fluid mass per unit volume, m_j the fluid mass flux and $\bar{q} = m_i n_i$ the fluid mass flux over boundary Γ with outward normal vector n_i. A finite element method is used to solve the field equations using a Newton-Raphson iterative scheme (Collin et al. 2006) to deal with material nonlinearity. A linearization of the classical part of the field equations around the updated configuration is required to find the test solution in the subsequent iteration. This linearization that approximates the constitutive behaviour for small increments $\delta[\cdot]$ around the updated configuration is given by the 7×7 consistent tangent stiffness matrix (CTSM) in Eq. (3) and relates the seven unknowns in (1) and (2) to σ_{ij}, m_i and $\dot{M}$.

The second gradient constitutive relation used here is elastic and isotropic, containing one parameter D[N] (Matsushima et al. 2002) implicitly controlling the width of possible localization bands.

$$\begin{bmatrix} C_{ijkl} & A_{ijl} & B_{ij} \\ D_{ikl} & E_{il} & G \\ H_{kl} & K_l & L \end{bmatrix} \begin{Bmatrix} \frac{\partial \delta u_k}{\partial x_l} \\ \frac{\partial \delta p}{\partial x_l} \\ \delta p \end{Bmatrix} = \begin{Bmatrix} \delta \sigma_{ij} \\ \delta m_i \\ \delta \dot{M} \end{Bmatrix} \tag{3}$$

3 Micro Scale Model

On the micro scale, a model for hydromechanical coupling (Frey et al. 2013) is used. A linear-elastic constitutive law is used for the solids, thereby concentrating possible degradation in the interfaces. Cohesive forces T_n and T_t between the solid grains act normally and tangentially to the interface orientation as a function of the history of relative displacement of the opposite interface boundaries Δ. Two separate damage laws are used to describe these forces (Fig. 1).

In addition to controlling the material damage the interfaces form a pore channel network that is fully saturated. Fluid transport is controlled by the conductivity of the channels formed by the interface openings and the fluid pressure distribution. Fluid pressure and pressure gradient within the interfaces lead to fluid normal

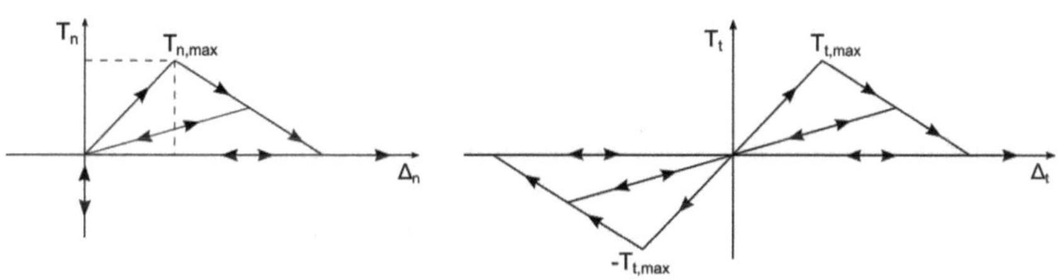

Fig. 1 Damage law for interface cohesive forces. Normal cohesive forces T_n and tangential cohesive forces T_t are a function of respectively relative normal displacement Δ_n and relative tangential displacement Δ_t of the opposite interface boundaries

forces and drag forces acting on the solids. In this way, hydromechanical coupling is controlled by the interface openings. Fluid compressibility k^w is taken into account by defining the fluid density ρ^w as a function of fluid pressure p:

$$\dot{\rho}^w = \frac{\rho^w}{k^w}\dot{p}, \quad \rho^w(0) = \rho_0^w \tag{4}$$

Laminar flow between smooth parallel plates is used as the model for hydraulic flow in the interfaces, which leads to a cubic relation between the interface hydraulic conductivity κ and interface normal opening Δ_n. Once the fluid problem is solved, the fluid effort on the grains is calculated and the mechanical system is solved, leading to new configuration with new channel conductivities. In this way the coupled problem is solved in an iterative way. Details on the micro model and a full development of the numerical procedures is given by Frey et al. (2013).

4 Micro-to-Macro Transition

Considering a balanced REV with periodic boundary conditions, the macro responses σ_{ij}^M, M^M and m_i^M are found by homogenization over REV domain V_Ω:

$$\sigma_{\mathrm{ij}}^{\mathrm{M}} = \frac{1}{\mathrm{V}_\Omega}\int\limits_\Omega \sigma_{ij}\, dV \tag{5}$$

$$M^M = \frac{1}{V}\int\limits_\Omega \rho^w dV \tag{6}$$

$$m_{\mathrm{i}}^{\mathrm{M}} = \frac{1}{\mathrm{V}_\Omega}\int\limits_\Omega m_i dV = \frac{1}{\mathrm{V}_\Omega}\int\limits_\Gamma \bar{q}x_j dS \tag{7}$$

The macro pressure p^M is defined as the average of the micro fluid pressure p. The macroscale constitutive behaviour is the incrementally nonlinear relation between the seven unknowns in Eqs. (1) and (2) and the seven REV response variables of Eqs. (5–7), given by the 7 × 7 CTSM in Eq. (3). For purely mechanical problems first order computational homogenization by static condensation (Kouznetsova et al. 2001) is often used to derive the CTSM. An extension of this method for the specific HM-coupled micromechanical model presented above is used here (Eijnden et al. 2013). This method contains a two-step homogenization that first reduces the fluid degrees of freedom from the micro scale global system of equations and second condenses of the remaining equations into (3).

5 Strain Localization in Biaxial Compression

Biaxial compression is a commonly-used modelling exercise to study localization problems. Here a biaxial test on a 0.5 m × 1.0 m sample with drainage on top and bottom under a compressive loading rate of 10^{-8} m/s is modelled using a 10 × 20 element mesh. A second gradient parameter D is chosen such that mesh independent results are obtained for the element size and classical constitutive relation that are used here. A 1 mm × 1 mm microscale REV with a periodic microstructure consisting of 4 grains is chosen (Fig. 2). A minimum permeability is guaranteed by defining a minimum hydraulic opening of the interfaces of 2.0 μm. The normalized reaction force on the sample and the fluid pressure in the center of the sample during compression are given in Fig. 2. To demonstrate the accuracy and efficiency of the computation homogenization for hydromechanical coupling, the simulation is done twice using respectively computational homogenization and numerical perturbation to derive the CTSM.

Figure 3a shows the normalized convergence criterion; the convergence obtained with computational homogenization and numerical perturbation are of the same quality, which demonstrates the consistency of the computational homogenization.

Although computation time efficiency is simulation-specific and influenced by many factors, it is the authors experience that, at least for small microstructures, computational homogenization is more efficient than numerical perturbation. Strain localization in a single band (Fig. 3a) takes place as an effect of evolving damage of the interfaces. After 1.5 % axial compression, no solutions can be found at the microlevel, where complete decohesion of the interfaces takes place in several integration points of the corner elements (Fig. 3b). Because most of the deformation of the microstructure is through sliding and normal opening

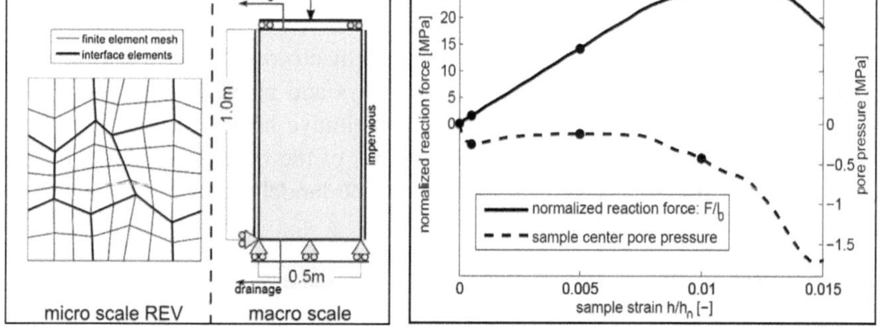

Fig. 2 **a** *Left* Microstructure with interfaces and macroscale boundary conditions. *Right* Reaction force to loading and pore pressure at sample center

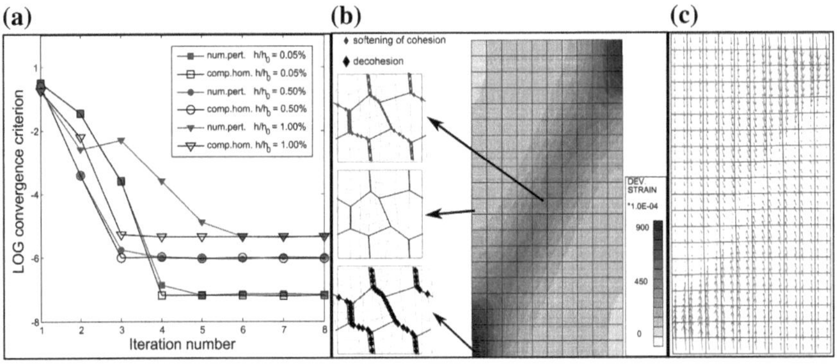

Fig. 3 **a** Convergence of the macroscale Newton-Raphson scheme. **b** Macro deviatoric strain and deformed microstructures. Macro strains are coupled with interface damage and decohesion. **c** Fluid flow at end of test, strongly related to the volumetric strain rate

of interfaces, the macro scale behaviour is mainly dilatant. This dilatancy leads to a pore volume increase in the localization band. As can be seen from Fig. 3c the fluid flow tends to be controlled by pore volume change, following the strain localization in the band.

6 Concluding Remarks

Modelling localization in a biaxial compression test with the presented FE^2 method demonstrates the suitability of the double scale model for hydromechanical coupling. The use of computational homogenization for hydromechanical problems proves to be an efficient technique for homogenization from microstructure to macro continuum. Although the complexity of the microstructural REV remains low and the micromechanical constitutive laws are simplistic, the model is able to capture complex nonlinear material behaviour and allows controlling anisotropy via the microstructure. However, the macro response remains a superposition of the micromechanical constitutive laws and physical phenomena need to be incorporated in the micromechanical constitutive laws in order to appear in the macroscopic response. Further developments of the microscale model including plastic and frictional behaviour will allow the modelling of a more advanced macro response.

Acknowledgments The support from the French national radioactive waste management agency (Andra) is gratefully acknowledged.

References

Collin F, Chambon R, Charlier R (2006) A finite element method for poro mechanical modelling of geotechnical problems using local second gradient models. Int J Numer Meth Eng 65(11):1749–1772

Eijnden Bvd, Collin F, Bésuelle P, Chambon R (2013) A double scale modelling approach for hydromechanical coupling. In: Proceedings of the third international symposium on computational geomechanics (COMGEO III), Kraków

Frey J, Chambon R, Dascalu C (2013) A two-scale poromechanical model for cohesive rocks. Acta Geotech 8(2):1–18

Kouznetsova V, Brekelmans W, Baaijens F (2001) An approach to micro-macro modeling of heterogeneous materials. Comput Mech 27(1):37–48

Matsushima T, Chambon R, Caillerie D (2002) Large strain finite element analysis of a local second gradient model: application to localization. Int J Numer Meth Eng 5(4):499–521

Mindlin R (1964) Micro-structure in linear elasticity. Arch Ration Mech Anal 16(1):51–78

Numerical Simulation and Length Scale Calibration of Micropolar Hypoplastic Model

Jia Lin and Wei Wu

Abstract A new micropolar hypoplastic model is developed with a complex tensor formulation. The model is implemented in a finite element code to simulate plane strain tests. The relationship between shear band width and internal length is studied, which can be used to calibrate of the length scale in the micropolar hypoplastic model.

1 Micropolar Hypoplastic Model

Granular materials consist of discrete grains. In spite of their discrete nature, granular materials can be reasonably well described by continuum theories. In general, conventional continuum theory without a characteristic length is sufficient for most engineering problems, where the domain size of interest is much larger than the mean grain size. However, if the domain size is of the order of the mean grain size, the underlying boundary-value problem shows scale dependence to some extent. A good example is the formation of shear band in granular materials (Mühlhaus and Vardoulakis 1987). There are several approaches to endow the constitutive equation with a characteristic length, e.g. micropolar theory, strain gradient theory and nonlocal continuum.

Hypoplastic constitutive equations for granular materials based on nonlinear tensorial functions are of current research interest. The original model did not contain a characteristic length (Wu and Kolymbas 1990). Micropolar hypoplastic constitutive model was proposed in Tejchman and Bauer (1996). A weakness of this model is that the constitutive equation between moment stress, curvature and their rates is obtained by simple analogy to the original stress-strain constitutive equation. Further justifications of this model are given by Huang et al. (2002, 2014).

J. Lin (✉) · W. Wu
Institute of Geotechnical Engineering, University of Natural Resources and Life Sciences, Vienna, Austria
e-mail: jia.lin@boku.ac.at

K.-T. Chau and J. Zhao (eds.), *Bifurcation and Degradation of Geomaterials in the New Millennium*, Springer Series in Geomechanics and Geoengineering,
DOI 10.1007/978-3-319-13506-9_10

Recently, an alternative approach was suggested by Goddard (2010) based on the tensor representation in complex numbers (Xiao 1998). In the present paper we attempt to establish the additional constitutive equation with the help of complex tensors. Consider the hypoplastic constitutive equation,

$$\dot{\sigma} = \mathrm{L} : \dot{\varepsilon} + \mathbf{N}\ \|\dot{\varepsilon}\|$$

where $\dot{\sigma}$ is the stress rate tensor, $\dot{\varepsilon}$ is the strain rate tensor, L is a linear function of the stress tensor and **N** is a nonlinear function of the stress tensor. We extend the stress and strain rate tensors to general tensors by adding the couple stress **m** and curvature rate $\dot{\kappa}$ to the imaginary part, in unify the units in the general tensors, a characteristic length scale need to be added in the general form.

$$\sigma' = \sigma + i\frac{\mathbf{m}}{l}, \dot{\varepsilon}' = \dot{\varepsilon} + il\dot{\kappa}$$

in which σ' and $\dot{\varepsilon}'$ are general stress and strain rate tensors and l is the characteristic length.

Therefore the general Jaumann stress rate tensor is:

$$\dot{\sigma}' = \dot{\sigma} + i\frac{\dot{\mathbf{m}}}{l}$$

The norm of the transformed strain rate is:

$$\left\|\dot{\varepsilon}'\right\| = \sqrt{\dot{\varepsilon} : \dot{\varepsilon} + l^2\dot{\kappa} : \dot{\kappa}}$$

Hence, the conversional hypoplastic constitutive equations can be easy changed into complex form:

$$\dot{\sigma}' = \dot{\sigma} + i\frac{\dot{\mathbf{m}}}{l} = \mathrm{L}(\sigma') : \dot{\varepsilon}' + \mathbf{N}(\sigma')\left\|\dot{\varepsilon}'\right\|$$

The ensuing equations are largely simplified by assuming that the tensors of stress, strain and their rates are symmetric, the tensors of couple stress, curvature and their rates are anti-symmetric. The equation has a real part and an imaginary part, therefore it can be separated into a stress rate equation and a couple stress rate equation.

$$\dot{\sigma} = \mathbf{H}_1(\sigma, \dot{\varepsilon}, \mathbf{m}, \dot{\kappa}, l)$$

$$\dot{\mathbf{m}} = \mathbf{H}_2(\sigma, \dot{\varepsilon}, \mathbf{m}, \dot{\kappa}, l)$$

This model is simpler than the existing micropolar hypoplastic model (Tejchman and Bauer 1996). There is only one additional parameter (the characteristic length) to be determined.

The complex tensor formulation is not restricted to the hypoplastic model. It can also be applied to other tensor formed constitutive models to obtain micropolar constitutive models.

Fig. 1 FEM simulations of plane strain tests, internal length 0.1, 0.25, 0.5, 0.75, 1.0 mm

2 FEM Implementation

This new model is implemented to a 2d FEM element in ABAQUS, each node has three degree of freedom (two translation and one rotation). Plane strain tests are simulated. The model has a height of 100 mm and widths of 25 mm. Elements have a friction angle of 35° and a Young's module of 20 MPa. The mesh size is 0.1 by 0.1 mm. However, it will be shown that the mesh size do not have big effects on the simulation results.

For isotropic loading, the bottom nodes are fixed in the vertical direction. The other three sides of the model is isotropically loaded with a pressure of 100 kPa. The moment on all the boundary nodes are set to be zero. In the biaxial loading, a constant displacement is applied on the nodes on the top of model.

The simulation results for different internal lengths are shown in Fig. 1. Gray scales are used to show the nodal rotation in radian.

It can be seen from the results that the rotation within the shear band can be simulated with the new micropolar hypoplastic model. The shear band width depends on the internal length as expected.

3 Mesh Independency

For classical continuum models, due to the lack of characteristic length, the result of FEM simulation depends on the size of mesh. In a micropolar continuum, the result of numerical simulation should be independent on the mesh size. This also holds for the model proposed above. In order to prove this, simulations with the same characteristic length but different mesh size are carried out. Mesh size 0.05 and 01 mm are used for fine and coarse mesh. The results of both simulations are shown in Fig. 2. It can be seen that both simulations predict the same shear band

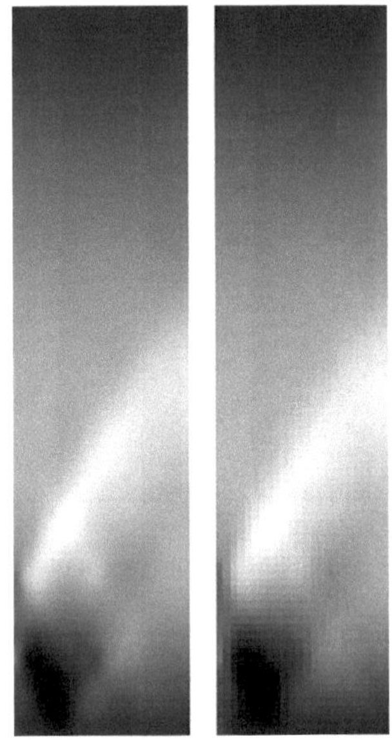

Fig. 2 FEM simulation results with fine mesh (*left*) and coarse mesh (*right*)

width for the same characteristic length. Hence, the FEM results of the micropolar hypoplastic model do not depend on the mesh size.

4 Length Scale Calibration

With the simulation results in Fig. 1, the length scale of the new micropolar hypoplastic model can be calibrated, as shown in Fig. 3.

A linear relationship between the internal length and shear band width is shown by the FEM results. Theoretically, the shear band width should be zero if the internal length is zero. However this cannot be realized either in numerical calculations or in physical experiments. Hence, according to the simulations, the shear band width has a starting value (about 4 mm), which is the shear band width when

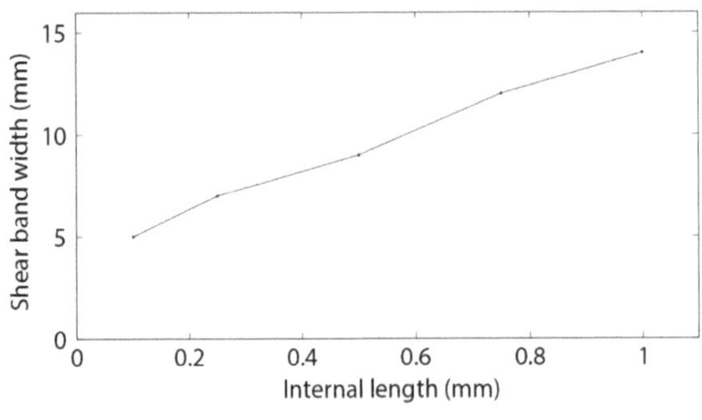

Fig. 3 Relationship between internal length and shear band width

characteristic length approaches zero. Taking away the stating value, the shear band width is about ten times the internal length. Hence, the internal length can be determined with:

$$l = (l_s - 4\ \text{mm})/10$$

where l_s is the shear band width.

5 Conclusions

A new micropolar hypoplastic model is proposed in this paper, which is obtained with the help of complex tensor formulations. The model is simple and only has one additional material parameter. By implementing this model to FEM and simulating plane strain tests, the relationship between internal length and shear band width is discussed, which can be used for the calibration of the length scale in the micropolar hypoplastic model.

References

Goddard JD (2010) Granular hypoplasticity with Cosserat effects. AIP Conf Proc 1227:323–332

Huang W, Nübel K, Bauer E (2002) Polar extension of a hypoplastic model for granular materials with shear localization. Mech Mater 34(9):563–576

Huang W, Sloan SW, Sheng D (2014) Analysis of plane Couette shear test of granular media in a Cosserat continuum approach. Mech Mater 69(1):106–115

Mühlhaus HB, Vardoulakis I (1987) The thickness of shear bands in granular materials. Géotech 37(3):271–283

Tejchman J, Bauer E (1996) Numerical simulation of shear band formation with a polar hypoplastic constitutive model. Comput Geotech 19(3):221–244

Wu W, Kolymbas D (1990) Numerical testing of the stability criterion for hypoplastic constitutive equations. Mech Mater 9(3):245–253

Xiao H (1998) On symmetries and anisotropies of classical and micropolar linear elasticities: a new method based upon a complex vector basis and some systematic results. J Elast 49(2):129–162

Micromechanical Analysis of Second Order Work in Granular Media

F. Nicot, N. Hadda, F. Bourrier, L. Sibille, A. Tordesillas and F. Darve

Abstract This paper examines the evolution of instabilities in granular materials from a microscopic point of view, using the discrete element method. The relationship between the unstable loading directions that result in negative values for the macroscopic expression of the second order work and the terms in the microscopic expression of the second order work was investigated. Good agreement was found between the microscopic and macroscopic expressions of the second order work for the two three-dimensional specimens studied, in both the elastic and plastic tensorial zones. The vanishing of the microscopic and macroscopic second order work is shown to coincide in both specimens.

1 Introduction

The second order work (Hill 1958) can be either expressed through the spatial description (Eulerian formalism) by assuming small deformations and neglecting geometrical aspects (Nicot and Darve 2007; Nicot et al. 2011) as $W_2 = \int_V \delta\sigma_{ij}\, \delta\varepsilon_{ij}\, dV$, $\overline{\overline{\sigma}}$ the Cauchy stress tensor, and $\overline{\overline{\varepsilon}}$ is the strain tensor.

F. Nicot (✉)
IRSTEA Geomechanics Group ETNA, Grenoble, France
e-mail: francois.nicot@irstea.fr

N. Hadda
Department of Civil Engineering, University of Calgary, Calgary, Canada

F. Bourrier
Irstea—Centre de Grenoble, BP 76, 38402 Saint-Martin d'Hères, France

L. Sibille · F. Darve
UJF-INPG-CNRS, Laboratoire Sols Solides Structures Risques, Grenoble, France

A. Tordesillas
Department of Mathematics and Statistics, University of Melbourne, Victoria, Australia

K.-T. Chau and J. Zhao (eds.), *Bifurcation and Degradation of Geomaterials in the New Millennium*, Springer Series in Geomechanics and Geoengineering,
DOI 10.1007/978-3-319-13506-9_11

It was shown (Nicot et al. 2012) that the second order work can be expressed in terms of the microscopic variables accounting for the microstructure of the material. Considering a granular assembly composed of N particles 'p' ($1 \le p \le N$), the second order work from microscopic variables writes

$$W_2^{\mu} = \sum_{p,q} \delta f_i^c \; \delta l_i^c + \sum_{p \in V} \delta f_i^p \; \delta x_i^p \tag{1}$$

where $\bar{l}^c$ is the branch vector relating the centers of contacting particles p and q, $\bar{f}^c$ the inter-particle contact force, $\bar{x}^p$ the position of particle 'p' and $\bar{f}^p$ the resultant force applied to the particle 'p'.

The numerical simulations performed herein are kinematically controlled. The strain rate is kept low, and inertial effects can hardly occur. The whole analysis can be done in quasi-static regime and the contribution of the term $\sum_{p \in V} \delta f_i^p \; \delta x_i^p$ becomes negligible (Hadda et al. 2013). Equation (1) therefore reduces to

$$W_2^{\mu} = \sum_{p,q} \delta f_i^c \; \delta l_i^c \tag{2}$$

when significant particle rearrangements occur (i.e., sliding between particles, contact loss and contact creation), the second term may no longer be negligible and may be an important contributor to W_2^{μ}. In this manuscript, however, we mainly focus on the comparison between macro and micro second order works and on the numerical validation of Eq. (2).

2 The Discrete Element Model

The vanishing of the second order work was proven to be related (among others) to both porosity and confining pressure of the assembly. In order to cover as many cases as possible and generalize the results found in this paper independently from such parameters, numerical simulations were conducted on two three-dimensional specimens of densely (porosity = 0.38) and loosely (porosity = 0.42) compacted assemblies of particles S_1 and S_2 respectively. We used the software 'Yade' (Šmilauer et al. 2010) based on the discrete element method (Cundall and Strack 1979).

The motion of the particles is governed by the elongation of both a normal and a tangential spring created at the initial contact point of each pairwise overlapping particles. In the normal direction, the contact constitutive relation is linear elastic and it assumes that the normal component f_n^c of the contact force is linearly related to the particle overlap through the normal contact stiffness denoted by k_n. The tangential component f_t^c of the contact force involves similarly the tangential contact stiffness denoted by k_t and the tangential relative displacement, which can be calculated by integrating the relative tangential velocity v_t in the contact plane during the lifetime of the overlap. The friction is incorporated at the contact level by the Coulomb friction law, thus f_t^c must obey the constraint $\left| f_t^c \right| \le \mu \, f_n^c$, where μ is the coefficient of friction. When this limit is reached, the tangential relative motion is regarded as

Table 1 Physical and mechanical parameters of the dense and loose specimens S1 and S2

	k_n/D (MPa)	k_t/k_n	φ_c (°)	Density (kg/m^3) (ρ)	Void ratio (e)	Coordination number (z)	κ ($\kappa = k_n/D\,p$) (kPa)
S_1	356	0.42	35	3,000	0.636	4.43	1,780 ($p = 200$)
S_2	356	0.42	35	3,000	0.71	4.22	3,560 ($p = 100$)

sliding with a friction force μf_n^c directed opposite to the tangential relative velocity. The two specimens considered are cubical in shape and contain 10,000 spherical particles of uniform radius distribution ranging from 2 to 12 mm enclosed within six rigid frictionless walls. The parameters in the contact constitutive relation are chosen such that $k_n/D = 356$ (MPa) and $k_t/k_n = 0.42$, where D is the mean diameter of the two particles in contact. Inter-particle friction angle ϕ_c is set to 35°. The characteristics and mechanical parameters of both specimens are detailed in Table 1 Both assemblies were compacted from initially sparse arrangements of particles to an isotropic state by increasing particles sizes until the desired isotropic pressure ($\sigma_1 = \sigma_3 = 200$ kPa for S_1 and $\sigma_1 = \sigma_3 = 100$ kPa for S_2) is reached. They are then subjected to an axisymmetric drained triaxial compression.

The evolution of both deviatoric stress $q = \sigma_1 - \sigma_3$ and volumetric strain ε_v versus the axial strain ε_1 are shown in Figs. 1 and 2 respectively for both specimens.

3 Second Order Work from Macroscopic Variables

In order to compute the second order work from macroscopic variables, three stress states defined by their deviatoric stress ratio $\eta = \frac{q}{p}$ (where p is the mean pressure in the specimen) are considered (represented by the points (A_1, B_1, C_1) and (A_2, B_2, C_2) in Fig. 1 for dense and loose specimens respectively). These arbitrary stress states are chosen before the Mohr-Coulomb condition is reached and correspond to values of η (see Table 2) smaller than η_{peak} for S_1 and $\eta_{plateau}$ for S_2 as illustrated in Fig. 1. In particular, A_1 and A_2 correspond to the isotropic state for each specimen.

The stress states defined above will constitute the initial states from which strain probes are performed. These strain probes consist in a series of strain loading increments $\Delta\bar{\varepsilon}$ defined in the Rendulic plane of strain increments by their direction α_ε (Fig. 1; varying here from 0 to 360° with a step of 5°) and their norm $\|\Delta\bar{\varepsilon}\|$ (set here equal to 5×10^{-5}). Once the stress response $\Delta\bar{\sigma}$ is computed for each strain loading increment, the macroscopic normalized second order work is computed as $W_2^n = \Delta\bar{\varepsilon} \cdot \Delta\bar{\sigma} / \|\Delta\bar{\varepsilon}\| \cdot \|\Delta\bar{\sigma}\|$ (Laouafa and Darve 2002), for all investigated strain directions α_ε and for all considered stress states. Figure 3 shows circular diagrams of second order work, plotted in the Rendulic plane of strain increments for the deviatoric stress ratios corresponding to (A_1, B_1, C_1) for S_1 and (A_2, B_2, C_2) for S_2.

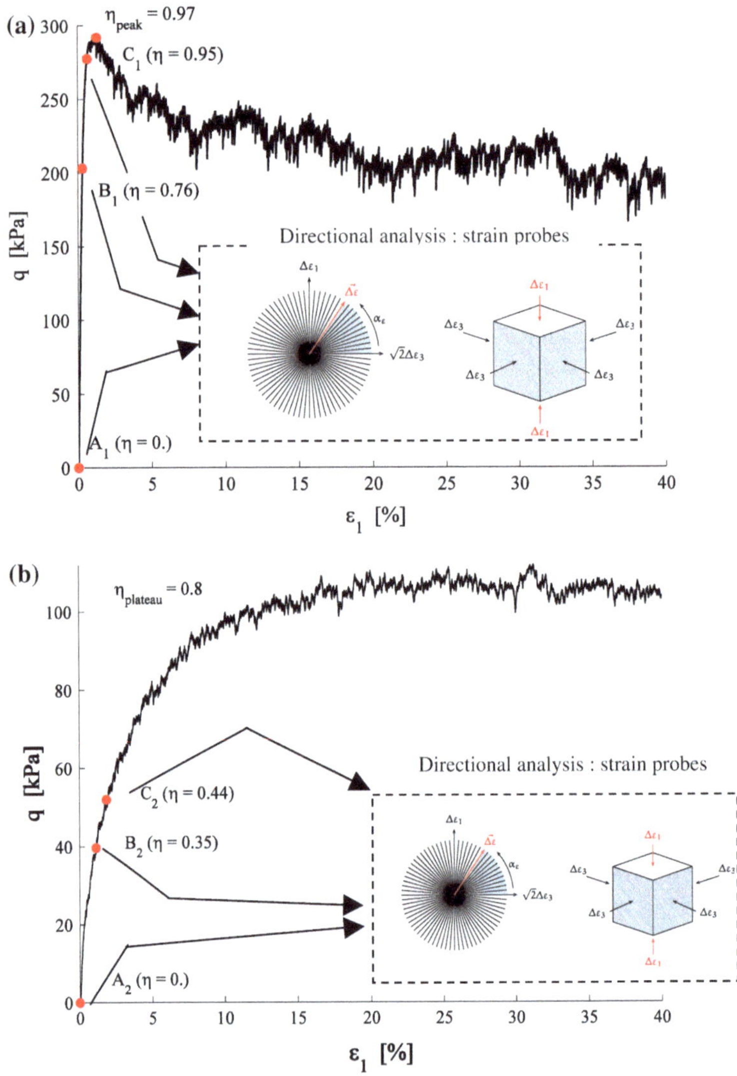

Fig. 1 Deviatoric stress in terms of axial strain at a confining pressure of 200 kPa for S1 (**a**), and 100 kPa for S2 (**b**)

For the readability of these representations, a constant $\ell = 0.5$ is added to the normalized value of W_2^n. Consequently, when W_2^n is negative the plot is inside the dashed circle of radius $r = \ell$, whereas the plot is outside the dashed circle for positive values of W_2^n. Cone of unstable strain directions (illustrated by the gray hatched zone and grouping the incremental strain directions along which the

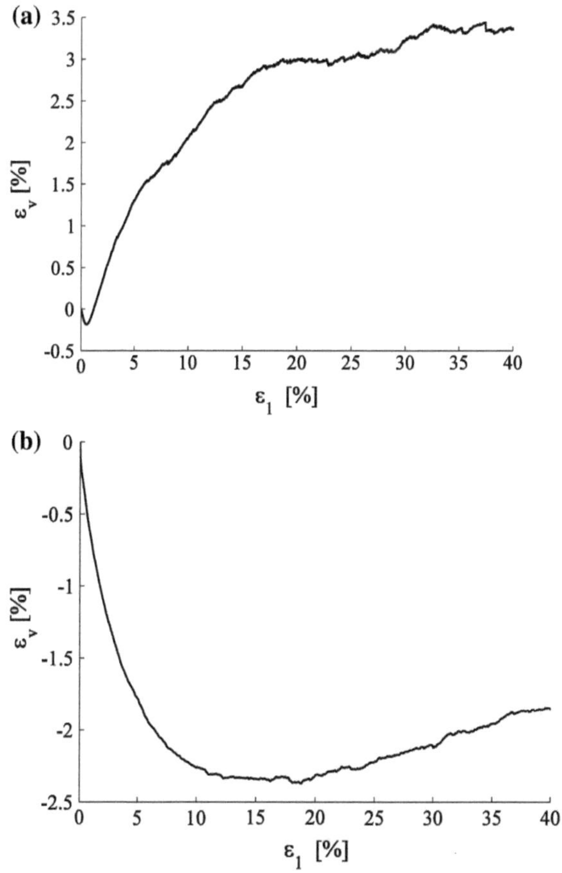

Fig. 2 Volumetric strain in terms of axial strain at a confining pressure of 200 kPa for S_1 (**a**), and 100 kPa for S_2 (**b**)

Table 2 Deviatoric stress ratio η corresponding to the stress states A1, B1 and C1 for S1 and A2, B2 and C2 for S2

	Specimen S_1			Specimen S_2		
	A_1	B_1	C_1	A_2	B_2	C_2
η	0	0.76	0.95	0	0.35	0.44

second order work takes negative values) are met in the second quadrant of the axisymmetric strain increment plane defined by $(-\sqrt{2}\Delta\varepsilon_3, \Delta\varepsilon_1)$ at deviatoric stress ratios $\eta = 0.95$ and $\eta = 0.44$ for S_1 and S_2 respectively.

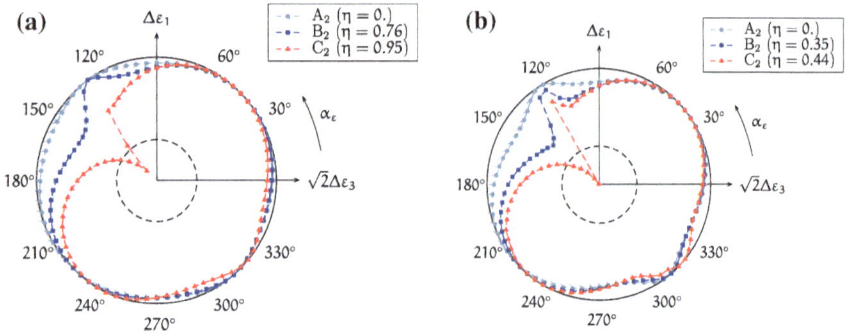

Fig. 3 Circular diagrams of the normalized second order work for dense specimen (**a**) and loose specimen (**b**)

4 Second Order Work from Microscopic Variables

The second order work can also be computed according to Eq. (2) from contact forces $\bar{f}^c$ and position vectors of particles $\bar{x}^p$. Figure 4 shows the evolution of the second order work (not normalized) from both macroscopic and microscopic variables at $\eta = 0.95$ and $\eta = 0.44$ for S1 and S2 respectively in terms of the direction of the strain loading increment α_ε.

A good agreement is found between the microscopic and macroscopic expressions of the second order work for the two cases examined, in the elastic tensorial zone as well as in the plastic tensorial zone[1] (the zone in light grey in Fig. 4), including the cone of instability (the dashed zone).

Similar results are found for other deviatoric stress states (not displayed here), showing that discrete element method agrees well with the theoretical development. As a starting point in the microscopic investigation, interest should be focused on the evolution of the number of contacts with negative values of $\Delta f_i^c \Delta l_i^c$ along strain probes and their spatial distribution in the specimen (Hadda et al. 2013).

5 Conclusions

Using discrete element simulations, the relationship between the unstable loading directions (characterized by negative values of the macroscopic expression of the second order work) and the elementary terms constituting the microscopic

[1] The plastic tensorial zone groups all directions along which the response to the strain probe has a predominant plastic component compared to elastic one.

Fig. 4 Comparison between the second order works from both macroscopic and microscopic variables for dense (**a**) and loose specimen (**b**)

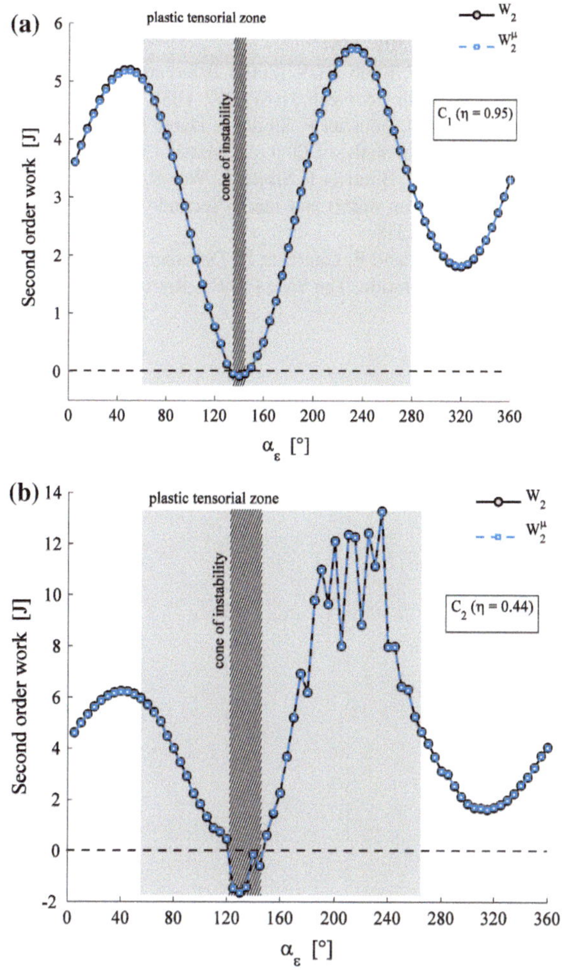

expression of the second order work (i.e. $\Delta f_i^c \Delta l_i^c$) was investigated. It was confirmed that the microscopic and macroscopic expressions of second order work are equivalent over a representative elementary volume, undergoing quasi-static deformation.

References

Cundall PA, Strack ODL (1979) A discrete numerical model for granular assemblies. Geotech 29(1):47–65

Hadda N, Nicot F, Bourrier F, Sibille L, Radjai F, Darve F (2013) Micromechanical analysis of second-order work in granular media. Granular Matter 15(2):221–235

Hill R (1958) A general theory of uniqueness and stability in elastic-plastic solids. J Mech Phys Solids 6(3):236–249

Laouafa F, Darve F (2002) Modelling of slope failure by a material instability mechanism. Comp Geotech 29(4):301–325

Nicot F, Darve F, Koah HDV (2007) Bifurcation and second-order work in geomaterials. Int J Num Anal Meth Geomech 31(8):1007–1032

Nicot F, Hadda N, Bourrier F, Sibille L, Darve F (2011) Failure mechanisms in granular media: a discrete element analysis. Granular Matter 13(3):255–260

Nicot F, Hadda N, Bourrier F, Sibille L, Wan R, Darve F (2012) Inertia effects as a possible missing link between micro and macro second-order work in granular media. Int J Solids Struct 49(10):1252–1258

Šmilauer V, Catalano E, Chareyre B, Dorofeenko S, Duriez J, Gladky A, Kozicki J, et al (2010) Yade documentation, The Yade Project. Retrieved from http://yade-dem.org/doc/

Stability and Bifurcation in Reinforced Boreholes

Euripides Papamichos

Abstract Rock nailing of a borehole is considered as a means to reinforce the rock and increase its borehole failure strength. The technique is modeled using Cox's original shear-lag method. In the continuum sense, nail reinforcement is viewed as a body force that acts as a confinement. The finite element formulation of the reinforced elastoplastic borehole is presented and results from an elastic analysis with stiff nails are compared against existing analytical solutions for elastic rocks with rigid nails. The formulation couples the mechanical behavior of the rock with the nail-rock contact law equation that have to be solved in a coupled manner. The bifurcation condition for borehole failure in a Cosserat elastoplastic rock is subsequently extended to a reinforced borehole for stability analysis.

1 Introduction

Nail reinforcement of a free surface or a borehole has been analyzed by Papamichos and Vardoulakis (2012) and Papamichos (2011), respectively. Elastic solutions with infinitely stiff nails were presented where the action of nails was modeled in a continuum sense through a distributed body force. In the following, the distributed body force approach is used to analyze elasto-plastic boreholes and investigate their stability. The formulation for analysis with the Finite Element Method (FEM) analysis is presented and it is followed with verification results for an elastic borehole with rigid nails where analytical solutions exist. Borehole

E. Papamichos (✉)
Department of Civil Engineering, Aristotle University of Thessaloniki, 54124 Thessaloniki, Greece
e-mail: epapamic@civil.auth.gr

E. Papamichos
SINTEF Petroleum Research, 7465 Trondheim, Norway

K.-T. Chau and J. Zhao (eds.), *Bifurcation and Degradation of Geomaterials in the New Millennium*, Springer Series in Geomechanics and Geoengineering,
DOI 10.1007/978-3-319-13506-9_12

stability is analyzed through the bifurcation condition for borehole failure. The satisfaction of the bifurcation condition in boreholes means that at that state shear band localizations or spalling failures are possible. The formulation of the bifurcation condition is performed for a Cosserat elastoplastic continuum which has internal length and thus allows for wave length selection of the warping of the borehole and hole size effect on the critical load.

2 Formulation

A vertical borehole of radius r_i is considered under initial vertical σ_V and horizontal σ_H in situ stresses. The radial stress at the borehole is reduced by $\sigma_{ri} > 0$ to simulate drilling of the borehole. In the vertical direction, plane strain is assumed since a long wellbore is considered. The wellbore is reinforced with radial nails of length L_n in a regular pattern with nail density per unit borehole surface m_n. Figure 1 shows a horizontal section of the considered borehole. The reinforcement creates a reinforced region I of radius $r_n = r_i + L_n$ and an unreinforced region II without nails. According to Papamichos (2011), the action of the radial nails corresponds to a distributed radial body force f_r given as

$$f_r = \frac{m_n r_i}{r} \frac{dN_n}{dr} \tag{1}$$

where N_n is the force on the nail. Following Cox (1952), it is assumed that

$$dN_n \big/ dr = M(u_n - u_r) \quad \text{for } r_i \leq r \leq r_n \tag{2}$$

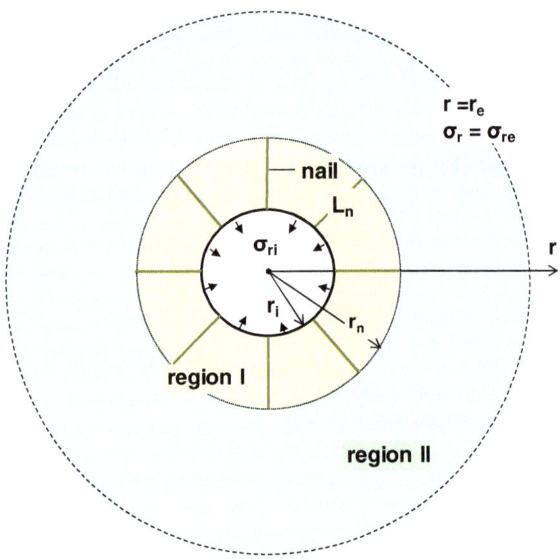

Fig. 1 Horizontal section of a vertical wellbore of radius r_i reinforced with radial nails of length L_n for a reinforced zone of radius r_n

where u_n is the displacement of the nail, u_r the radial displacement of the rock and M a contact law parameter with dimension of stress. According to this law, the added force transmitted from the nail to the rock and vice versa due to their stiffness difference, is linearly related to their relative displacement. For elastic nails, the nail force and displacement are related as

$$N_n = E_n A_n du_n / dr \quad \text{for } r_i \leq r \leq r_n \tag{3}$$

where E_n is the nail Young's modulus and A_n the nail's cross-sectional area.

2.1 Primary Loading

In the primary loading path the deformations are axisymmetric and thus shear stresses and Cosserat couple stresses vanish. Thus the solution can be significantly simplified and become 1 − d with dependence of the quantities only along the radial direction. An external boundary at radius r_e is assumed, with $r_e \gg r_i$, where an external stress $\sigma_{re} = \sigma_H$ is applied. The finite element formulation is based on the variational equation within a volume V with boundary ∂V, which in axisymmetric problems and plane strain conditions can be written in the cylindrical coordinate system (r, θ, z) as

$$\int_V (\delta\varepsilon_{rr}\sigma_{rr} + \delta\varepsilon_{\theta\theta}\sigma_{\theta\theta})dV - \int_V \delta u_r f_r dV = \int_{\partial V_\sigma} \delta u_r t_r dS \tag{4}$$

where σ_{rr} is the radial and $\sigma_{\theta\theta}$ the tangential stress, dual in energy to the radial ε_{rr} and tangential $\varepsilon_{\theta\theta}$ strains, respectively and t_r is a prescribed traction on a part ∂V_σ of the boundary ∂V. Moreover, $\delta\varepsilon_{rr}$, $\delta\varepsilon_{\theta\theta}$ and δu_r are variations of the radial and tangential strains and the radial displacement u_r, respectively. Equation (4) can be simplified and written in matrix form as

$$\int_{r_i}^{r_e} \{\delta\varepsilon\}^T\{\sigma\} r dr - \int_{r_i}^{r_e} \delta u_r f_r r dr = -\delta u_r|_{r=r_i}\sigma_{ri} r_i + \delta u_r|_{r=r_e}\sigma_{re} r_e \tag{5}$$

where the strain and stress variation vectors are defined, respectively, as $\{\delta\varepsilon\}^T = \{\delta\varepsilon_{rr}, \delta\varepsilon_{\theta\theta}\}$ and $\{\sigma\}^T = \{\sigma_{rr}, \sigma_{\theta\theta}\}$.

Since the static equations for the mechanical behavior are coupled to the deformation of the nail, the contact law equation Eq. (2) is required to solve the problem of a reinforced continuum for the unknown rock and nail displacements. The variational form of Eq. (2) can be obtained in the standard Galerkin manner as

$$\int_{r_i}^{r_n} \delta u_{n,r} N_n dr + \int_{r_i}^{r_n} \delta u_n M(u_n - u_r) dr = 0 \tag{6}$$

where δu_n is the variation of nail displacement and account has be taken for the zero nail force boundary conditions at the hole and at r_n.

In an elastoplastic analysis, incremental and iterative methods are used to solve these equations, where the total external loads are added in increments step by step. Using the incremental elastoplastic constitutive relations for the solid $\{d\sigma\} = [C^{ep}]\{d\varepsilon\}$, the nail Eqs. (3) and (1) for the body force, Eqs. (5) and(6) at the $(m+1)$th loading step can be written as

$$\begin{aligned}
&\int_{r_i}^{r_e} \{\delta\varepsilon\}^T \left[C^{ep}\right] \{d\varepsilon\} r dr + \int_{r_i}^{r_e} \delta u_r m_n r_i M (du_r - du_n) dr \\
&={}^{m+1} \left[\delta u_r \big|_{r=r_i} \sigma_{ri} r_i + \delta u_r \big|_{r=r_e} \sigma_{re} r_e\right] \\
&\quad - \int_{r_i}^{r_e} \{\delta\varepsilon\}^{T\,m} \{\sigma\} r dr + \int_{r_i}^{r_e} \delta u_r^m f_r r dr \\
&\quad - \int_{r_i}^{r_n} \delta u_n M (du_r - du_n) dr + \int_{r_i}^{r_e} \delta u_{n,r} E_n A_n du_{n,r} dr \\
&= - \int_{r_i}^{r_n} \delta u_{n,r}^m N_n dr - \int_{r_i}^{r_n} \delta u_n^m f_r r / (m_n r_i) dr
\end{aligned} \tag{7}$$

Figure 2 compares numerical results for loading of an elastic reinforced borehole with the analytical solution for rigid nails (Papamichos 2011). In the numerical solution stiff nails with En = 2,000 GPa were used in order to be able to compare the results. The rock is a soft sandstone with Young's modulus E = 0.5 GPa.

2.2 Hole Failure

The FEM formulation of the bifurcation condition for lateral borehole failure (Papamichos 2010) in a Cosserat elastoplastic rock is extended here to a nail reinforced elastoplastic borehole. The condition is formulated on the basis that in addition to the trivial solution of cylindrical convergence of the hole during the primary loading path, there exists another non-trivial warping solution that fulfils homogeneous boundary conditions. Thus, the bifurcation problem for the borehole failure can be formulated with the homogeneous and incremental form of the virtual work equation for a Cosserat continuum in a nail reinforced medium

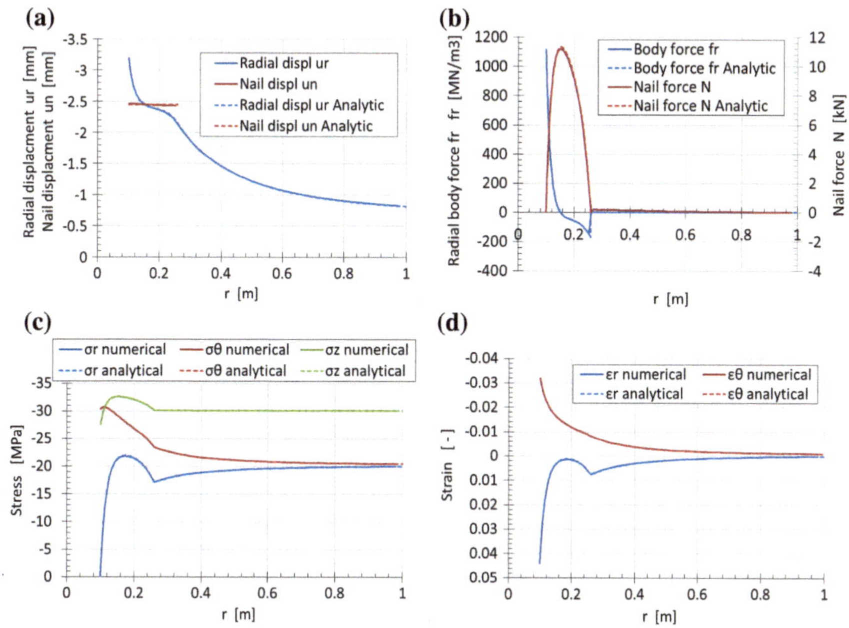

Fig. 2 Reinforced borehole **a** Radial displacement and nail displacement, **b** Body force and nail force, **c** Radial, tangential and axial stress profiles, and **d** Radial and tangential stress profiles. Compression is negative

$$
\begin{aligned}
&\int\limits_V \{\delta d\hat{\varepsilon}\}^T \{d\hat{\sigma}\} dV - \int\limits_V \{\delta d\hat{u}\}^T \left\{d\hat{f}\right\} dV = 0 \\
&\int\limits_{r_i}^{r_n} \delta d\hat{u}_{n,r} d\hat{N}_n dr + \int\limits_{r_i}^{r_n} \delta d\hat{u}_n M \left(d\hat{u}_n - d\hat{u}_r\right) dr = 0
\end{aligned} \tag{8}
$$

where $\{d\hat{\sigma}\}$ and $\{d\hat{\varepsilon}\}$ are the generalized stress and strain increment vectors, respectively, that correspond to the non-trivial bifurcation solution (Papamichos 2010). The bifurcation condition for lateral borehole failure is obtained by solving Eq. (8) for the non-trivial, plane-strain, displacement and micro-rotation increment field

$$
\begin{aligned}
&d\hat{u}_r(r,\theta) = V_r(r)\cos m\theta, \quad d\hat{u}_\theta(r,\theta) = V_\theta(r)\sin m\theta, \quad d\hat{u}_z = 0, \\
&d\hat{\omega}_z^c(r,\theta) = W_z(r)\sin m\theta, \quad d\hat{\omega}_r^c = 0, \quad d\hat{\omega}_\theta^c = 0, \quad d\hat{u}_n = V_n(r)\cos m\theta
\end{aligned} \tag{9}
$$

where $m = 1, 2, 3, \ldots$ is the wavenumber of the warping mode. Following the analysis in Papamichos (2010) for axisymmetric problems submitted to non-symmetrical loadings, Eq. (8) becomes

$$\begin{bmatrix} [K_1] & -[K_2] \\ -[K_3] & [K_4] \end{bmatrix} \begin{Bmatrix} \{dU\} \\ \{dU_n\} \end{Bmatrix} = 0$$

$$[K_1] = \int_{r_i}^{r_e} \left(\left[B^s\right]^T [D] \left[B^s\right] + \left[B^c\right]^T [D] \left[B^c\right] \right) r dr$$

$$+ \int_{r_i}^{r_e} m_n r_i M \{\overline{N}\} \{\overline{N}\}^T dr$$

$$[K_2] = \int_{r_i}^{r_e} m_n r_i M \{\overline{N}\} \{N\}^T dr, \tag{10}$$

$$[K_3] = \int_{r_i}^{r_n} M \{N\} \{\overline{N}\}^T dr$$

$$[K_4] = \int_{r_i}^{r_n} \left(E_n A_n \{N_{,r}\} \{N_{,r}\}^T + M \{N\} \{N\}^T \right) dr$$

where $\{dU\}$ and $\{dU_n\}$ are, respectively, the rock and nail displacement increment vectors at the nodal points. Expressions for the strain-displacement matrices $[B^s]$ and $[B^c]$ and the constitutive matrix $[D]$ are given in Papamichos (2010) where only the first term of the [K] matrix is present. In the remaining terms due to the nail reinforcement it holds

$$\{N\}^T = [N_i], \quad \{N_{,r}\}^T = \left[N_{i,r}\right], \quad \{\overline{N}\}^T = \left[N_i \; 0 \; 0\right] \tag{11}$$

Equation (10) results in an eigenvalue problem for the critical bifurcation loads σ_{ri} and σ_{re}. Within a finite element discretization scheme, the solution to this problem is obtained by requiring that the global stiffness matrix becomes singular, i.e.

$$\det \begin{vmatrix} [K_1] & -[K_2] \\ -[K_3] & [K_4] \end{vmatrix} = 0 \tag{12}$$

which is the bifurcation condition for lateral instability.

3 Conclusions

The formulation of the bifurcation problem in reinforced boreholes is presented to study the enhanced borehole stability due to reinforcement. Nail reinforcement is modeled through a distributed body force and involves an additional equation necessary for the calculation of the nail displacement, which is coupled with the rock displacement. The bifurcation condition for a Cosserat reinforced solid is an extension of the condition for unreinforced boreholes.

Acknowledgments This work has been performed as a part of the research project "Formation reinforcement for wellbore stability and sand control" funded by the PETROMAKS programme of the Research Council of Norway and BP Norway (Project number 215667/E30).

References

Cox HL (1952) The elasticity and strength of paper and other fibrous materials. Br J Appl Phys 3:72

Papamichos E (2010) Analysis of borehole failure modes and pore pressure effects. Comput Geotech 37(1–2):141–152

Papamichos E (2011) Stability of nail reinforced wellbores. In: Bonelli S et al (ed) Advances in bifurcation and degradation in geomaterials. Proceedings of the 9th international workshop on bifurcation and degradation in geomechanics. Springer, Porquerolles, France, pp 93–99

Papamichos E, Vardoulakis I (2012) Rock nail reinforcement of a free surface. Int J Numer Anal Meth Geomech 36(2):203–218

Shear Banding in Torsion Shear Tests on Cross-Anisotropic Deposits of Fine Nevada Sand

Poul V. Lade, Eugene J. Van Dyck and Nina M. Rodriguez

Abstract A series of torsion shear experiments was performed on large hollow cylinder specimens of Fine Nevada sand with major principal stress directions relative to vertical, α, varying from 0° to 90° and with the intermediate principal stress, $\sigma 2$, varying from $\sigma 3$ to $\sigma 1$ as indicated by $b = (\sigma 2 - \sigma 3)/(\sigma 1 - \sigma 3)$. The Fine Nevada sand was deposited by dry pluviation, thus producing a sand fabric with horizontal bedding planes and cross-anisotropic characteristics. The various stress conditions were achieved by varying the pressures inside and outside the hollow cylinder specimen relative to the shear stress and the vertical deviator stress according to a pre-calculated pattern. All stresses and all strains were determined from careful measurements so that analysis of the soil behavior could be made reliably. The soil behavior was determined for a pattern of combinations of α varying with increments of 22.5° from 0° to 90° and b varying with increments of 0.25 from 0.0 to 1.0. Thus, 25 test locations were established, but many tests were repeated to study the consistency of the results. The friction angles varied considerably with α and b, thus indicating the importance of the intermediate principal stress and the principal stress directions relative to the horizontal bedding planes. The observed shear bands essentially followed the expected directions, but due to the cross-anisotropy, shear bands were also observed in the direction of the major principal stress in regions with high b-values. The strength variation was also influenced by the flexibility of the boundaries in these regions.

P.V. Lade (✉)
Department of Civil Engineering, The Catholic University of America, Washington, DC 20064, USA
e-mail: lade@cua.edu

E.J. Van Dyck
Schnabel Services Inc., Glen Allen, VA 23059, USA

N.M. Rodriguez
Department of the Navy, Naval Sea Systems Command 05C, Washington Navy Yard, Washington, DC 20376, USA

K.-T. Chau and J. Zhao (eds.), *Bifurcation and Degradation of Geomaterials in the New Millennium*, Springer Series in Geomechanics and Geoengineering,
DOI 10.1007/978-3-319-13506-9_13

1 Introduction

The effects of principal stress direction relative to the bedding planes and the effects of the relative magnitude of the intermediate principal stress on the direction of shear banding in cross-anisotropic sand deposits were studied in a torsion shear apparatus. Large hollow cylinder specimens were prepared by dry pluviation of Fine Silica sand, which creates cross-anisotropic deposits similar to those found in situ. To study the variation of shear strength and direction of shear banding for all directions of the major principal stress relative to vertical, α, and all relative values of the intermediate principal stress, as expressed by $b = (\sigma_2 - \sigma_3)/(\sigma_1 - \sigma_3)$, a systematic program of drained torsion shear experiments were performed at each of the 25 intersection points of $b = 0.0$, 0.25, 0.50, 0.75, and 1.00 and $\alpha = 0°$, 22.5°, 45.0°, 67.5°, and 90.0°. The tests performed in this experimental program had different, varying internal and external pressures and therefore, were able to cover constant intermediate principal stress ratios as expressed by b, principal stress directions, as expressed by α, and mean normal stress, as expressed by σ_m (=101 kPa).

2 Preparation of Hollow Cylinder Specimens

All torsion shear tests were performed on Fine Nevada sand, which is composed of subangular to subrounded grains consisting mainly of quartz (98 %). The properties of this sand are as follows: Mean diameter, $D_{50} = 0.23$ mm; coefficient of uniformity, 2.08; coefficient of curvature, 1.05; specific gravity, 2.65; maximum void ratio, 0.771; and minimum void ratio, 0.507.

The boundaries of the hollow cylinder specimen consisted of custom molded inner and outer latex rubber membranes attached to stainless steel end rings. Hollow cylinder specimens with horizontal bedding planes were prepared using the pluviation and saturation techniques described by Lade et al. (2014). All specimens had inner and outer radii of 18.0 and 22.0 cm, respectively, and the wall thickness was therefore 2.0 cm. The height of the hollow cylinder was 40 cm. A void ratio, $e = 0.53$ was targeted for each specimen. This corresponds to a relative density of 91 % for the Fine Nevada sand.

3 Shear Bands in Cross-Anisotropic Sand

The shear band directions were measured at the end of each test while the specimen was held on a vacuum. Usually, the shear bands are oriented such that the normal to the shear band is contained in the wall, as indicated on the photograph in Fig. 1. All but the experiments at $b = 0.0$ have resulted in peak failure caused by development of shear bands. The experiments with $b = 0.0$ exhibited smooth peak failure with shear banding developing in the softening regime.

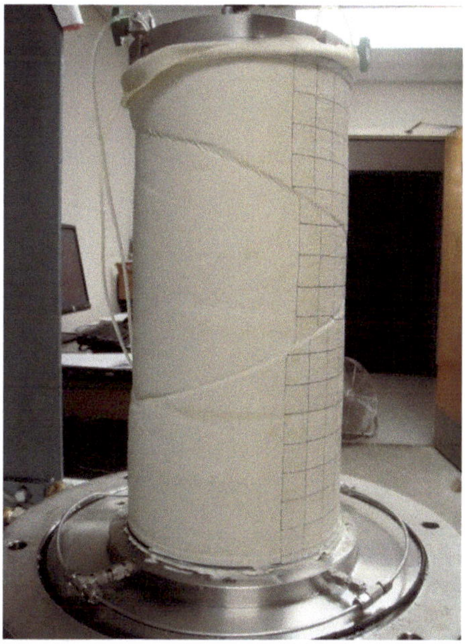

Fig. 1 Hollow cylinder specimen after failure under stress conditions with $b = 0.25$ and inclination of major principal stress relative to vertical at $\alpha = 90°$

Figure 2 show comparisons of measured shear band directions, β, relative to the σ_1-directions with the theoretical directions calculated from expressions proposed by Coulomb, Roscoe (1970) and Arthur et al. (1977) for the torsion shear tests with $\alpha = 0°$, 45° and 90°. The shear band inclinations typically vary within 5°–10° along its length around the hollow cylinder specimens due to small variations in void ratio. Due to experimental scatter, complete consistency between the experimental results and the theoretically predicted inclinations for the shear bands could not be expected. Within this scatter, the diagrams show that the shear band inclinations in the torsion shear tests appear to fit best with the variation proposed by the Coulomb direction, i.e. the shear bands tend to form angles of $\pm(45° - \phi/2)$ with the direction of the major principal stress, σ_1, or $\beta = \pm(45° + \phi/2)$ with the σ_1-plane, as shown in the diagrams. This is true for all α-angles and low to medium b-values.

For high b-values the hollow cylinder specimens showed more than one type of shear band pattern. This is due to the cross-anisotropic character of the sand deposit for which the shear strength is lower in the horizontal direction than in the vertical direction. Thus, shear failure would occur in the horizontal direction and across the wall for high b-values, even though the intermediate principal stress, σ_2, was smaller than the major principal stress, σ_1. This has previously been observed in true triaxial tests on cross-anisotropic sand deposits (Abelev and Lade 2003). For b-values greater than 0.6–0.7, horizontal shear failures were observed in all the hollow cylinder specimens, independent of the inclination, α, of the major principal stress. The cut-off value of b (=0.6–0.7) for the occurrence of "canyon" shear bands depends on the degree of cross-anisotropy of the sand deposit.

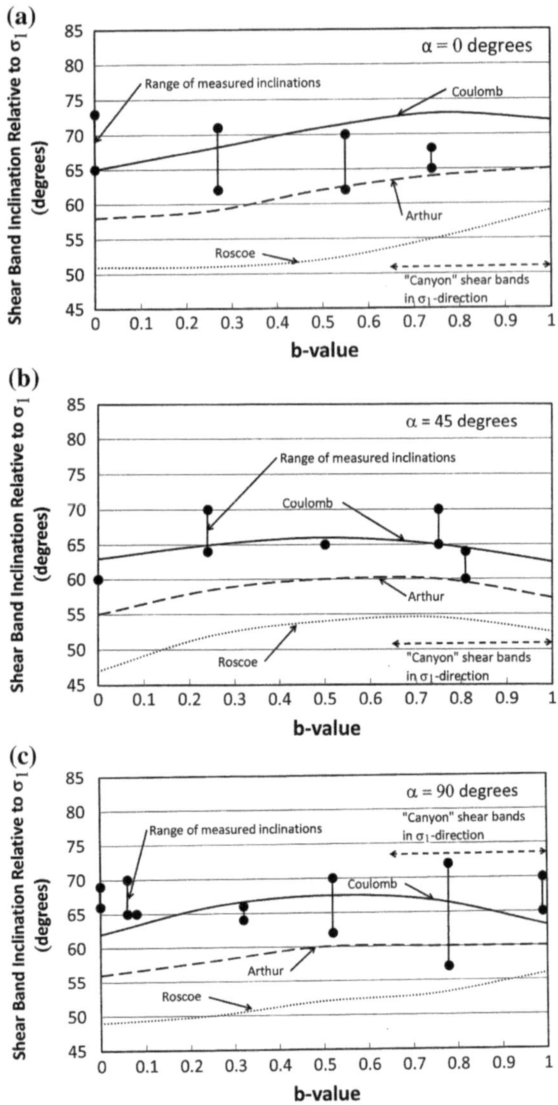

Fig. 2 Shear band inclination relative to σ_1-direction for torsion shear tests with **a** $\alpha = 0°$, **b** $\alpha = 45°$, and **c** $\alpha = 90°$

For these higher b-values the Coulomb inclination of the shear bands were observed in all experiments, but additional shear banding occurred across the wall thickness of the specimen.

Figure 3 shows two sketches of the shear band pattern for high b-values. In both cases the shear bands appear to be inclined in the σ_1-direction, but this is because extension type conditions occur with the minor principal stress perpendicular to the indicated σ_1-directions. The shear bands therefore appear as wide

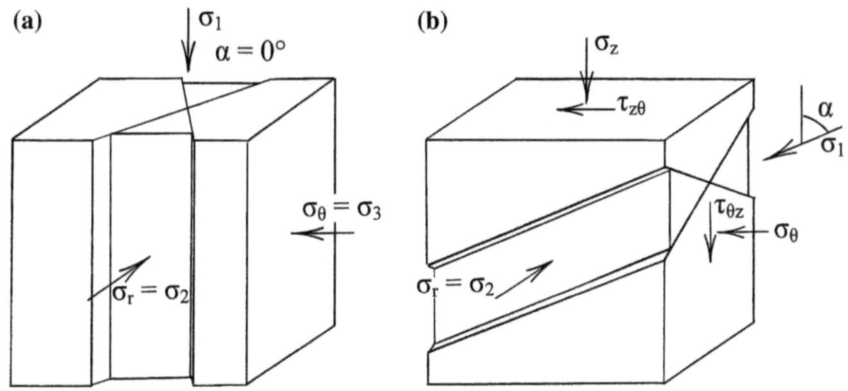

Fig. 3 Sketches of shear band pattern in torsion shear tests on cross-anisotropic sand deposits tested at high b-values

"canyons" on the outside surface of the hollow cylinder specimen rather than as relatively thin shear bands of the type shown in Fig. 1. Once the shear banding initiates in the horizontal direction, the shear strength in the σ_1-direction is also affected and it begins to decline.

Examples of shear banding at high b-values are shown in Fig. 4a–e. The photos correspond to α-values close to 0°, 22.5°, 45°, 67.5° and 90°. The actual values of α and b at the time of shear banding are given in the legends of the photos. It is clear that the reason these shear bands occur is that the surrounding flexible rubber membrane allows non-uniform deformations to occur. If the membranes had been rigid rather than flexible, then shear banding would have been impeded or prevented until higher stresses could have been reached.

Note that the vertical "canyon" shear band in Fig. 4a occurs due to shear banding in the radial direction, but no shearing occurs along the shear band in the vertical direction due to the stiff steel end rings. Neither does shear banding occur at angles of $45° \pm \phi/2$ relative to vertical, because the configuration and kinematic constraints of the hollow cylinder specimen does not allow this to occur. For the remaining tests in Fig. 4b–e, shearing along the "canyon" shear band created by the radial stress can occur because it is not inhibited by stiff ends. In the test in Fig. 4e with $\alpha = 90°$ and $b = 1.0$, a combination of "conventional" and "canyon" shear bands are observed, because both patterns are equally likely.

For values of b near zero, the shear banding in cross-anisotropic sand with horizontal bedding planes should be equally likely in the $\sigma_1 - \sigma_3$ plane and in the $\sigma_1 - \sigma_2$ plane. However, the kinematic conditions and the different inside and outside pressures determine the modes of developing shear banding. Shear bands developed in the wall for $\alpha = 0°$ with a zigzag pattern around the hollow cylinder

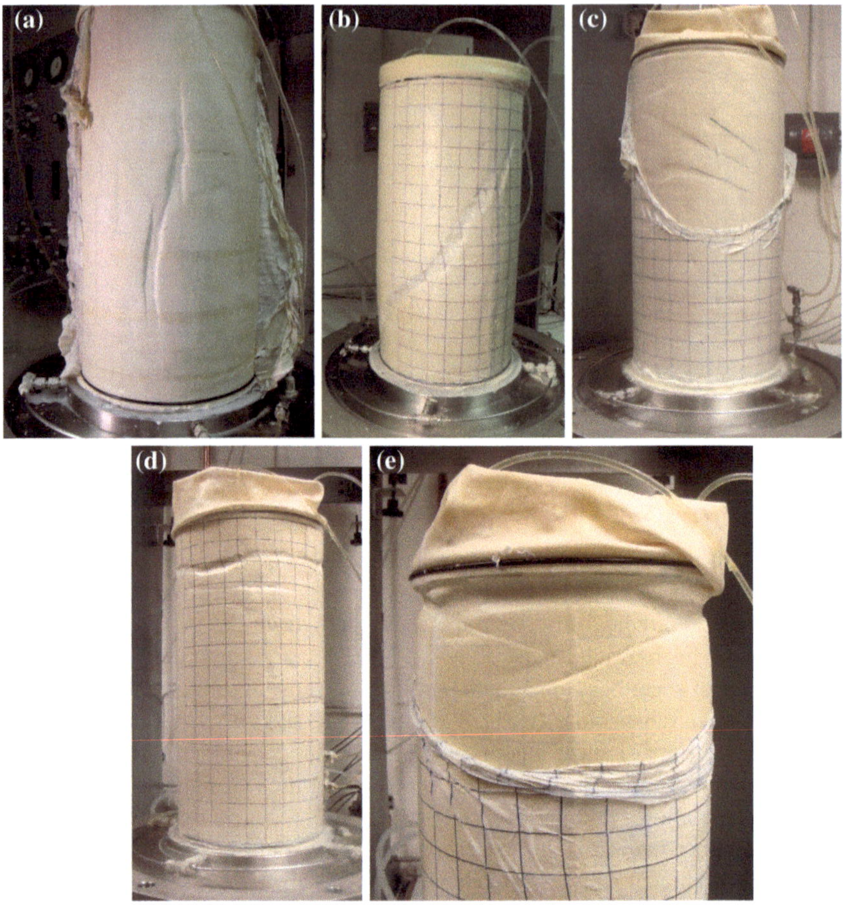

Fig. 4 "Canyon" shear band patterns oriented in direction of the major principal stress in hollow cylinder specimens with **a** $\alpha = 0°$ and $b = 0.79$, **b** $\alpha = 22.5°$ and $b = 0.76$, **c** $\alpha = 45°$ and $b = 0.75$, **d** $\alpha = 67.5°$ and $b = 1.0$, and **e** $\alpha = 90°$ and $b = 1.0$

specimen. For $b = 0$ and $\alpha = 67.5°$ and $90°$ the outside pressure was higher than the inside pressure and for these cases the hollow cylinders imploded, as exemplified in Fig. 5.

Thus, it appears that the loading scheme to create various conditions of α and b and the kinematic constraints in the torsion shear tests on hollow cylinder specimens impose certain limitations on the failure modes in the cross-anisotropic sand deposits and they are apparently affected by the boundary conditions provided in these tests.

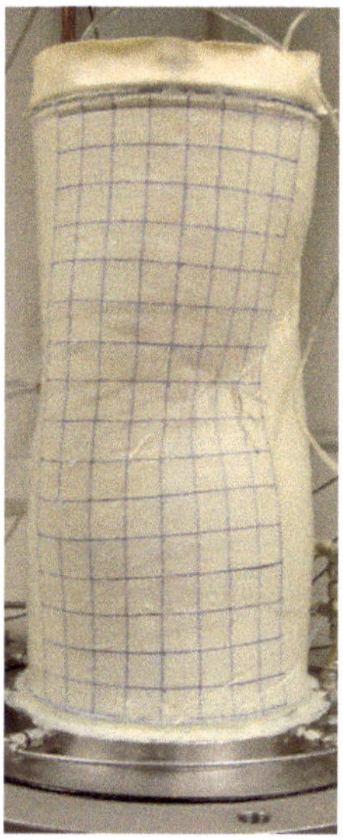

Fig. 5 Implosion of hollow cylinder specimen tested in torsion shear at $\alpha = 90°$ and $b = 0.0$

4 Conclusions

Experimental research is being performed to contribute to establishment of more realistic modeling of soil behavior. Many real soils, as they occur in situ, clearly exhibit cross-anisotropic behavior with a vertical axis of rotational symmetry. This real behavior is most often assumed to be isotropic, and a number of observed behavior patterns are therefore not predicted correctly. Torsion shear experiments have been performed on hollow cylinder specimens to determine the behavior and variation of the friction angle of dense, Fine Nevada sand deposited with cross-anisotropic fabric. Systematic variations of the intermediate principal stress and directions of the major principal stress have been employed in these experiments. Based on these results and previous knowledge it appears that six different factors play important roles in determining the behavior for a given sand: (1) initial void ratio, (2) minor principal stress, (3) intermediate principal stress, (4) cross-anisotropic fabric, (5) orientation of the major principal stress relative to the bedding planes, and (6) occurrence of shear bands.

Acknowledgments The research presented here was performed with support from the National Science Foundation under Grant No. CMMI-0757827. Grateful appreciation is expressed for this support.

References

Abelev AV, Lade PV (2003) Effects of cross-anisotropy on three dimensional behavior of sand. I: stress-strain behavior and shear banding. J Eng Mech 129(2):160–166
Arthur JRF, Dunstan T, Al-Ani QAJL, Assadi A (1977) Plastic deformation and failure in granular media. Geotechnique 27(1):53–74
Lade PV, Rodriguez NM, Van Dyck EJ (2014) Effects of principal stress directions on 3D failure conditions in cross-anisotropic sand. J Geotech Geoenvironmental Eng 140(2):04013001-1–04013001-12
Roscoe KH (1970) The influence of strains in soil mechanics. Geotechnique 20(2):129–170

Effects of Flooding on Crushable Sand

Carlos Ovalle, Christophe Dano, Pierre-Yves Hicher and Mónica Cisternas

Abstract We carried out compression tests on crushable sand to study the effect of creep and relaxation after flooding. We show that there is a consistent framework linking the breakage ratio with the compressibility, which could be useful to develop constitutive models.

1 Introduction

The design of safe and environmentally friendly rockfill dams requires ensuring mechanical and hydraulic stability. These structures are mainly built with a mix of coarse rock aggregates, which may be degraded by stress and hydraulic conditions during lifetime, causing particle crushing and therefore increasing the compressibility. It is well known that the mechanical behavior of crushable granular materials is a function of the properties of the granular packing and its water content, as well as mechanical and geometrical characteristics of individual grains (Lee and Coop 1995; Lade et al. 1996; Oldecop and Alonso 2003; Ovalle et al. 2013a, b; Ovalle et al. 2014). In fact, the source of grain crushing comes from the mechanical properties of grains and its loading conditions at inter-granular contacts. A close look to the microscopic scale (internal flaws) gives a physical explanation of this phenomenon, as described by classical brittle fracture mechanics (Irwin 1957) through the concept of fracture toughness K_{IC} (for Mode I), which depends on the relative humidity (Oldecop and Alonso 2007).

C. Ovalle (✉)
Department of Structural and Geotechnical Engineering,
Pontificia Universidad Católica de Chile, Santiago, Chile
e-mail: covalle@ing.puc.cl

C. Dano · P.-Y. Hicher · M. Cisternas
LUNAM University, Ecole Centrale de Nantes, CNRS UM, 6183 Nantes, France
e-mail: christophe.dano@ec-nantes.fr

K.-T. Chau and J. Zhao (eds.), *Bifurcation and Degradation of Geomaterials in the New Millennium*, Springer Series in Geomechanics and Geoengineering,
DOI 10.1007/978-3-319-13506-9_14

Many advances have been made in the understanding of the phenomenon of grain crushing. However, the role of the breakage ratio as a governing parameter for the effect of the water content on crushable soils is not clear. Therefore, in this paper we present an extensive experimental program of the effects of flooding in crushable sand. The main objective is to discuss the influence of stress path, water content and time, on the link between the mechanical behavior and the increment of the breakage ratio.

2 Experimental Work

We used angular sand obtained from the grinding of a quartzite shale rock from Trois Vallées quarry, in the north of France. Samples were prepared with uniform initial grain size distribution (GSD) sieved between 2 and 2.5 mm. To avoid crushing during preparation, loose samples of dry material were prepared in cylindrical moulds by pluviation at low height. As explained later, some samples were flooded with demineralized water after a phase of compression. For flooding, water was slowly added to dry samples through a pipe connected to its base, so as to maintain a constant water level slightly over the upper level of the sample. After each test, samples were dried and accurately sieved between 2.5 and 0.08 mm.

2.1 *Triaxial and Isotropic Compression Tests*

Loose samples of dry material were prepared at an average void ratio of 0.937 ± 0.035 (Tests 1–8) in a cylindrical mould of 110 mm high and 70 mm in diameter. Strain controlled CID triaxial compression test at confining pressures (σ'_3) of 0.4 and 0.8 MPa were performed. Volume changes of the sample were measured through the confinement water volume control system, which allows testing dry material. Some dry tests were flooded after the isotropic consolidation phase and others after a triaxial compression pushed up around 7–12 % of axial strain (ε_a) (see Tests 4, 5, 6 and 8 in Fig. 1). However, the methodology does not allow for the verification of the degree of saturation through Skempton's B parameter. First, the vertical strain increment was stopped and then flooding was carried out by imposing a water flow through the sample from the bottom to its top face. Then, a period of deviatoric stress ($q = (\sigma'_1 - \sigma'_3)$) relaxation of one hour was applied (i.e., at both constant ε_a and σ'_3). The final ε_a for all triaxial tests was 20 %. Figure 1 shows the stress-strain response in terms of the main stress ($p' = (\sigma'_1 + 2\sigma'_3)/3$) and the void ratio. Regardless the point of flooding, the material is more compressible after flooding. Moreover, all flooded samples at a given σ'_3 joined approximately the same volumetric response and almost the same breakage ratio (i.e., the same final GSD).

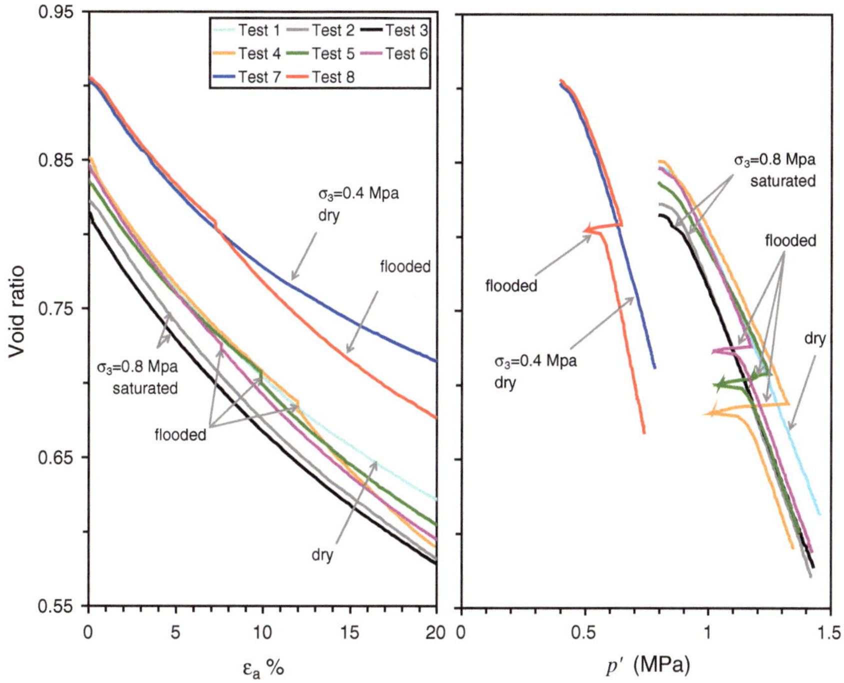

Fig. 1 Stress-strain curves after triaxial tests

Similarly, dry and flooded drained isotropic compression tests (with stress control) at σ'_3 of 0.4 and 0.8 MPa were carried out using the same sample preparation methodology and with an average void ratio was 0.980 ± 0.088.

2.2 *Oedometric Tests*

Loose samples of dry material were prepared at an average void ratio of 1.018 ± 0.05 in a cylindrical oedometric device of 19 mm high and 70 mm of diameter (Tests 14–25). Stress controlled oedometric compression tests were performed using effective vertical stress levels (σ'_v) from 0.15 to 2.10 MPa. 1 h of strain stabilization was allowed for each stress step. Tests were run for three sample conditions: dry, saturated and flooded. In this case, a saturated material stands for a sample which was initially flooded, i.e., before the initial loading stage of $\sigma'_v = 0.15$ MPa, even if there is no verification of Skempton's B parameter. Figure 2a presents the oedometric compression curves, where it can be seen that a good repeatability was obtained. Hence, two main compression curves can be identified, one for dry conditions and the other for saturated and flooded conditions. For example, tests 23, 24 and 25 were flooded at σ'_v of 0.40, 0.85 and 1.5 MPa,

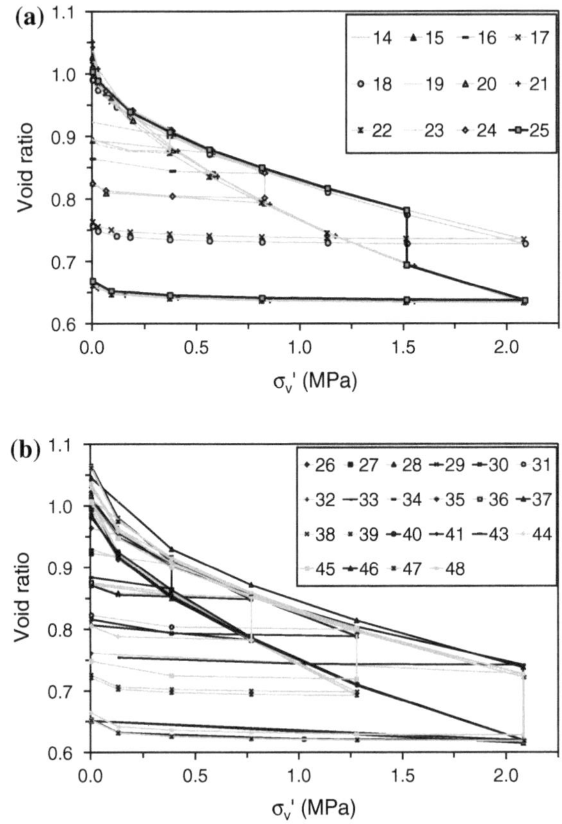

Fig. 2 Compression curves for oedometric tests (**a**) Tests 14–25 (**b**) Tests 26–48

respectively. It can be seen that those tests followed the dry compression curve and after flooding, they joined the behavior of the saturated material. Next, after flooding and 1 h of creep in Test 25, σ'_v was increased up to 2.10 MPa and the compression curve followed the saturated material response. It has also been observed that after flooding in Test 25, the material reaches almost the same GSD that the saturated sample (Test 22), with more grain crushing compared to the dry case (Test 18).

2.3 Creep in Oedometric Tests

In order to study the creep behavior of the material, a second series of oedometric tests was carried out. Using the same aforementioned methodology, several loose samples at a void ratio of 1.019 ± 0.046 were prepared (Test 26–48, see Fig. 2b). This time, load was kept during 24 h of creep at each stress level. As well as for 1 h of creep in oedometric compression, we observed that regardless the point of flooding, the oedometric compression curve is unique for saturated and flooded samples.

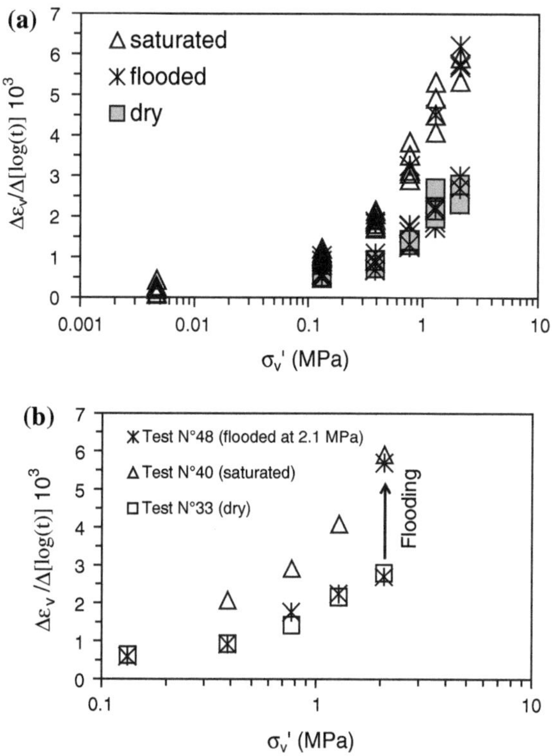

Fig. 3 Creep compressibility

Figure 3a presents the time dependent compressibility index defined as $\Delta\varepsilon_v/\Delta[\log(t)]$ (for t in minutes and measured from 1 min to 24 h). It is clear that dry samples have a lower index and flooded ones joined the index of saturated samples after flooding. This last statement can be clearly appreciated in Fig. 3b for three representative tests at $\sigma'_{v\ max} = 2.1$ MPa.

2.4 Effect of Water in the Fracture Toughness of Rock Samples

In order to study the effect of water at the source of the particle crushing phenomenon (i.e., at the individual grain scale), we have carried out crushing tests on core rock samples. We used the cracked chevron notched Brazilian disc (CCNBD) method to obtain K_{IC}, as suggested by Fowell (1995). This method gives consistent fracture toughness values and allows testing rock anisotropy since the crack orientations can be easily arranged. The notch cut along the core diameter causes crack propagation to start at the tip of the notch and to proceed radially outwards in a stable fashion until the point where the fracture toughness is obtained.

Table 1 K_{IC} of CCNBD specimens with (a) orthogonal and (b) parallel core orientation

Rock aggregate	Sample N°	Condition	K_{IC} MPa m$^{0.5}$
(a)			
1	1	Saturated	0.58
	2		0.59
	3	Dry	1.48
	4		1.57
	5		1.49
2	1	Saturated	0.42
	2	Dry	0.54
	3		0.63
(b)			
1	1	Saturated	0.23
	2	Dry	0.22
2	1	Saturated	0.16
	2		0.11
	3	Dry	0.21
	4		0.22
	5		0.19
	6		0.23

We machined rock cores of D = 52 mm in diameter, following both parallel and orthogonal orientations with respect to the rock anisotropies composed by the sheeted structure of the shale rock. We have conducted tests on air-dried samples, as well as samples submerged under water during 24 h before testing. Core samples were numbered according to two coarse rock aggregates from they were machined. Due to the heterogeneity of the shale rock and in order to clearly identify the effect of water in K_{IC}, in this paper we only compare cores coming from the same aggregate. Table 1 summarize the experimental results, which are in the same order of magnitude of typical values for shale rocks reported in the literature of around $K_{IC} = 0.5$–1.0 MPa m$^{0.5}$ (Ashby and Jones 2006). For orthogonal orientated cores coming from aggregate N°1 we obtained an average reduction of approximately 60 % from dry to saturated specimens. Similarly, the reduction for orthogonal orientated cores from aggregate N°2 was about 20–30 %. On the other hand, in parallel orientated samples K_{IC} drops significantly compared to the previous case, due to the induced chevron notch on the weakest plane of the rock. Moreover, the difference between dry and saturated specimens is less important compared to orthogonal orientated cores.

3 Conclusions

The experimental data presented in this paper show that saturated and flooded crushable sand samples are more compressible than dry samples, mainly due to the increment of grain crushing in the presence of water. This phenomenon could be explained due to corrosive attacks of water at the micro-cracks scale inside the particles. As shown experimentally, this phenomenon implies a reduction of particle strength, and accordingly of K_{IC}, in saturated rock aggregates compared to dry ones.

Our results also show that, for given initial conditions (GSD and density) and compression state (stress path and p'_{max}), the compression curve in the void ratio—p′ plane of flooded and saturated samples is virtually the same, regardless the loading-flooding sequence. It is well known that, at a given stress state, the more polydisperse the GSD is, the higher the packing density will be (Biarez and Hicher 1997). Therefore, the uniqueness of the compression curve of saturated and flooded sand tests can be explained because they have reached the same GSD, which has been possible due to grain crushing.

Figure 4 presents the breakage ratio (B_r) after each test (using the definition of Einav 2007) as a function of the plastic work (W^p) for all tests presented in this paper (Tests 1–48). It is clear that dry materials (squares symbols) require more W^p in order to reach a given B_r, compared to saturated and flooded cases, which can be considered as equivalent. Consequently, for a material with a given initial density and initial GSD, the amount of particle breakage depends only on the plastic work and the water content.

Figure 4 also shows that the relation between B_r and the lowest void ratio attained for all tests presented here is essentially unique and it does not depend on the initial density, the loading condition (stress path, creep or relaxation) nor the test condition (dry, saturated or flooded). However, it should depend on intrinsic parameters as particle shape and initial GSD. This finding is relevant for constitutive modelling, since the prediction of the final GSD could be possible based on calibration tests in only one stress path. Moreover, constitutive models based on a variable critical void ratio as a function of grain crushing, as the pioneering work

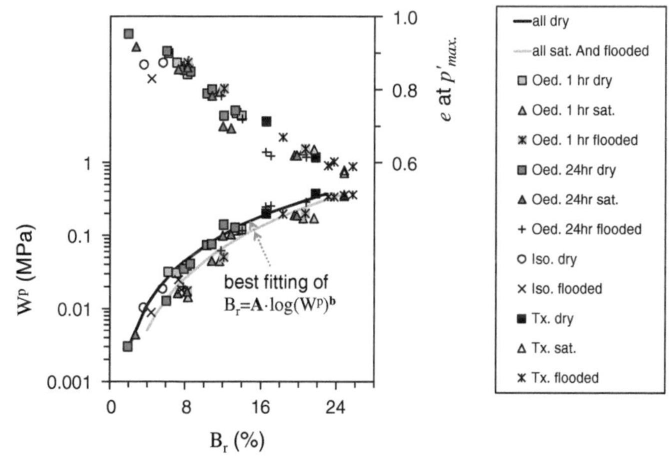

Fig. 4 Void ratio at maximum p′ and plastic work as a function of B_r for all tests

of Daouadji et al. (2001), could be enhanced in order to model the effect of water on grain crushing.

Moreover, the uniqueness of the mechanical response for flooded and saturated samples is also observed in the creep behaviour. The creep compressibility index increases with the mean stress, and for a given stress its magnitude is higher for saturated sand samples compared to dry ones. After flooding, this index increases to the same value obtained in saturated samples, analogous to the results on both the compression curve and the breakage ratio. As reported in the literature, this could be explained due to the increment of crack propagation velocity inside the grains in the presence of water (Oldecop and Alonso 2007).

References

Ashby M, Jones D (2006) Engineering materials 1, 3rd edn. Elsevier, Oxford

Biarez J, Hicher PY (1997) Influence de la granulométrie et de son évolution par ruptures de grains sur le comportement mécanique de matériaux granulaires. Revue Française de Génie Civil 1(4):607–631

Daouadji A, Hicher PY, Rahma A (2001) An elastoplastic model for granular materials taking into account grain breakage. Eur J Mech A-Solid 20:113–137

Einav I (2007) Breakage mechanics—part I: theory. J Mech Phys Solids 55(6):1274–1297

Fowell R (1995) Suggested method for determining Mode I fracture toughness using cracked chevron notched brazilian disc (CCNBD) specimens. Int J Rock Mech Min Sci Geomech Abstr 32(I):57–64

Irwin G (1957) Analysis of stresses and strains near the end of a crack traversing a plate. J Appl Mech 24:361–364

Lade P, Yamamuro J, Bopp P (1996) Significance of particle crushing in granular materials. J Geotech Eng 122(4):309–316

Lee I, Coop M (1995) The intrinsic behaviour of a decomposed granite soil. Géotechnique 45(1):117–130

Oldecop L, Alonso E (2003) Suction effects on rockfill compressibility. Géotechnique 53(2):289–292

Oldecop L, Alonso E (2007) Theoretical investigation of the time dependent behavior of rockfill. Géotechnique 57(3):289–301

Ovalle C, Dano C, Hicher PY (2013a) Experimental data highlighting the role of surface fracture energy in quasi-static confined comminution. Int J Fract 182(1):123–130

Ovalle C, Dano C, Hicher P-Y, Cisternas M (2014) An experimental framework for evaluating the mechanical behavior of dry and wet crushable granular materials based on the particle breakage ratio. Can Geotech J. doi:10.1139/cgj-2014-0079

Ovalle C, Frossard E, Dano C, Hu W, Maiolino S, Hicher PY (2013b) The effect of size on the strength of coarse rock aggregates and large rockfill samples through experimental data. Acta Mech 225:2199–2216. doi:10.1007/s00707-014-1127-z

Numerical Analysis of 1D Seepage Failure Process of Sandy Materials by CWENO Method

Kosuke Tsujimura, Kazunori Fujisawa and Akira Murakami

Abstract This paper proposes a numerical method to simulate the seepage failure of sandy materials, known as sand boils or quicksand. The sand is largely deformed and the particles rapidly migrate with the seepage water after the seepage failure. In order to deal with the large deformation and the migration of the sand particles, an Eulerian formulation is applied to the governing equations, i.e., the equations for the conservation of mass and momentum of seepage water and sand, which yields a system of hyperbolic partial differential equations. To achieve stable and accurate numerical solutions for the hyperbolic system, the Central Weighted Essentially Non-Oscillatory (CWENO) method is employed herein. The numerical simulation of the 1D seepage failure of sand, induced by an upward seepage flow, is conducted and the computation is compared with analytical solutions. The results show that the proposed method can reproduce the behavior of sand particles and seepage water during seepage failure.

1 Introduction

The seepage failure of sand, known as sand boils or quicksand, occurs when the effective stress and the stiffness of the sand vanishes due to an increase in pore water pressure and its gradient. The critical hydraulic gradient has been known as the quick condition; it can predict whether a sand boil may occur or not. However, methods for predicting the behavior of sand after the seepage failure are currently limited. The sand fluidizes and behaves as a fluid during the sand boiling, which involves a multiphase flow consisting of sand particles and seepage water. This makes the phenomenon after the sand boiling difficult and complex. This paper provides a numerical method for computing the motion of sand and seepage water during sand boiling. Herein, an Eulerian formulation is adopted to derive the

K. Tsujimura · K. Fujisawa (✉) · A. Murakami
Graduate School of Agriculture, Kyoto University, Kyoto, Japan
e-mail: fujik@kais.kyoto-u.ac.jp

K.-T. Chau and J. Zhao (eds.), *Bifurcation and Degradation of Geomaterials in the New Millennium*, Springer Series in Geomechanics and Geoengineering,
DOI 10.1007/978-3-319-13506-9_15

governing equations, which form a hyperbolic system of partial differential equations and enable us to deal with the migration and the large deformation of fluidized sand. The CWENO method (e.g., Levy et al. 2002) is employed to easily solve the hyperbolic system without Riemann solvers, and the one-dimensional behavior of the sand during upward sand boiling is solved numerically and analytically.

2 Governing Equation

The mass conservation of both seepage water and sand and the equations of motion for them are adopted as the governing equations describing the failure process of sand induced by a seepage flow. An Eulerian formulation of the equations gives the following forms for the partial differential equations:

$$\frac{\partial(1-n)}{\partial t}+\frac{\partial(1-n)v_i}{\partial x_i}=0 \tag{1}$$

$$\frac{\partial(1-n)v_i}{\partial t}+\frac{\partial(1-n)v_iv_j}{\partial x_j}=-\frac{1}{\rho_s}\frac{\partial\sigma'_{ij}}{\partial x_j}-\frac{1-n}{\rho_s}\frac{\partial p}{\partial x_i}+(1-n)g_i+\frac{n^2\rho g}{k\rho_s}(q_i-v_i) \tag{2}$$

$$\frac{\partial\rho n}{\partial t}+\frac{\partial\rho n q_i}{\partial x_i}=0 \tag{3}$$

$$\frac{\partial\rho n q_i}{\partial t}+\frac{\partial\rho n q_i q_j}{\partial x_j}=-n\frac{\partial p}{\partial x_i}+\rho n g_i-\frac{n^2\rho g}{k}(q_i-v_i) \tag{4}$$

$$\frac{\partial\sigma'_{ij}}{\partial t}+\frac{\partial\sigma'_{ij}v_k}{\partial x_k}=E_{ijkl}\frac{1}{2}\left(\frac{\partial v_k}{\partial x_l}+\frac{\partial v_l}{\partial x_k}\right)+\sigma'_{ij}\frac{\partial v_k}{\partial x_k}+\omega_{ik}\sigma'_{kj}-\sigma'_{ik}\omega_{kj} \tag{5}$$

where n, v_i, q_i, σ'_{ij}, ω_{ij}, E_{ijkl}, p, ρ, ρ_s, k, g_i and x_i denote the porosity, the spatially-averaged velocity of the sand particles and the seepage water, the effective stress tensor, the spin tensor, the elastic modulus tensor, the water pressure, the densities of water and the sand grains, the hydraulic coefficient, the gravitational acceleration and the components of the rectangular coordinate system, respectively.

3 Numerical Methods

The CWENO scheme is based on the finite volume method originally proposed by Levy et al. (2002); it is employed to solve the above governing equations. To apply the CWENO scheme, Eqs. (1)–(5) are rewritten in the following form:

$$\frac{\partial \boldsymbol{u}}{\partial t}+\frac{\partial \boldsymbol{F}(\boldsymbol{u})}{\partial x_1}+\frac{\partial \boldsymbol{G}(\boldsymbol{u})}{\partial x_2}=\boldsymbol{S}(\boldsymbol{u}) \tag{6}$$

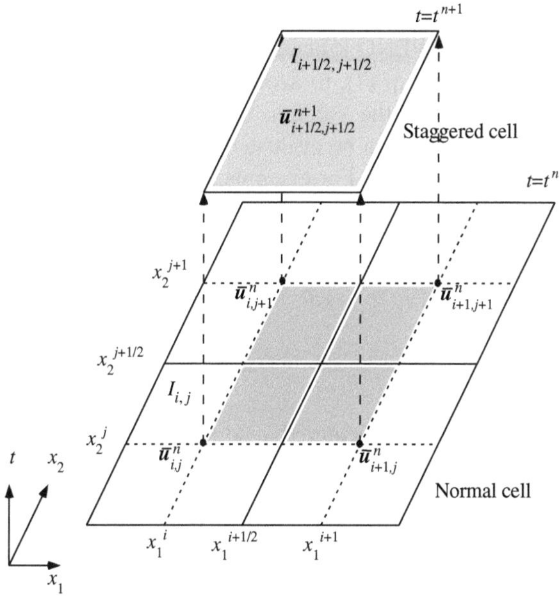

Fig. 1 Normal and staggered grids for computing the conserved variable

where $\boldsymbol{u}$, $\boldsymbol{F}(\boldsymbol{u})$, $\boldsymbol{G}(\boldsymbol{u})$ and $\boldsymbol{S}(\boldsymbol{u})$ denote the vectors of the conserved variables, the convective fluxes in the x_1 and x_2 directions and the source term, respectively.

For the numerical analysis presented later, a rectangular grid with mesh sizes of $h = \Delta x_1 = \Delta x_2$ in both directions is used for simplicity. Defining $I_{i,j}$ as the domain of a finite volume cell around (x_1^i, x_2^j), i.e., $[x_1^i - h/2, x_1^i + h/2] \times [x_2^j - h/2, x_2^j + h/2]$ and letting Δt and $\bar{\boldsymbol{u}}_{i,j}$ denote the interval of time steps and the spatial average of the conserved variable $\boldsymbol{u}$ over $I_{i,j}$, the integration of Eq. (6) over the staggered cell $I_{i+1/2,j+1/2}$, i.e., $\left[x_1^i, x_1^i + h\right] \times \left[x_2^j, x_2^j + h\right]$, and the interval of time t^n to $t^{n+1}(=t^n + \Delta t)$ results in the following equation (See Fig. 1):

$$\begin{aligned}\bar{\mathbf{u}}^{n+1}_{i+1/2,j+1/2} = \bar{\mathbf{u}}^{n}_{i+1/2,j+1/2} &- \frac{1}{h^2}\int_{t^n}^{t^{n+1}}\int_{x_2^j}^{x_2^{j+1}} \mathbf{F}(\mathbf{u}(x_1^{i+1}, x_2, t)) - \mathbf{F}(\mathbf{u}(x_1^{i}, x_2, t))\, dx_2 dt \\ &- \frac{1}{h^2}\int_{t^n}^{t^{n+1}}\int_{x_1^j}^{x_1^{j+1}} \mathbf{G}(\mathbf{u}(x_1, x_2^{j+1}, t)) - \mathbf{G}(\mathbf{u}(x_1, x_2^{j}, t))\, dx_1 dt \\ &+ \frac{1}{h^2}\int_{t^n}^{t^{n+1}}\int_{x_2^j}^{x_2^{j+1}}\int_{x_1^i}^{x_1^{i+1}} \mathbf{S}(\mathbf{u}(x_1, x_2, t))\, dx_1 dx_2 dt\end{aligned} \tag{7}$$

where $\bar{\mathbf{u}}^n_{i+1/2,j+1/2}$ denotes the spatial average of $\boldsymbol{u}$ over $I_{i+1/2,j+1/2}$ at the nth time step. The CWENO scheme is characterized by the integration over the staggered cell, seen in Eq. (7), to avoid the implementation of Riemann solvers. Using the smoothness of the solution on the segment $(x_1^i, x_2^j) \times [t^n, t^{n+1}]$, the time integral in Eq. (7) can be evaluated by a simple quadrature rule rather than solving a Riemann problem. For example, a fourth-order method can be achieved by using Simpson's rule as

$$\int_{t^n}^{t^{n+1}} f(\mathbf{u}(x_1^i, x_2^j, t)\, dt = \frac{\Delta t}{6}\Big\{f(\mathbf{u}(x_1^i, x_2^j, t^n)) + 4f(\mathbf{u}(x_1^i, x_2^j, t^n + \Delta t/2)) + f(\mathbf{u}(x_1^i, x_2^j, t^{n+1}))\Big\} \tag{8}$$

and the following centered quadrature rule in space for the integral in spaces:

$$\int_{x_1^i}^{x_1^{i+1}} f(x_1)\, dx_1 = \frac{h}{24}\Big\{-f(x_1^{i+2}) + 13f(x_1^{i+1}) + 13f(x_1^i) - f(x_1^{i-1})\Big\} \tag{9}$$

Equation (7) is the fundamental equation of the CWENO scheme for updating variable $\boldsymbol{u}$. Although the detailed procedure cannot be explained here due to space limitation, the CWENO scheme is briefly summarized into the following four steps:

1. Find an accurate evaluation for cell averages on staggered grid $\bar{\mathbf{u}}^{n+1}_{i+1/2,j+1/2}$ from the given cell averages $\bar{\mathbf{u}}^{n+1}_{i,j}$.
2. Compute the point value of the first derivatives of fluxes at the integer grid points. These sets of data are required in order to obtain the mid-values at t^n and t^{n+1} by solving a sequence of Cauchy problems.
3. Evaluate the mid-values using a Runge-Kutta scheme.
4. Update the staggered control volume averages by Eq. (7).

At the end, the solution is projected back to the center of each finite volume cell via an inverse reconstruction similar to Step 1. Details of each step can be found in Levy et al. (1999, 2002).

4 Numerical Analysis

The aim of this section is to numerically analyze the sand boiling induced by an upward seepage flow and to compute the behavior of the sand particles and the seepage water during the phenomenon. Figure 2 shows the finite volume cells discretizing a sand column subjected to a seepage flow, where the normal and the staggered cells are drawn with black and gray lines, respectively.

The hydrostatic distribution of water pressure was assumed as the initial condition for the seepage water, and the seepage velocity was zero. The sand was

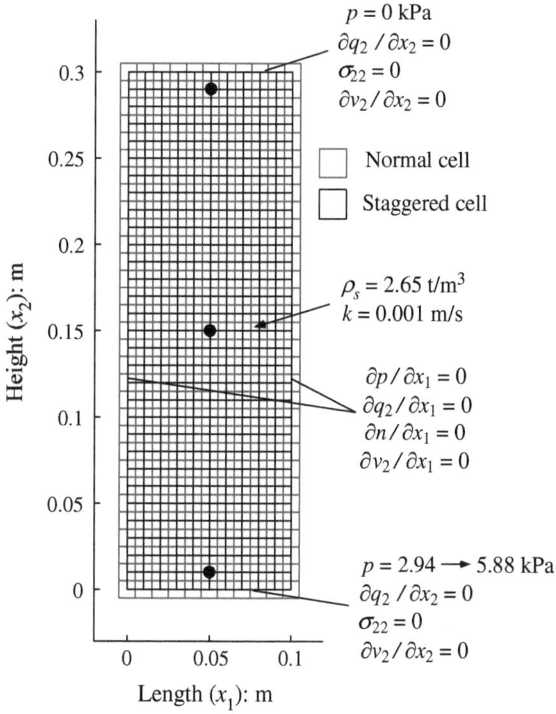

Fig. 2 Computational domain and boundary conditions

also assumed to be static, so that it initially had the earth pressure at rest. The sand boil was generated by gradually raising the water pressure at the bottom, while the water pressure at the top ($x_2 = 0.3$) was maintained. The water pressure at the bottom ($x_2 = 0$) was increased from the initial value of 2.94–5.88 (=5.85 × 1.005) kPa within 10 s (See Fig. 2). The bottom pressure of 5.85 kPa corresponded to the quick condition of the sand.

Figure 3 exhibits computed water pressure p and effective stress σ'_{22} at upper, middle and lower points of the sand column, which are indicated by black dots in Fig. 2. As the water pressure increases, in accordance with the given boundary condition, the effective stress at all these points decreases. Figure 4 shows the changes in velocities of seepage flow q_2 and sand particles v_2 at the upper point. From Fig. 4, it is seen that the velocities rapidly increase after 10 s when the water pressure reaches the quick condition. Assuming the incompressibility of water and the uniformity of the flows in the upward direction, the analytical solutions for the velocities of seepage water and sand during sand boiling can be obtained as follows by solving Eqs. (2) and (4) under the condition that the effective stress is zero.

$$
\begin{aligned}
v_2(t) &= g\varepsilon t - \frac{\alpha q_{2,0}\varepsilon}{\alpha+\beta}\left\{1 - \mathrm{e}^{-(\alpha+\beta)t}\right\}, \\
q_2(t) &= g\varepsilon t + q_{2,0} + \frac{\beta q_{2,0}\varepsilon}{\alpha+\beta}\left\{1 - \mathrm{e}^{-(\alpha+\beta)t}\right\}
\end{aligned}
\tag{10}
$$

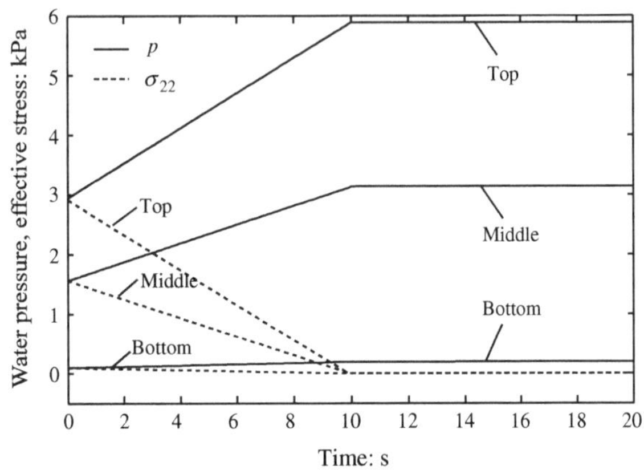

Fig. 3 Computed water pressure and effective stress

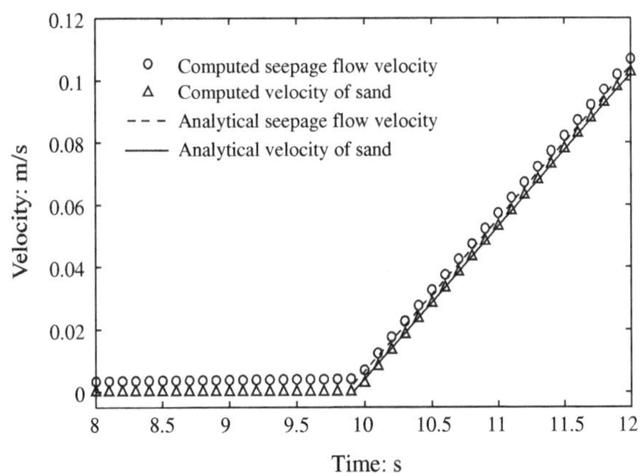

Fig. 4 Comparison between computed and analytical velocities of sand and seepage water

in which

$$\alpha = \frac{n^2 \rho g}{k\rho_s(1-n)}, \quad \beta = \frac{ng}{k}, \quad q_{2,0} = \frac{k}{n}(1-n)\left(\frac{\rho_s}{\rho} - 1\right) \tag{11}$$

ε in Eq. (10) denotes the ratio of the excess hydraulic gradient to the critical one of the quick condition; it should be noted that the t in Eq. (10) is the elapsed time after the sand boil has occurred. The value of ε is 0.005 in this analysis because the given value of the water pressure at the bottom was 5.88 kPa (=1.005 × 5.85 kPa), as mentioned above, which exceeds the critical value of 5.85 kPa by 0.5 %. The solid and dotted lines in Fig. 4 indicate the analytical solutions of velocities v_2 and q_2 calculated from Eq. (10), respectively. Figure 4 shows that the computed velocities are in good agreement with the analytical solutions.

References

Levy D, Puppo G, Russo G (2002) A fourth-order central WENO scheme for multidimentional hyperbolic systems of conservation laws. SIAM J Sci Comput 24(2):480–506

Levy D, Puppo G, Russo G (1999) Central WENO schemes for hyperbolic systems of conservation laws. Math Model Numer Anal 33(3):547–571

Mechanisms of Deformation in Porous Rocks at the Grain Scale

Martin Tjioe and Ronaldo I. Borja

Abstract This paper investigates two dominant pore-scale deformation mechanisms in porous rocks: crystal plasticity and micro-fracture generation. Modeling and simulation are conducted on the mesoscopic scale, defined as a scale smaller than the specimen but larger than the grain. Imperfections are introduced in the form of random voids and random crystal orientations. Depending on the imposed overall deformation, the overall stress response could vary dramatically with assumed pore-scale mechanisms.

1 Introduction

Processes such as carbon sequestration, geothermal energy generation, hydrocarbon extraction, and mining rely on a solid understanding of the mechanical response of rocks to avoid catastrophic failure and optimize yield of the end product. The responses of rocks to these activities are governed by micro-mechanisms that interact with one another to produce complex emergent behavior. This paper investigates the pore-scale deformation mechanisms in high-porosity rocks to better understand their impact on the overall macroscopic behavior. In particular, we focus on two dominant grain-scale mechanisms resulting from purely mechanical, isothermal loading: crystal plasticity and generation of micro-fractures. The investigation is conducted at the mesoscopic scale, defined as a scale smaller than the specimen but larger than the grain. The rationale for using the mesoscopic scale as opposed to the specimen scale is that different parts of the specimen may be undergoing different stages of deformation at any given time, thus obscuring the finer-scale mechanisms when viewed on the larger scale.

M. Tjioe · R.I. Borja (✉)
Department of Civil and Environmental Engineering, Stanford University, Stanford, CA 94305, USA
e-mail: borja@stanford.edu

K.-T. Chau and J. Zhao (eds.), *Bifurcation and Degradation of Geomaterials in the New Millennium*, Springer Series in Geomechanics and Geoengineering,
DOI 10.1007/978-3-319-13506-9_16

The finite element (FE) method is used to model grain-scale deformation. Imperfections in the form of voids and different crystal groups are introduced into the solid matrix. Crystal plasticity is investigated using a rate-independent theory that allows systematic tracking of slip activation sequences, thus effectively dealing with linearly dependent slip systems. Micro-fracture generation is modeled using a strong discontinuity approach based on the assumed enhanced strain (AES) method. This method allows static condensation of slip degrees of freedom on the element level, thus preserving the size and structure of the global FE matrix equations.

2 Crystal Plasticity

Crystal plasticity is a result of permanent rearrangement of the crystal atoms, and can manifest itself in the form of either dislocation or mechanical twins (Tjioe et al. 2012). Crystal plasticity arises when the shear stress on each crystallographic plane exceeds the shear capacity. A slip tensor is formed from a unit vector normal to a crystallographic plane, and a slip direction vector tangent to this plane. The total plastic strain is then obtained from the contribution of all the plastic slips on the active planes. One major aspect of the implementation of crystal plasticity theory is the linear dependence of the active slip systems. A face-centered cubic (f.c.c.) crystal lattice, for instance, has eight glide planes and each plane has three slip directions, resulting in 24 potentially active slip systems. However, the plastic strain tensor, being a second order symmetric tensor that has to satisfy the constraint of isochoric deformation, has only five independent components. This implies that many of the slip systems are redundant as the deformation progresses, creating a computational difficulty. This is tackled through the 'ultimate algorithm' (Borja and Wren 1993; Borja and Rahmani 2012). By following the sequence of slip system activation, the algorithm is able to identify the next active slip system and determine whether it is linearly independent from other active slip systems. Systematic identification of the linearly independent active slip systems is an important attribute of the ultimate algorithm.

3 Micro-fracture Generation

Micro-fractures within a solid grain result in slip or displacement jump within the grain. Micro-fractures are often observed in porous rocks near imperfections such as voids or inclusions. These imperfections attract local stress concentrations and cause the surrounding grains to develop localized deformation in the form of displacement jumps.

When a fracture is closed and the stress normal to the fracture is compressive, the shear stress is balanced by both cohesive and frictional resistance. However, as slip develops, the cohesive force typically diminishes and the resistance becomes purely frictional, a phenomenon called 'slip weakening.' In the simulations, a linear slip-weakening model is employed, in which the cohesion decreases linearly with slip. Once the slip reaches a certain critical value, the cohesion vanishes completely.

To numerically capture the development of a micro-fracture, the initial cohesion is set equal to the shear stress at the point of initiation of the micro-fracture. Activation of the primary crystallographic slip system then triggers the initiation of the micro-fracture. Once this slip system activates, the normal vector to the crystallographic plane and the tangential slip direction are used to define the orientation of the micro-fracture.

The assumed enhanced strain (AES) method (Borja and Regueiro 2001; Foster et al. 2007) is utilized in the finite element implementation, where the source of plasticity comes solely from the discontinuity as the host element unloads elastically. Each element has two additional degrees of freedom in addition to the displacement unknowns, namely tangential slip and normal slip. The AES method allows the two additional degrees of freedom to be statically condensed on the element level without altering the size and structure of the global finite element matrix equation. This is an advantage of the algorithm as it leads to a lower computational cost and easier implementation.

4 Numerical Simulations

We investigate the pore-scale deformation of a mesoscopic volume of limestone. The volume is a cube consisting of two types of inhomogeneity: two crystal groups with different orientations, and randomly positioned voids (Fig. 1). Cubical voids with different sizes are embedded into the solid to give a total overall porosity of 4.1 %. Even though limestone consists of calcite and aragonite, which have trigonal-hexagonal scalenohedral and orthorhombic crystal structures, it is assumed to be made up of f.c.c. crystals in this simulation.

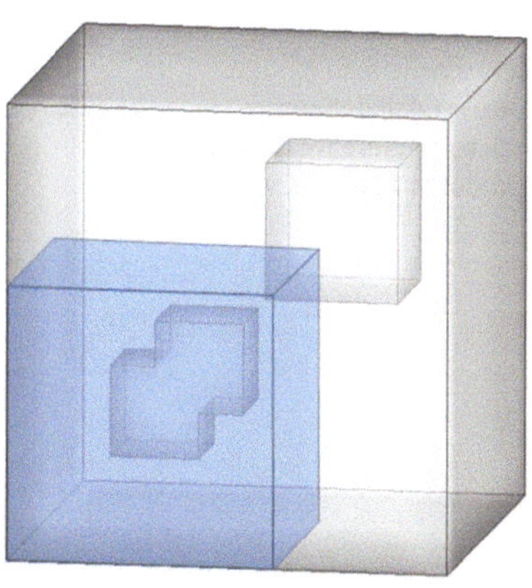

Fig. 1 Randomly-positioned cubical voids are shown in the mesoscopic volume. Two crystal groups are also shown where one crystal group is shown shaded in *blue* while the remaining part belongs to another crystal group

The elastic parameters used in the simulations include Young's modulus and Poisson's ratio, which are equal to 20,000 MPa and 0.30, respectively. The crystal yield stress is the maximum shear stress that can be reached before a slip system activates. In the simulations, all the slip systems are assumed to have a crystal yield stress of 30 MPa. For micro-fracture generation, the coefficient of friction dictating the amount of frictional resistance is set to the value of 0.75; the cohesion on the surface of discontinuity is assumed to vanish in one time step.

There are many ways in which the stress response at the grain scale can be homogenized to obtain the mesoscopic overall stress. The computational homogenization technique we employ calculates the mesoscopic stress by taking a volume average of the stresses at each integration point in the FE calculations.

In the first simulation, we subject the mesoscopic volume to a homogeneous deformation involving both shear and tension. With every time increment, the displacement on the surface is increased linearly. The plot of the overall von Mises stress against the volumetric strain for the mesoscopic volume is shown in Fig. 2. The initiations of localization within the volume for both mechanisms occur at the same time step. It can be observed that the overall stress-strain curves deviate from linearity in the limit of elastic response. For crystal plasticity, this deviation is due to the fact that the shear stresses at the slip systems cannot go beyond the crystal yield stress, and it is only significant at a later stage of deformation. For micro-fracture generation, a drop in the stress level is observed after localization, attributed to significant slip weakening in the fractured elements as the shear stresses drop to zero due to tensile normal stress on the fracture plane.

In the second simulation, we subject the mesoscopic volume to a homogeneous deformation involving compression in one direction and extension by an amount 20 % of the value of compression in the other two directions. The plot of the overall von Mises stress against the volumetric strain for the volume is shown in Fig. 3. For the crystal plasticity simulation, we observe the same plateau as in the

Fig. 2 Plot of the overall von Mises stress-volumetric strain curve when the mesoscopic volume is subjected to shear and tensile loading. It is observed that the micro-fracture mechanism produces a softer response

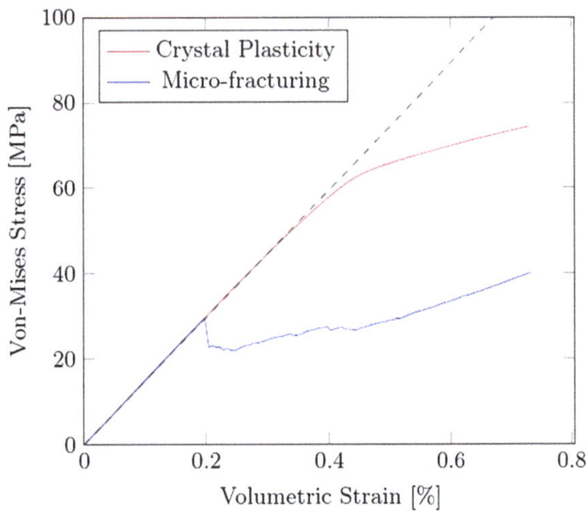

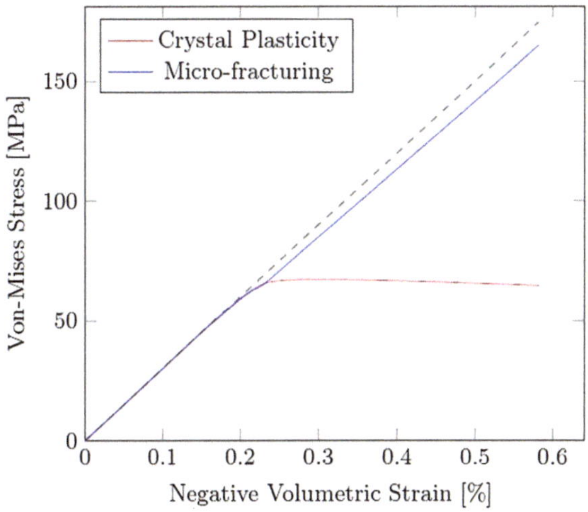

Fig. 3 Plot of the overall von Mises stress-volumetric strain curve when the mesoscopic element is subjected to compressive loading. For the micro-fracturing simulation, an increasing normal stress, and thus frictional resistance, on the fracture surface allows increasingly larger von Mises stress to develop

previous simulation because the shear stresses are constrained to be lower than the yield stress once plastic deformation occurs. For the micro-fracturing simulation, the shear stress on the fracture plane is predominantly governed by the frictional resistance on the surface, which continues to increase as a result of the increasing normal stress on the surface. This implies that increasingly larger shear stress can be accommodated on the fracture plane, which translates into increasing overall von Mises stress in the mesoscopic volume.

5 Conclusions

In this paper, we have presented the numerical simulations of two micro-mechanisms in porous rocks, namely, crystal plasticity and micro-fracture generation. For shear and tensile loading, micro-fractures produce a softer mesoscopic response than crystal plasticity. For compressive loading, however, the increasing frictional resistance on the micro-fractures allows the mesoscopic stress to continue to increase.

Other important micro-mechanisms such as pore collapse or diffusional mass transfer have not been considered, and when acting together they are expected to producc a complex mesoscopic response dissimilar to the investigation of the individual mechanisms presented in this paper.

Acknowledgments This material is based upon work supported by the U.S. Department of Energy, Office of Science, Office of Basic Energy Sciences, Geosciences Research Program, under Award Number DE-FG02-03ER15454. Support for the first author is provided in part by the Stanford Graduate Fellowship (SGF).

References

Borja RI, Wren JR (1993) Discrete micromechanics of elastoplastic crystals. Int J Numer Meth Eng 36(22):3815–3840

Borja RI, Rahmani H (2012) Computational aspects of elasto-plastic deformation in polycrystalline solids. J Appl Mech 79(3):031024

Borja RI, Regueiro RA (2001) Strain localization of frictional materials exhibiting displacement jumps. Comput Meth Appl Mech Eng 190(20–21):2555–2580

Foster CD, Borja RI, Regueiro RA (2007) Embedded strong discontinuity finite elements for fractured geomaterials with variable friction. Int J Numer Meth Eng 72(5):549–581

Tjioe M, Rahmani H, Borja RI (2012) Pore-scale deformation in high-porosity rocks. ARMA/NARMS 12-670

The Influence of Fines Content on the Onset of Instability and Critical State Line of Silty Sand

T.-K. Nguyen, N. Benahmed, P.-Y. Hicher and M. Nicolas

Abstract This communication presents the results of an experimental study in the laboratory on the mechanical behavior of silty sand through conventional compression triaxial tests, under both drained and undrained conditions. We particularly put an emphasis on the effect of fines content on the evolution of the maximum and the minimum void ratio, the onset of instability and the critical state line. Several series of monotonic triaxial tests were carried out on sand/silt mixtures, with fines content ranging from 0 to 20 %. In some series, the samples were reconstituted at the same initial global void ratio e_i while in other ones, the samples were reconstituted at the same initial intergranular void ratio $e_i{}^g$ to simulate the mechanical behaviour of natural material subjected to erosion phenomenon such suffusion where the fines are removed and the sand matrix kept constant. The results from this study revealed that the fines content affects both the liquefaction potential of silty sand and the position of the critical state line. Moreover, the results showed that the effect of fines depends strongly on the density parameter used for the interpretation, void ratio or intergranular void ratio.

T.-K. Nguyen · N. Benahmed (✉)
IRSTEA d'Aix en Provence, Unité de Recherche Ouvrages Hydrauliques, 3275 Route de Cézanne CS 40061, 13182 Aix en Provence cedex 5, France
e-mail: nadia.benahmed@irstea.fr

P.-Y. Hicher
Institut de Recherche en Génie Civil et Mécanique GeM, UMR CNRS 6183, Ecole Centrale de Nantes, BP 92101, 44321 Nantes cedex 3, France

M. Nicolas
IUSTI, UMR CNRS 6595, Aix-Marseille Université, 5 rue Enrico Fermi, 13453 Marseille cedex 13, France

K.-T. Chau and J. Zhao (eds.), *Bifurcation and Degradation of Geomaterials in the New Millennium*, Springer Series in Geomechanics and Geoengineering,
DOI 10.1007/978-3-319-13506-9_17

1 Introduction

Since several last decades, the instability of granular materials has taken a lot of interest of geotechnical researches. While this phenomenon on the clean sand were deeply investigated and well known already, the instability of the silty sand, which is widely encountered in nature, unfortunately, is not clear yet. However, it is established now that the effect of fines content is an important parameter when evaluating the mechanical behaviour of silty sands.

It is considered that when the low plastic fines content (FC) is below than a certain value, which is called threshold fines content Fcth, the mechanical behavior of material is dominated by coarse grains and the material can be named as silty sand. In contrast, when the FC is above this limit, the mechanical response is governed by the fines particles and the material is sandy silt.

In analyzing silty sand behavior, the concept of intergranular void ratio is largely used which is an extension of classical void ratio. This state parameter considers the volume of fines content as voids. Thevanayagam (1998) proposed a simple formula to calculated the intergranular void ratio at a certain state from the void ratio at corresponding state as following:

$$e^g = \frac{e + Fc/100}{1 - Fc/100} \tag{1}$$

In this study, the concepts of intergranular void ratio and classical void ratio, were both used in analyzing the effect of fines content on silty sand behaviour.

2 Experimental Procedure

2.1 Materials Used

The natural Camargue silty sand was collected from the Camargue flood protection dikes. The clean sand used is uniform with angular and rounded particles shape, a uniformity coefficient $C_U = 2.40$, and an effective diameter $D_{50} = 0.314$ mm. The minimum and maximum void ratio is equal to 0.610 and 1.100, respectively. The natural fines are low plastic fines with a plasticity index Ip = 8. The specific gravity of both sand and silt particles is $\rho = 2.67$ g/cm^3. The host sand and host silt were blended together to obtain the different mixtures with many desired target fines content (by weight) varying from 0, 5, 10, 15 and 20 %. The grains size distribution of all materials used in this study are presented in Fig. 1.

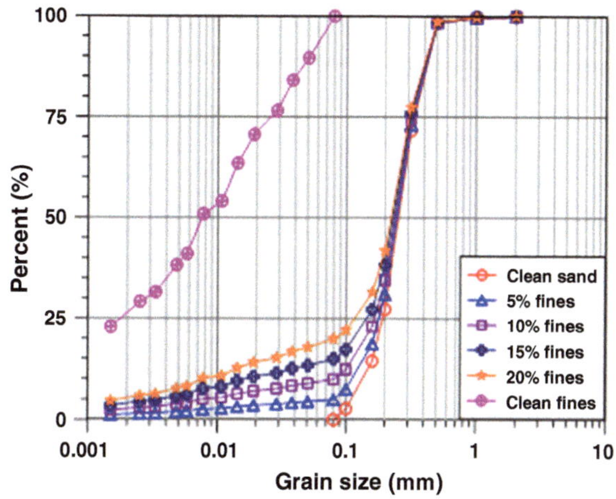

Fig. 1 Grain size distribution of materials used in this study

2.2 Specimens Reconstitution and Test Procedure

The moist tamping method was used to prepare all the specimens in this study to avoid segregation between coarse grains and fines particles and to get homogeneous samples. The oven-dried mixture of sand with fines was divided into seven equal parts and each part was mixed to a moisture content of 2 %, placed into a split mould and then compacted at predetermined height to achieve the target density. The nominal dimension of the specimens was 140 mm in height and 70 mm in diameter.

A fully computed controlled Dynamic Triaxial System from Wykeham Farrance was used. All the tests were conducted in strain controlled mode with constant rate of 1 % per minute.

3 Experimental Results and Discussion

3.1 Effect of Fines on Evolution of Min and Max of Void Ratio

It's worth to note that the value of maximum and minimum void ratio depends strongly on the procedure used, especially with high fines content. However, there is currently no standard to determine these parameters that is applicable for entire

Fig. 2 Evolution of maximum and minimum void ratio with fines content

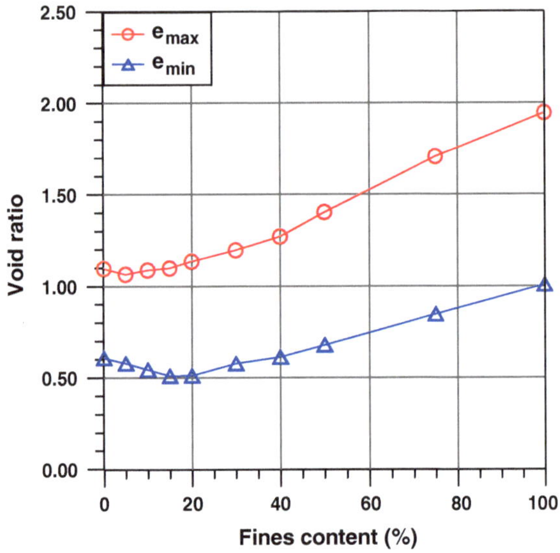

range of fines content. In literature, different authors used different methods, it can be either ASTM procedure (Amini and Qi 2000; Huang et al. 2004; Stamatopoulos 2010; Xenaki and Athanasopoulos 2003), Proctor test, pouring material into the mold by funnel or some kinds of combination (Papadopoulou and Tika 2008; Polito and Martin 2001; Yang et al. 2006). In this study, we used both French standard's procedure and Japanese standard's one to get consistent value. The evolution of maximum and minimum void ratio is reported in Fig. 2.

Based on physical meaning of threshold fines content, which is the maximum fines amount that can be contained in the void space while maintaining a contiguous sand skeleton (Polito and Martin 2001), the evolution of minimum void ratio suggests that its value in our case is about 20 %.

3.2 Undrained Behaviour of Sand with Fines

Four series of triaxial undrained tests were performed to investigate the effect of fines content on the behaviour of silty sand. In two first series, the initial overall void ratio was kept constant which means reducing the amount of sand when increasing the amount of fines. In each one, the specimens had the same initial overall void ratios, $e_i = 1.05$ and 1.00, corresponding to density indexes of 0.10 and 0.20, respectively, with the percentage of fines varied of 0, 5, 10, 15 and 20 %. The results of the first series are presented in Fig. 3.

In two other series, the specimens were prepared at a constant initial intergranular void ratio by keeping the same amount of sand and adding the fines particles progressively. The tests were carried out with two initial intergranular void ratios

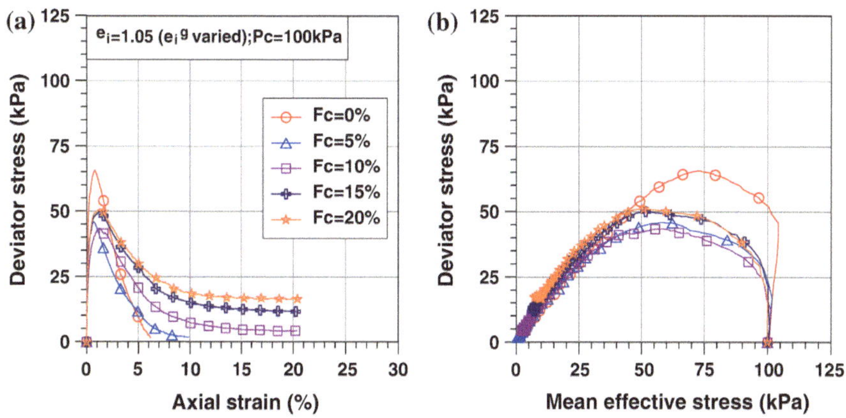

Fig. 3 Effects of fines content on undrained response of silty sand-series $e_i = 1.05$: **a** $q - \varepsilon_a$; **b** $q - p'$

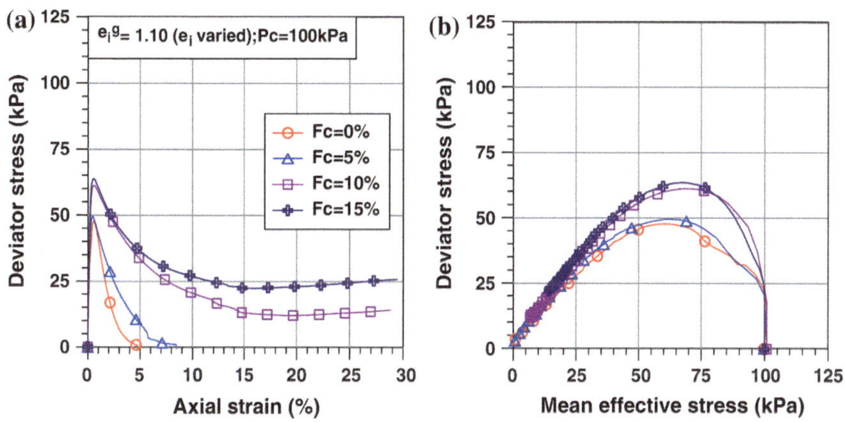

Fig. 4 Effects of fines content on undrained response of silty sand-series $e_i{}^g = 1.10$: **a** $q - \varepsilon_a$; **b** $q - p'$

$e_i{}^g = 1.20$ and 1.10, which correspond to density indexes of −0.20 and 0 respectively for host sand while they are higher for other mixtures. The fines content was varied by 0, 5, 10 and 15 %. The results are presented in Fig. 4.

In order to clearly highlight the evolution of the peak shear strength, the latter of all the four series is reported in Fig. 5, by two groups separately, as a function of the fines content. We observed that in two first series (case 1), the peak strength decreases significantly as the fines content increases up to about 5 % and then varies slightly with a further increase of fines content. As for the two latter series (case 2), the peak shear strength was insensitive to fines content up to 5 % and then increases with further increasing of fines content up to 15 %.

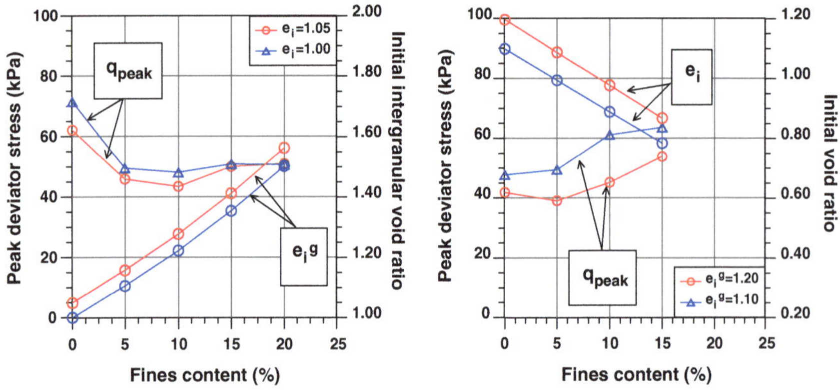

Fig. 5 Effect of fines content on peak deviator stress—two series at the initial void ratio e_i constant (*left*) and two series at the initial intergranular void ratio e_i^g constant (*right*)

We supposed that when the fines content is very low, i.e. below 5 % in this case, the fines particles are almost inactive. They fall into the voids between the sand particles without contributing to soil strength. Therefore, the peak resistance drops noticeably in case 1 because of the loosening of sand skeleton and does not change in case 2 because the skeleton remains unchanged. With a further increase of fines content, some of the fines particles start to participate in the force chain of the structure, hence the peak resistance decreases less significantly in case 1 and increases slightly in case 2.

3.3 Drained Behavior of Sand with Fines

Three series of drained tests were performed with the confining pressure of 100, 200 and 400 kPa. In each series, the specimens were prepared at the same initial intergranular void ratio e_i^g but with different percentages of fines: 5, 10 and 15 %. The Fig. 6 presents a comparison of three tests of the series carried out at confining pressure of 100 kPa.

Although there is no clear trend of the influence of fines on shear strength, as was seen in the case of the undrained tests, a slight increase of shear strength can be seen with the increase of fines. Nevertheless, the volumetric strains seem to be very sensitive to the percentage of fines. Indeed, the mixture of sand and fines is even less contractive as fines content increases. This trend is quite obvious since the specimens have the same amount of sand, and the addition of fines leads to denser, and hence, less contractive specimens.

All the tests were conducted up to 30 % axial strain where they reached to a very stable state in terms of both volumetric strain and deviator stress. The Fig. 7 presents the critical state lines obtained for different mixtures as well as those obtained on clean sand.

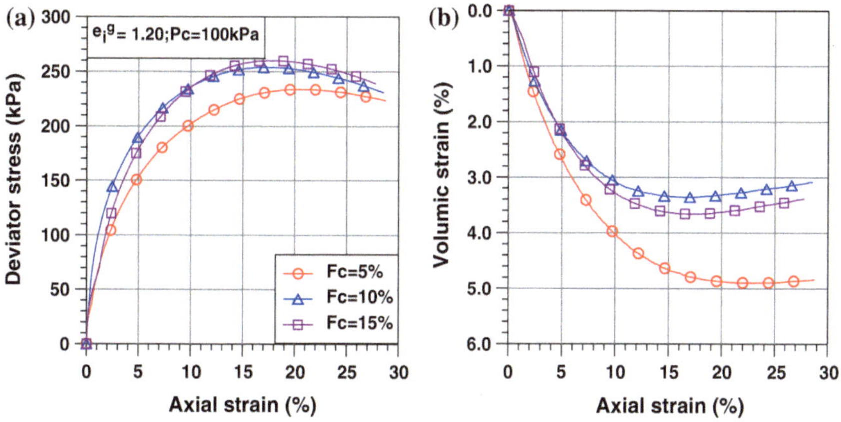

Fig. 6 Effects of fines content on drained response of silty sand: **a** q $-$ ε_a; **b** ε_v $-$ ε_a

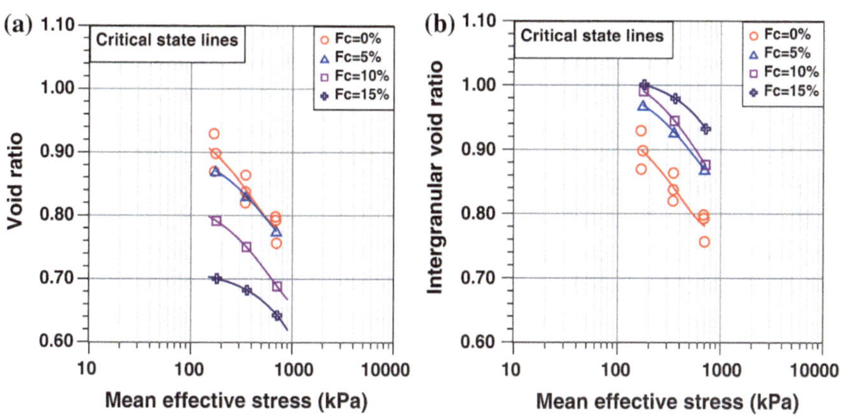

Fig. 7 Critical state line of different mixtures in two spaces: **a** e $-$ logp′; **b** e* $-$ logp′

In classical space e $-$ logp′ (Fig. 7a), it is interesting to note that each mixture creates its own critical state line, which moved quickly downward as the fines content increased up to 15 %. In Fig. 7b, the intergranular void ratio is used as the state parameter instead of classical void ratio. In this case, the trend is reversed with a slight upward movement of critical state line as the fines content increased. Similar results were reported by Papadopoulou and Tika (2008).

Since the internal friction angle at critical state remained unchanged, the movement of critical state lines means that at a given void ratio, the shear resistance at the critical state decreased remarkably as the fines content was increased up to 15 %. In contrast, at a given intergranular void ratio, the shear resistance at the critical state increases softly as fines content is increased up to 15 %. It means that, similarly to the peak strength, the material resistance at the critical state is rather controlled by sand skeleton whereas the fines particles play less important role.

Furthermore, we found that the fines particle have an effect on critical state even its percentage is as small as 5 %. It may be explained by the fact that the structure has changed during shearing and has become denser at the end of drained test leading to an increase of active fines particles amount.

4 Conclusion

An experimental study was conducted to determine the mechanical behaviour of silty sand using monotonic triaxial compression tests under drained and undrained conditions. The effect of fines content on mechanical response was investigated by using different sand/fines mixtures with different percentages of fines.

The results of undrained tests showed that the effects of fines content on the peak shear strength depend on the density parameter used in the interpretation, initial classical void ratio or initial intergranular void ratio. It's necessary to point out that the evaluations presented in current study were performed on specimens having the same initial density parameter (void ratio or intergranular void ratio) but not the same post-consolidation density parameter. We noted that the densification of specimens during the fabrication procedure increases as the fines content is high. Therefore, the evolutions of the peak strength presented here include the effect of density. Hence, the evaluation based on post-consolidation density indexes may provide different results and will be presented elsewhere in detail.

The results of drained tests revealed that the presence of fines particles affect the position of the critical state line. This is coherent with the fact that each mixture creates a new material.

References

Amini F, Qi GZ (2000) Liquefaction testing of stratified silty sands. J Geotech Geoenvironmental Eng 126(3):208–217

Huang YT, Huang AB, Kuo YC, Tsai MD (2004) A laboratory study on the undrained strength of a silty sand from Central Western Taiwan. Soil Dyn Earthq Eng 24(9–10): 733–743

Papadopoulou A, Tika T (2008) The effect of fines on critical state and liquefaction resistance characteristics of non-plastic silty sands. Soils Found 48(5):713–725

Polito CP, Martin JR II (2001) Effects of nonplastic fines on the liquefaction resistance of sands. J Geotech Geoenvironmental Eng 127(5):408–415

Stamatopoulos C (2010) An experimental study of the liquefaction strength of silty sands in terms of the state parameter. Soil Dyn Earthq Eng 30(8):662–678

Thevanayagam S (1998) Effect of fines and confining stress on undrained shear strength of silty sands. J Geotech Geoenvironmental Eng 124(6):479–491

Xenaki VC, Athanasopoulos GA (2003) Liquefaction resistance of sand-silt mixtures: an experimental investigation of the effect of fines. Soil Dyn Earthq Eng 23(3):183–194

Yang SL, Lacasse S, Sandven R (2006) Determination of the transitional fines content of mixtures of sand and non-plastic fines. Geotech Test J 29(2):102–107

A Multiscale Investigation of Strain Localization in Cohesionless Sand

Ning Guo and Jidong Zhao

Abstract We develop a hybrid finite-discrete element method (FEM/DEM) multiscale framework to model strain localization in cohesionless sand. This framework takes advantage of a hierarchical solution structure based on the two numerical techniques, by modeling the material as a continuum at large using FEM and deriving the material point response of the FEM mesh from a DEM assembly attached to each integration point. In doing so, the phenomenological assumptions on the macroscopic constitutive law in conventional continuum modeling can be naturally bypassed. We demonstrate the predictive capability of the model by simulating a biaxial compression test on sand where strain localization is observed. The influences of different boundary conditions on the shear band pattern are examined.

1 Introduction

Strain localization is one of the major failure mechanisms for granular materials and is considered as a precursor of many catastrophic geo-hazards in civil engineering such as slope instability, debris flow and failure of foundation. Modeling of strain localization has been an active area for both engineers and theorists using either continuum-based methods such as finite element method (FEM) (Gao and Zhao 2013; Tejchman and Wu 2009) or discrete element method (DEM) (Fu and Dafalias 2011; Ishihara and Oda 1998). However, the phenomenon of strain localization in granular media is still not well understood due partially to the two facts: First, the continuum-based methods neglect the discrete nature of granular media

N. Guo (✉) · J. Zhao
Department of Civil and Environmental Engineering, The Hong Kong University of Science and Technology, Clear Water Bay, Kowloon, Hong Kong
e-mail: ceguo@ust.hk

J. Zhao
e-mail: jzhao@ust.hk

K.-T. Chau and J. Zhao (eds.), *Bifurcation and Degradation of Geomaterials in the New Millennium*, Springer Series in Geomechanics and Geoengineering,
DOI 10.1007/978-3-319-13506-9_18

while the macroscopic response of the material is indeed encoded in its microscopic structural character, which makes the continuum-based method inept to explain the underlying mechanism for the macroscopic observations and being frequently criticized as phenomenological. Second, strain localization is indeed a boundary value problem (BVP) even for a sample in laboratory test, which typically involves large number of particles which cannot be effectively handled by most discrete methods due to computational cost. To address these issues, herein we present a hybrid method to study strain localization in granular media, aiming to retain the advantages of both FEM and DEM and meantime to avoid their respective limits as mentioned.

In the hybrid FEM/DEM model, the two methods are coupled in a hierarchical manner where FEM is used to solve the global BVP and a DEM assembly is attached at each Gauss integration point of the FEM mesh serving as representative volume element (RVE) (see our previous work, Guo and Zhao 2013). The local material responses are captured by the embedded DEM simulations and are provided for global solution by the FEM, while the FEM provides displacement/strain conditions as boundary conditions for each DEM at each time step. In so doing, the phenomenological nature in conventional constitutive modeling can be totally bypassed, whilst the computational efficiency of FEM in solving BVPs is retained. More importantly, the framework provides microscopic insights at the particle level thanks to the local DEM simulations, which can potentially shed lights on important physical mechanisms accounting for some macroscopic observations including the initiation and development of strain localization. The predictive capability of the hybrid multiscale framework is demonstrated by a simulation on biaxial compression tests, where different strain localization patterns are observed when an initially uniform specimen is subjected to different boundary conditions.

2 Approach

2.1 Formulations

The hybrid model is implemented by coupling two open-source codes—Escript (Gross et al. 2007) and YADE (Šmilauer et al. 2010), for FEM and DEM computations respectively. The entire problem domain is first discretized into a proper FEM mesh. The FEM solution procedure follows the conventional displacement driven formulation in solving the global weak-form governing equation for quasi-static loading in the absence of body force. For a typical nonlinear problem as in dealing with granular media, Newton-Raphson iterative method is usually adopted to solve the nonlinear equation system which requires the tangent operator $\boldsymbol{D}$ to assemble the tangent matrix. Different from the conventional FEM solution procedure where $\boldsymbol{D}$ is derived from an ad hoc constitutive assumption (e.g. the elasto-plastic modulus), the current hybrid FEM/DEM model obtains the tangent operators from the embedded discrete element assemblies.

We estimate the tangent operator from the elastic modulus $\mathbb{D}$ of a DEM packing, which can be analytically derived based on the assumption of uniform strain field (Kruyt and Rothenburg 1998):

$$\mathbb{D} = \frac{1}{V} \sum_{N_c} \left(k_n \boldsymbol{n}^c \otimes \boldsymbol{d}^c \otimes \boldsymbol{n}^c \otimes \boldsymbol{d}^c + k_t \boldsymbol{t}^c \otimes \boldsymbol{d}^c \otimes \boldsymbol{t}^c \otimes \boldsymbol{d}^c \right) \tag{1}$$

where V is the total volume of the DEM assembly, N_c is the number of contacts in the volume, k_n and k_t are the normal and tangential contact stiffnesses, $\boldsymbol{n}^c$ and $\boldsymbol{t}^c$ are the unit vectors in the outward normal and tangential directions at a contact, respectively. $\boldsymbol{d}^c$ is the branch vector connecting the centroids of the two contacted particles. This tangent operator is then used to assemble the tangent matrix for the FEM solver.

Besides the tangent operator, the stress tensor is also homogenized from the DEM packing following the Love's formula $\boldsymbol{\sigma} = \sum_{N_c} \boldsymbol{d}^c \otimes \boldsymbol{f}^c / V$ where $\boldsymbol{f}^c$ is the contact force. For 2D case, the mean and deviatoric stress can be calculated $p = \mathrm{tr}\boldsymbol{\sigma}/2$, $q = \sqrt{\boldsymbol{s} : \boldsymbol{s}/2}$ ($\boldsymbol{s} = \boldsymbol{\sigma} - p\boldsymbol{\delta}$). Similarly for the local strain measure, we take the symmetric part of the DEM deformation $\nabla \boldsymbol{u}$ (the anti-symmetric part accounts for rigid body rotation) $\boldsymbol{\varepsilon} = (\nabla \boldsymbol{u} + \nabla \boldsymbol{u}^T)/2$.

2.2 DEM Model and RVE Calibration

We use a linear contact law in the DEM and the Coulomb's friction criterion to describe the frictional behavior at each interparticle contact. Three input parameters are required for DEM model: the Young's modulus E_c, Poisson's ratio ν_c and the friction angle φ. The normal and tangential stiffnesses are determined by $k_n = E_c r$ and $k_t = \nu_c k_n$ where $r = 2r_1 r_2/(r_1 + r_2)$ is the common radius of the two contacted particles. A threshold for the tangential force is imposed by the friction angle such that $|f_t^c| \leq f_n^c \tan \varphi$. Note that k_n and k_t will be used in Eq. (1) to calculate the tangent operator. A simple local non-viscous damping force is also added to dissipate kinetic energy $\boldsymbol{f}^{damp} = -\alpha f^{resid} \boldsymbol{v}/v$ where α is the damping ratio, $\boldsymbol{f}^{resid}$ is the residual force (unbalanced force) on the particle, $\boldsymbol{v}/v$ returns the direction of the particle velocity. All the parameters relevant to the DEM model are summarized in Table 1.

After careful calibration, we determine to use a RVE containing 400 circular disks with a particle size distribution (PSD) ranging from 3 to 7 mm such that its mechanical response is in qualitative agreement with laboratory tests on sand. The gravity force is not considered in the DEM model and periodic boundary is used in both directions of the RVE. Initially, the RVE is isotropically consolidated to a mean pressure $p_0 = 100$ kPa with an initial void ratio of 0.177.

Table 1 Parameters for the DEM model

Radii (mm)	Density (kg/m^3)	E_c (MPa)	ν_c	φ (rad)	α
3 ~ 7	2,650	600	0.8	0.5	0.1

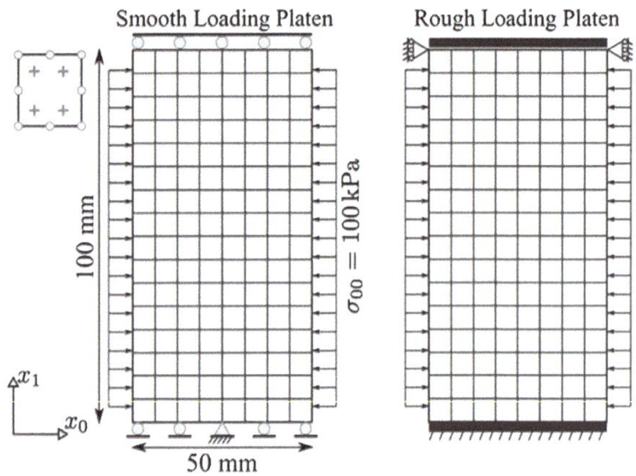

Fig. 1 FEM mesh and boundary conditions for the biaxial compression tests. *Left* smooth boundary case; *right* rough boundary case. *Top left* 8-node 4-integration point serendipity element used for the FEM mesh in both cases

3 Results and Discussion

We demonstrate the predictive capacity of the hybrid FEM/DEM model described above by using it to simulate a biaxial compression test. Particular focus is placed here on the initiation and development of strain localization and the influence of boundary condition. The specimen has a dimension of 50 mm in width and 100 mm in height. First, it is discretized into a FEM mesh with 8×16 quadrilateral elements. We use the 8-node serendipity element with 4 Gauss integration point each. In total, 512 RVEs are used for the whole domain and they all possess the same initial condition, which leads to a uniform sample. Drained condition is imposed to the specimen, by maintaining a constant lateral confining pressure (i.e. $\sigma_{00} = 100$ kPa) during the compression. The FEM mesh with quadrilateral element and the boundary conditions for the specimen are shown in Fig. 1.

The resultant normal stresses on the loading platen are shown in Fig. 2 for both cases of boundary conditions. The result from a pure DEM test on the RVE is also

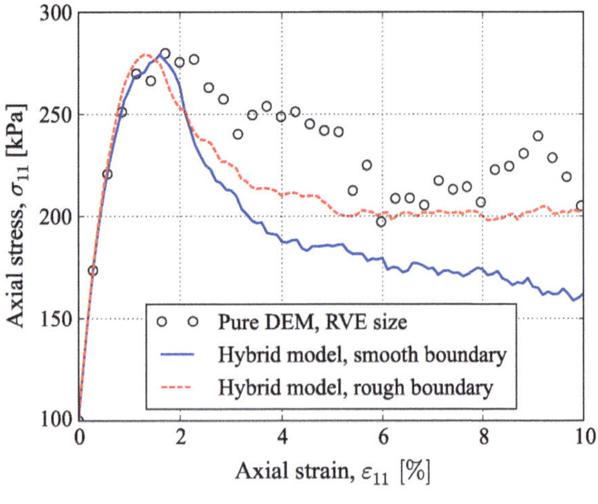

Fig. 2 Resultant normal stress on the loading platen during the compression tests

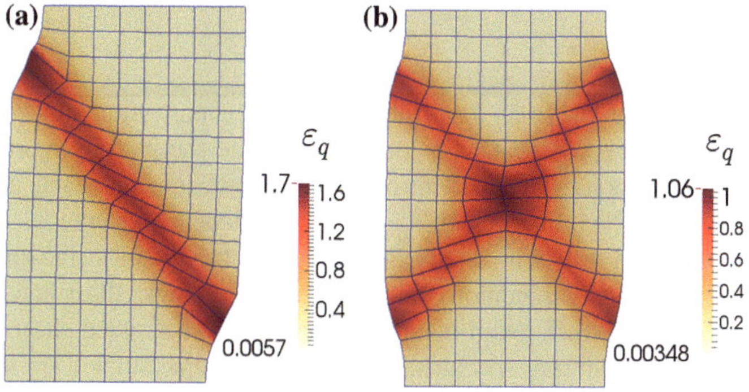

Fig. 3 Accumulated deviatoric strain contour showing strain localization: **a** *smooth* boundary; **b** *rough* boundary

presented for comparison. As seen from the figure, the hybrid model successfully predicts the typical response of dense sand subject to the biaxial compression as observed in laboratory and other conventional FEM simulations. For the smooth boundary case, the pre-peak response from the hybrid model matches the pure DEM test very well, while its post-peak response is much softer than that from the pure DEM test, which is attributable to the global strain localization as to be discussed later. For the rough boundary case, the initial stiffness is slightly higher than that in the smooth boundary condition due to the deformation of the specimen is restricted by the rough loading platen. The residual stress in this case (200 kPa) is also higher than that in the smooth case (160 kPa). However, the specimen attains almost the same peak strength in both cases (around 280 kPa).

The accumulated deviatoric strain contours at the final state (i.e. $\varepsilon_{11} = 10\,\%$) are presented in Fig. 3 which clearly shows that strain localization occurs in both cases. For the smooth boundary condition, a dominant single shear band is observed, while in the case using rough boundary, a pattern with crossed double shear bands is observed. These observations are consistent with the results presented by Gao and Zhao (2013) and the laboratory tests by Tatsuoka et al. (1990). Note that as the specimen is initially uniform and the loading condition is symmetric, no strain localization can be predicted by most conventional models when smooth boundary is used in conventional FEM tests, due to the lacking of symmetry breaker. Many studies have introduced artificial imperfections or random fields in order to trigger strain localization. While in the current hybrid model, the material response is captured by the RVE which is disordered and contains inherent heterogeneity in nature. This heterogeneity, though considerably small when the RVE is prepared to be isotropic, leads to a non-coaxial behavior of the granular media which initiate the strain localization (c.f. Gao and Zhao 2013).

4 Conclusions

We present a study on the strain localization in biaxial compression tests with a novel hybrid FEM/DEM approach. This modeling technique successfully reproduces the granular media responses without introducing any phenomenological constitutive relation for FEM. The influences of boundary conditions are examined. The use of rough loading platen results in a stiffer response than the use of a smooth one. And with a smooth boundary condition, only one dominant single shear band is developed; while in the case of rough boundary condition, two crossed shear bands are seen in the initially uniform specimen. Further microscopic studies are required to reveal the underlying mechanisms of the initiation of shear band in granular media.

Acknowledgments This work was supported by RGC HK (Grant No. 623609).

References

Fu P, Dafalias YF (2011) Fabric evolution within shear bands of granular materials and its relation to critical state theory. Int J Numer Anal Meth Geomech 35:1918–1948
Gao Z, Zhao J (2013) Strain localization and fabric evolution in sand. Int J Solids Struct 50:3634–3648
Gross L, Bourgouin L, Hale A, Muhlhaus H (2007) Interface modelling in incompressible media using level sets in escript. Phys Earth Planet Inter 163:23–34
Guo N, Zhao J (2013) A hierarchical model for cross-scale simulation of granular media. AIP Conf Proc 1542:1222–1225
Ishihara K, Oda M (1998) Rolling resistance at contacts in simulation of shear band development by DEM. J Eng Mech 124:285–292
Kruyt N, Rothenburg L (1998) Statistical theories for the elastic moduli of two-dimensional assemblies of granular materials. Int J Eng Sci 36:1127–1142
Šmilauer V, Catalano E, Chareyre B, Dorofeenko S, Duriez J, Gladky A, Kozicki J, Modenese C, Scholtès L, Sibille L, Stránský J, Thoeni K (2010) Yade reference documentation. The Yade Project
Tatsuoka F, Nakamura S, Huang CC, Tani K (1990) Strength anisotropy and shear band direction in plane strain tests of sand. Soils Found 30:35–54
Tejchman J, Wu W (2009) Non-coaxiality and stress-dilatancy rule in granular materials: FE investigation within micro-polar hypoplasticity. Int J Numer Anal Meth Geomech 33:117–142

The Effect of Rotational Degrees of Freedom on the Formation of Deformation Patterns in Granular Materials Using Digital Image Correlation

Maxim Esin, Arcady V. Dyskin and Elena Pasternak

Abstract The shear band formation and evolution is a predominant mechanism of deformation patterning in granular materials. Independent rotations of separate particles can affect the pattern formation by adding the effect of rotational degrees of freedom to the mechanism of instability. Measurement of independent grain rotations is technically challenging. We use a special experimental technique based on digital image correlation in order to recover both displacement and independent rotation fields in 2D physical models of granular material. In the model the particles are represented by smooth steel disks with speckles painted on them to enable the rotation reconstruction. Both mono- and polydispersed particle assemblies are used. We show that the average values of the angles of disk rotations are insignificantly different from zero. It means that the rotations are microscopic (at the scale of the grain size) and do not proliferate to the macroscopic scale. However the particle rotations exhibit mesoscopic clustering: monodispersed assemblies produce vertical columns of particles rotating the same direction; polydispersed assemblies 2D form clusters of particles with alternating rotations.

1 Introduction

Shear band formation is the main mechanism of instability, localisation of deformation and eventually failure of granular materials and its investigation is important for predicting failures of natural and engineering structures. Experimental studies of shear band formation conducted to date use a variety of granular materials to

M. Esin (✉) · E. Pasternak
Deep Exploration Technologies Cooperative Research Centre, School of Mechanical and Chemical Engineering, University of Western Australia, Perth, Australia
e-mail: esinmy@gmail.com

A.V. Dyskin
Deep Exploration Technologies Cooperative Research Centre, School of Civil, Environmental and Mining Engineering, University of Western Australia, Perth, Australia

K.-T. Chau and J. Zhao (eds.), *Bifurcation and Degradation of Geomaterials in the New Millennium*, Springer Series in Geomechanics and Geoengineering, DOI 10.1007/978-3-319-13506-9_19

observe the deformation and characteristics of this localised phenomenon. The use of sand is very common and it is the granular material observed and analysed in most experiments conducted (Gudehus and Nubel 2004; Rechenmacher et al. 2010; Andrade et al. 2010; Hall et al. 2010). A special 2D experiment using coins as particles was conducted in Tamura and Yamada (1996) in order to visualize the localized deformation and shear band formation. In this paper we develop further the approach proposed in Tamura and Yamada (1996) and include the measurements of particle rotations. Furthermore, we discover the behavior of the polydispersed assembly of the particles. It should be noted that although the physics of polydispersed granular materials was studied before Tordesillas et al. (2008, 2012) nobody analyzed the particle rotations of the polydispersed assembly.

2 Physical Model

The model consists of a transparent PMMA frame and a slider to effect pressure on the assembly of the disks. Studying the monodispersed assembly, it consists of a 10×11 array of smooth disks with speckles painted on them to enable the rotation reconstruction using the DIC photogrammetric technique. All disks in the physical model where initially in contact with all the neighbouring disks forming hexagonal packing. The disks were made of steel and have thickness of 1.6 mm. The diameter of disks is constant and equal to 23.5 mm, the mass of each disk is 5.45 g. This model is similar to that of a model used in a previous experiment conducted in Tamura and Yamada (1996) where Japanese 10 yen coins (thickness is 1.5 mm, diameter is 23.5 mm, mass is 4.5 g) were used.

For the polydispersed assemblies we used 3 different numbers of disks randomly packed in the experimental setup. The assembly consists of 19, 38 and 33 disks with diameters of 20, 23.5 and 27 mm respectively.

3 The Experiment

The experimental setup with the monodispersed assembly of the disks is shown in Fig. 1. The experiment consisted of moving the slider by rotating the hand crank, measuring of the spring length to determine applied load and taking the photos to recover both displacement and independent rotation fields of the disks using the digital image correlation. We used software (The VIC-2D System) and set the measurement areas. VIC-2D uses correlation algorithms for obtaining full-field displacement and strain data. In-plane movements can be determined for every point within the measurement area, as well as the Lagrangian strain tensor. At the moments of re-compaction of the disks a high-speed camera instead of a regular camera is used in order to capture the rapid processes associated with the shear band formation and re-compaction.

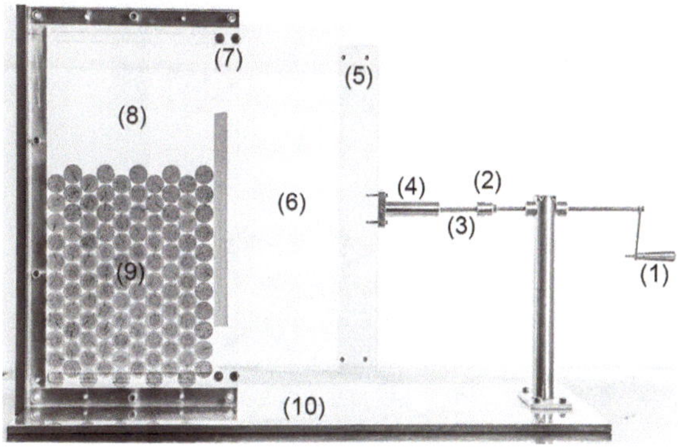

Fig. 1 The experimental apparatus: *1* hand crank, *2* compression block *3* compression spring, *4* hollow cylinder, *5* slide mount, *6* slider, *7* roller bearings, *8* perplex plates, *9* physical model, *10* base

4 Experimental Results for Monodispersed Assembly

Figure 2 shows the force versus slider displacement plots obtained from the experiments with a monodispersed assembly. They represent several consecutive tests each covering three re-compactions of the assembly. It should be noted that the second re-compaction sometimes does not look similar to the first and third one. Figure 3a, b show stages of the first re-compaction, Fig. 3c, d show the lattice-like pattern observed during the second re-compaction. This difference in patterns can be explained by small imperfections in the disks, their variable friction with the walls, small inclinations of the slider or other factors, which require further investigation. The shear bands observed always form at 30°, which is consistent with the hexagonal packing of the disks.

The localised deformation and rotations observed throughout the experimentation are qualitatively similar to that of the experiments conducted in Tamura and Yamada (1996). The shear bands observed throughout experimentation and those in Tamura and Yamada (1996) look identical. This can be attributed to the similarities of the physical models.

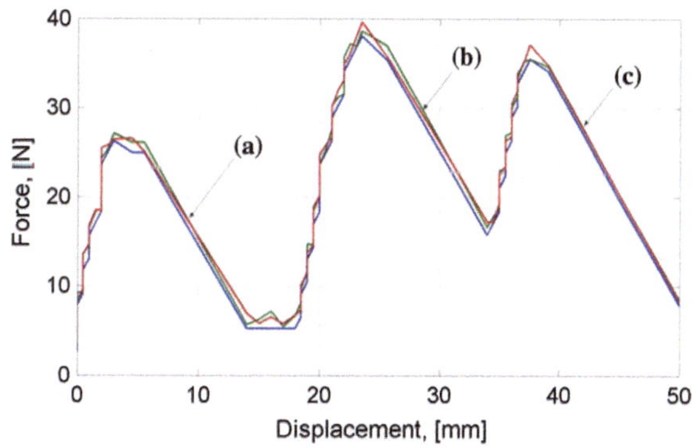

Fig. 2 The force versus slider displacement graphs for three trials for the monodispersed assembly. Shear band formation during: **a** the first re-compaction, **b** the second re-compaction, **c** the third re-compaction

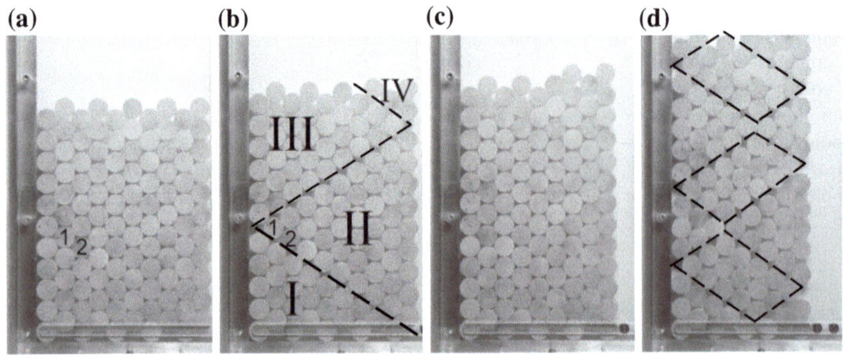

Fig. 3 Stages during the re-compactions: **a** initial model before first re-compaction, **b** shear band formation, **c** initial model before second re-compaction, **d** shear band formation (shear bands are represented by *dashed lines*)

As can be seen from Fig. 3 there are four separate regions (I, II, III and IV) formed during the shear band formation. Analysing the disk rotations it can be seen that in each region the resultant angles of the disk rotations have the same sign in each vertical column, but alternate signs between the columns. The values of angles are similar. The only exception is the region I where constraints on displacements and additional friction with the walls of the experimental apparatus play a considerable role. As a result, most of the disks included of the region I rotate to small angles. Also for each side of shear bands between regions II, III and IV it is seen that the signs of the disk rotations change when we are going from one region to another. The rotation angles averaged over each of the four regions are found to be statistically insignificantly different from zero. This suggests that the shear band formation is accompanied by formation of oriented rotation patterns at the scale intermediate between the particle size and the region size.

Figure 4 shows the rotation angle versus slider displacement for two disks. The solid blue line (a) shows the rotation angle of the disk #1, the dotted green line (b) shows the rotation angle of the disk #2 (Fig. 3a). Fluctuations of rotation angles are due to the noise associated with the digital image correlation.

Fig. 4 The angles of disk rotations versus displacement graphs: **a** the rotation angle of the disk #1, **b** the rotation angle of the disk #2

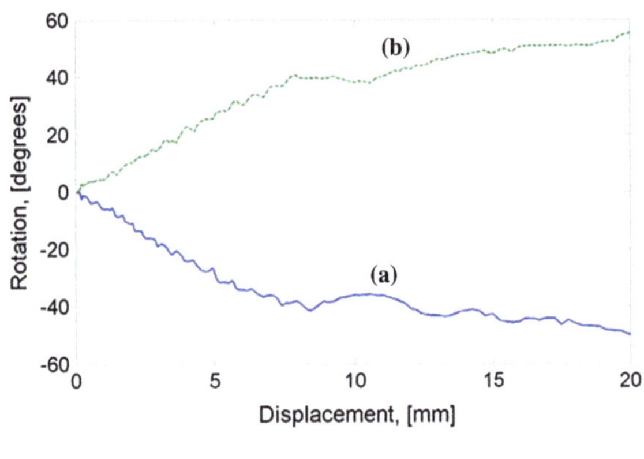

5 Experimental Results for the Polydispersed Assembly

Figure 5 shows the force versus slider displacement graph for a polydispersed assembly. As with the monodispersed assembly these graphs were obtained from several consecutive tests. The most important feature of polydispersed assemblies is that their behaviour depends on the packing. Random polydispersed assemblies can be packed with quite different volume fractions (Jaeger et al. 1996). Obviously the mechanical behaviour will be different as well. In the tests we observed two types of the re-compactions of the polydispersed assembly: constant gradual re-compactions without significant drops in the force (a) and recompactions with significant drops in the force similar to the graph for the monodispersed assembly (b). However, in polydispersed assemblies we did not observe shear bands similar to monodispersed assemblies.

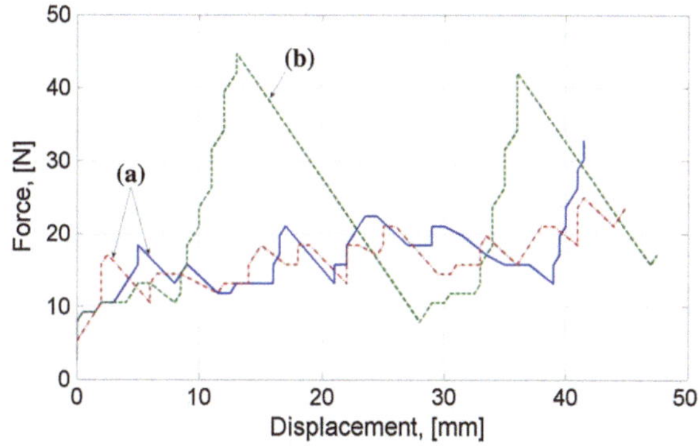

Fig. 5 Force versus slider displacement for three trials for the polydispersed assembly: **a** constant gradual re-compactions without significant drops in the force; **b** re-compactions with significant drops in the force

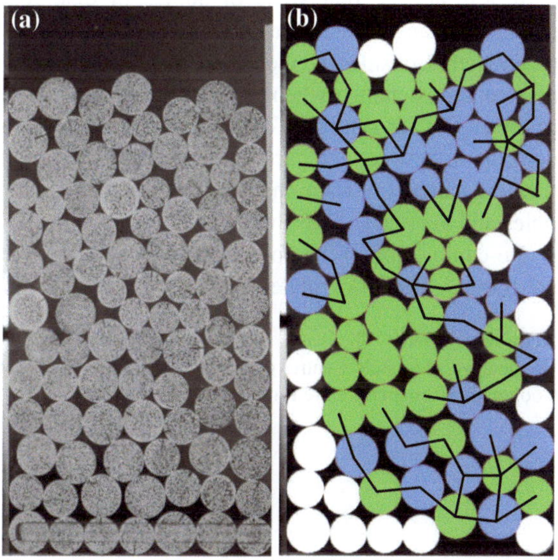

Fig. 6 Original (**a**) and final positions with the qualitative picture of the rotation angles (**b**) of the disks in the polydispersed assembly

Figure 6 shows the original and final positions of the disks and their rotations after the re-compactions of the polydispersed particle assembly in our physical experiment. Here we do not show the exact values of the angles, but rather represent the qualitative picture, which is more interesting at the moment.

Green colour indicates the counterclockwise rotation, blue colour—clockwise rotation, white colour—non-rotating disks i.e. the disks with rotation angles within the measurement error. As can be seen during the experiment with the polydispersed assembly, the disks rotate in clusters rather than individually. To show the clusters we connect the neighbouring contacting disks with opposite signs of rotations by black lines. It can be seen that one of the cluster passes through almost the entire assembly. The maximum length of this cluster is 200 mm, the number of links is 15.

6 Conclusion

The results observed throughout experimentation with the monodispersed assembly were comparable to that of the similar experiment conducted in Tamura and Yamada (1996). In particular, the shear band formations we observed were similar. In both cases the assembly is divided into separate similar regions during the first re-compaction.

We experimentally investigated angles of disk rotations during re-compactions of both mono- and polydispersed assemblies. It was found that the average values of the angles of disk rotations are insignificantly different from zero. It means that the rotations are microscopic (at the scale of the grain size) and do not proliferate to the macroscopic scale.

It was noticed that signs of angles of disk rotations in the monodispersed assembly change in each region when we pass from one vertical column of disks to another. Considering the whole monodispersed assembly, signs of angles of disk rotations change when we pass from one region to another. The magnitudes of the angles are similar in columns within each region. It is evident that there is a correlation between the particle rotations within the columns. This corresponds to forming patterns of rotations at an intermediate scale between the scale of the particles and the scale of the shear band.

Re-compaction of the polydispersed assembly is different: a number of clusters are formed. The disks rotate in clusters rather than individually.

Acknowledgment The work has been supported by the Deep Exploration Technologies Cooperative Research Centre whose activities are funded by the Australian Government's Cooperative Research Centre Programme. This is DET CRC Document 2014/351.

References

Andrade JE, Avilaa CF, Hall SA, Lenoir N, Viggiani G (2010) From grain kinematics to continuum mechanics. J Mech Phys Solids 59(2):237–250

Gudehus G, Nubel K (2004) Evolution of shear bands in sand. Géotechnique 54(3):187–201

Hall SA, Wood DM, Ibraim E, Viggiani G (2010) Localised deformation patterning in 2D granular materials revealed by digital image correlation. Granular Matter 12(1):1–14

Jaeger HM, Nagel SR, Behringer RP (1996) The physics of granular materials. Phys Today 49(4):32–38

Rechenmacher A, Abedi S, Chupin O (2010) Evolution of force chains in shear bands in sands. Géotechnique 60(5):343–351

Tamura T, Yamada Y (1996) A rigid-plastic analysis of granular materials. Soils Found 36(3):113–121

Tordesillas A, Muthuswamy M, Walsh SDC (2008) Mesoscale measures of nonaffine deformation in dense granular assemblies. J Eng Mech 134(12):1095–1113

Tordesillas A, Walker DM, Froyland G, Zhang J, Behringer RP (2012) Transition dynamics and magic-number-like behavior of frictional granular clusters. Phys Rev E 86:011306

Stick-Slip Motion and the Associated Frictional Instability Caused by Vertical Oscillations

Iuliia Karachevtseva, Arcady V. Dyskin and Elena Pasternak

Abstract We investigate the influence of harmonic normal force on the frictional sliding. Frictional sliding demonstrates stick-slip motion even when the friction coefficient is constant. Applying vertical oscillations with a proper phase shift with respect to the beginning of the stick slip process creates a time window of decreasing stick-slip velocities. Reducing the amplitude of the harmonic normal force delays the time of the appearance of the window but extends its duration. More surprisingly, the reduction in the amplitude of the normal force oscillation leads to decreasing the stick-slip velocity within the time window.

1 Introduction

Sliding over pre-existing fractures and interfaces is one of the forms of instability in geomaterials. Sliding is often accompanied by stick-slip—a spontaneous jerking motion between two contacting bodies, sliding over each over. Stick-slip motion is an undesirable effect in engineering as it can lead to excessive energy consumption and premature wear. In the Earth's crust stick-slip in fault sliding is associated with the occurrence of earthquakes (Byerlee 1970).

It is usually assumed that the mechanism of stick-slip motion lies in intermittent change between static and kinetic friction (Popp and Rudolph 2004).

I. Karachevtseva (✉) · E. Pasternak
Deep Exploration Technologies Cooperative Research Centre, School of Mechanical and Chemical Engineering, The University of Western Australia, Perth, Australia
e-mail: juliso22@gmail.com

E. Pasternak
e-mail: elena.pasternak@uwa.edu.au

A.V. Dyskin
Deep Exploration Technologies Cooperative Research Centre, School of Civil, Environmental and Mining Engineering, The University of Western Australia, Perth, Australia
e-mail: arcady.dyskin@uwa.edu.au

K.-T. Chau and J. Zhao (eds.), *Bifurcation and Degradation of Geomaterials in the New Millennium*, Springer Series in Geomechanics and Geoengineering,
DOI 10.1007/978-3-319-13506-9_20

We however showed that the stick-slip type of oscillations could be observed even with a constant friction coefficient if a block of a mass M is driven with a constant velocity V_0 through a spring of stiffness k (Karachevtseva et al. 2014). In this case frequency Ω of the stick-slip oscillations of sliding velocity is $\Omega^2 = k/M$. Two types of sliding can be observed in this situation:

Steady slip with the slip velocity $V = V_0$;
Stick-slip ($V \neq V_0$ and oscillating with frequency Ω).

The first type is an analogue of fault creep in the Earth's crust, the second type represents the earthquake cycle. Both types of sliding are usually investigated using a spring-block model introduced by Burridge and Knopoff (Turcotte 1992).

In this paper we consider the effect of normal vibrations on the mechanism of sliding. In the Earth crust such vibrations can be induced by either tidal stresses (very low frequencies) or seismic waves produced by a nearby instability (high frequencies). At lower scale such vibrations can be either technogenic or even intentionally induced in order to reduce friction (as for instance in cutting, Popov (2010)).

When tidal normal stress is negative (directs inward to a fault plane), the total normal pressure on the fault would increase and keep the system stationary (stick phase). In case of reduction of normal force the friction will decrease and allow fault slipping (slip phase). The amplitudes of tidal stress on a seismic fault depend on such factors as fault type, geographical location, orientation and tidal forces produced by sun and moon (Xu et al. 2011).

2 A Single-Degree-of-Freedom Model of Frictional Sliding

Consider a single degree of freedom block-spring model driven by a spring whose other end is attached to a driver moving with a constant velocity (Fig. 1). The system consists of mass M, spring of stiffness k and a driver that moves with the constant velocity V_0. Now we apply vertical oscillations to the block by generating harmonic force $F(t) = A\sin\omega t$. In nature the sources of vertical vibration can be waves emitted seismic vibration and tides.

The system of equations which represents the motion of the block reads:

$$\begin{cases} \dot{V} = T - \mathrm{sgn}(V)\mu(N - A\sin\omega t) \\ \dot{T} = k(V_0 - V) \end{cases} \tag{1}$$

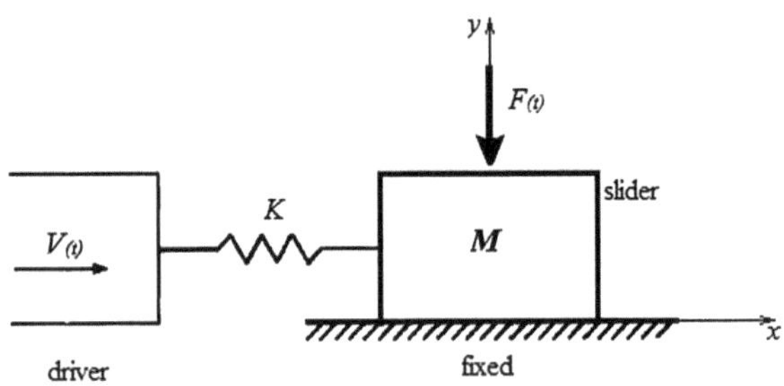

Fig. 1 The single-degree-of-freedom block-spring model

where $\mathrm{sgn}(V) = -1$ for $V < 0$, 0 for $V = 0$ and 1 for $V > 0$. The sign function represents the fact that friction always acts against velocity.

In order to represent the system of Eq. (1) in dimensionless form, it is convenient to introduce a dimensionless time t^*:

$$t^* = t\omega_0\,,\ \omega_0^2 = \frac{k}{M} \tag{2}$$

where ω_0 is the eigen frequency of the block-spring system, M is the block mass and k is the spring stiffness.

The equations in the dimensionless form read:

$$\begin{cases} \dot{V} = T - \mu(1 - A\sin\omega t) \\ \dot{T} = 1 - V \end{cases} \tag{3}$$

where the dot represents the derivative with respect to the dimensionless time t^*.

We consider the following initial conditions:

$$V(0) = 0,\ T(0) = 0 \tag{4}$$

3 Stick-Slip Regime

Figures 2 and 3 represent the behaviour of the system (dimensionless velocity and force vs. dimensionless time). First, we can see that the system exhibits the stick-slip type sliding even with constant friction coefficient owing to pure self-excitation. The oscillating velocity shows a phase shift of $\varphi = 0.3$ with respect to the oscillations of the force. Second, imposing normal vibrations, $F(t) = A\sin\omega t$ on the different frequencies does not lead to noticeable change in the behaviour of the system. Thus, the stick-slip is stable with respect to normal oscillations.

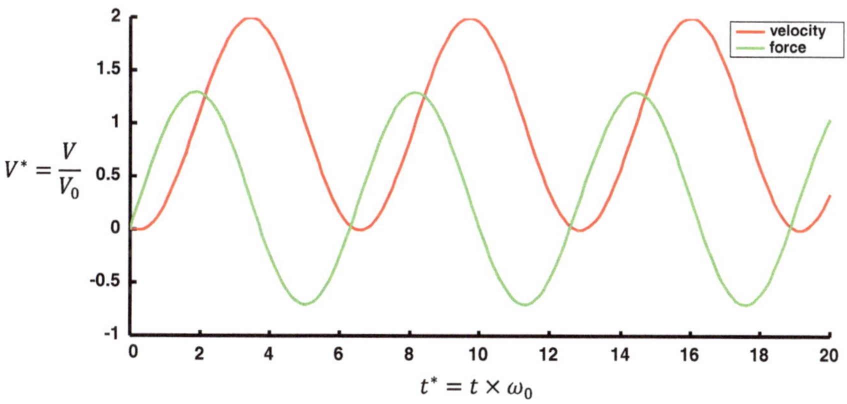

Fig. 2 Stick-slip sliding in the absence of vibration

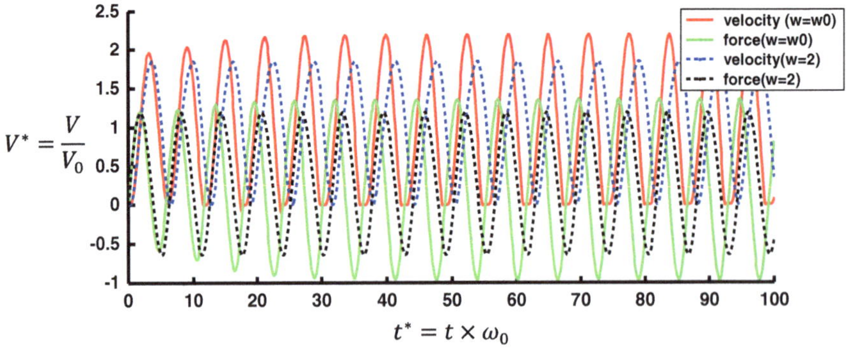

Fig. 3 Stick-slip sliding in the presence of vibration

4 Effect of Phase Shift

We now consider the effect of phase shift in the normal vibrations. We impose the vibrations the eigen frequency of the stick-slip motion ($\omega = \omega_0$). The normal force includes the phase of stick-slip $\varphi = 0.3$:

$$F(t) = A\sin(\omega t + \phi) \tag{5}$$

As can be seen from Fig. 4 the phase shift equal to the phase shift of the velocity does not have a noticeable influence on the system either. A completely different picture is observed when we add the phase shift $\varphi_1 = \frac{\pi}{4}$ to the imposed vibrations:

$$F(t) = A\sin(\omega t + \phi + \frac{\pi}{4}) \tag{6}$$

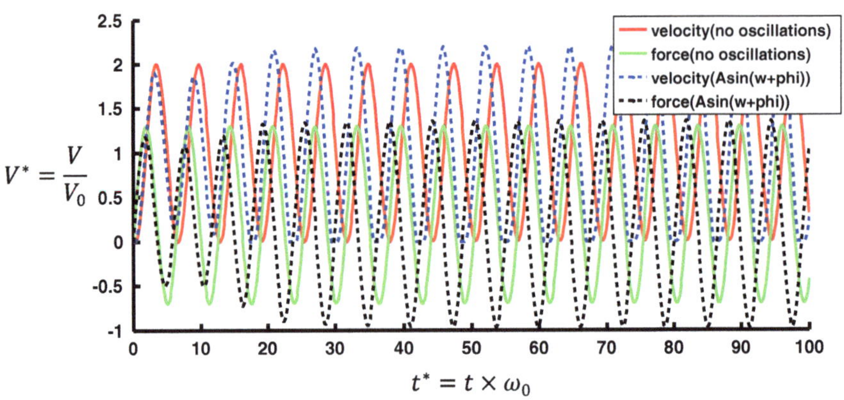

Fig. 4 Stick-slip sliding in the absence and presence of vibration (including the phase shift of stick-slip motion)

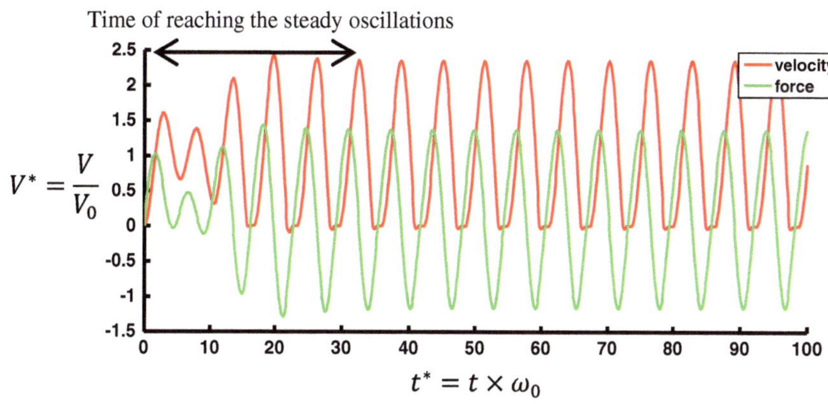

Fig. 5 The zone of reduced vibrations for $A = 1$

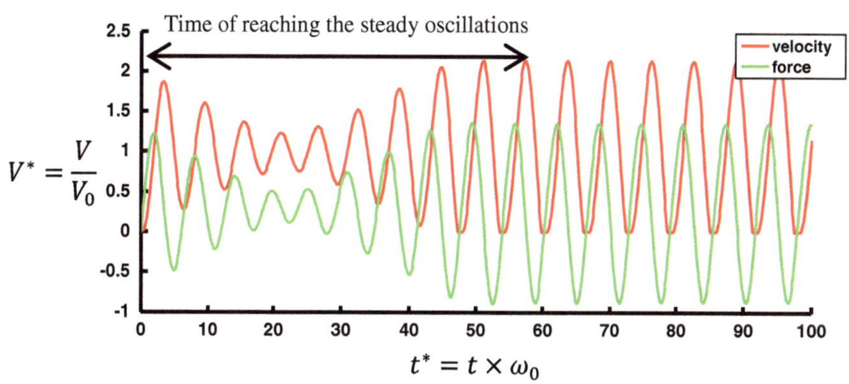

Fig. 6 The zone of reduced vibrations for $A = 0.3$

The oscillator produces different responses depending on the amplitude parameter A, for which the following values are chosen: $A = 0.1$, $A = 0.3$, $A = 1$. Figures 5, 6 and 7 show that the addition the phase shift $(\varphi + \frac{\pi}{4})$ lead to formation the zone of reduced vibration. It is seen that the zone of reduced vibrations extends with the *decrease* of amplitude (the smaller the amplitude of oscillations the longer the zone of reduced vibration). Moreover, it is moves from the origin with the decrease of amplitude. This phenomenon opens a way to control stick slip vibrations through the application of forced normal vibrations on the eigen frequency of the system with an appropriate choice of the phase shift.

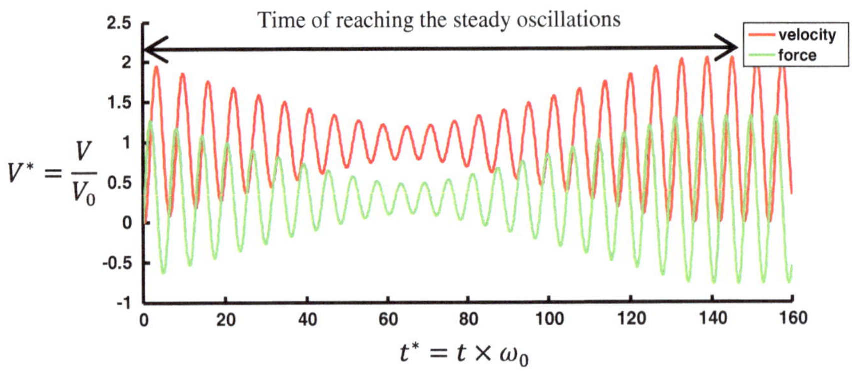

Fig. 7 The zone of reduced vibrations for $A = 0.1$

5 Conclusion

A single degree of freedom block-spring model with a constant friction coefficient was studied to investigate the influence of harmonic normal force on the frictional sliding. We imposed vertical oscillations on the eigen frequency of the single-degree-of-freedom block-spring system. The stick-slip motion can occur in the absence of normal vibrations even when the friction coefficient is constant. The effect of the imposed harmonic normal vibrations on the eigen frequency depends upon the phase shift with respect to the oscillations of the force of the original block-spring system, i.e. the system without the normal vibrations. There exist a phase shift that leads to the formation of a zone of reduced stick-slip velocity. The time of the beginning and the extent of this zone depend upon the amplitude of imposed vibrations: the *lower* the amplitude higher the stick-slip reduction and also the later this zone starts and the longer it lasts. Obviously as the amplitude tends to zero the time of the beginning of the zone of reduced stick-slip tends to infinity. That is why at zero amplitude this zone is not observed.

Acknowledgments The work has been supported by the Deep Exploration Technologies Cooperative Research Centre whose activities are funded by the Australian Government's Cooperative Research Centre Programme. This is DET CRC Document 2013/350.

References

Byerlee JD (1970) The mechanics of stick-slip. Tectonophysics 9(5):475–486

Karachevtseva Iu, Dyskin AV, Pasternak E (2014) The cyclic loading as a result of the stick-slip motion, Adv Mater Res 891–892:878–883

Popp K, Rudolph M (2004) Vibration control to avoid stick-slip motion. J Vib Control 10:1585–1600

Popov LV (2010) Friction under the influence of ultrasonic vibrations. Contact mechanics and friction. Springer (Eds.), Berlin, pp 285–299

Turcotte DL (1992) Fractals and chaos in geology and geophysics. Cambridge University Press, Cambridge

Xu YJ, Wu XP, Yan CH, Huang Y, Wang Y (2011) The tidal coulomb failure stresses on various kinds of seismic fault. Chin J Geophys 54(2):197–206

Investigation of Direct Shear Interface Test Using Micro-polar Continuum Approach

B. Ebrahimian and E. Bauer

Abstract This paper investigates the influence of the length scale of a cohesionless granular soil specimen on the evolution of shear resistance in the direct shear interface test. Numerical simulations are carried out for two different specimen sizes using a micro-polar hypoplastic material model and the finite element method in the updated Lagrange frame. Due to the presence of the lateral rigid boundaries of shear box, the deformation and stress fields become significantly inhomogeneous. Correspondingly, the mobilized shear resistance and the shear band thickness are not constant along the interface. It is shown that, for the case of medium rough wall, some shear localization occurs along part of the interface between the sand specimen and the rough bottom surface and some shear localization takes place within the sand specimen. The evolution of the average mobilized friction angle obtained from the interface shear test is also influenced by the scaling effect of the testing device particularly at the beginning of shearing.

1 Introduction

For studying the mechanical behavior of granular soils in contact with a rough wall, the so-called direct shear interface test is a practical standard laboratory test widely used in geotechnical engineering, e.g. Boulon (1989). As the boundary conditions of the interface shear box apparatus usually does not correctly reflect

B. Ebrahimian (✉)
School of Civil Engineering, Faculty of Engineering, University of Tehran, Tehran, Iran
e-mail: bebrahimian@ut.ac.ir; ebrahimian.babak@gmail.com

B. Ebrahimian
The Highest Prestigious Scientific and Professional National Foundation, Iran's National Elites Foundation (INEF), Tehran, Iran

E. Bauer
Institute of Applied Mechanics, Graz University of Technology, Graz, Austria
e-mail: erich.bauer@tugraz.at

K.-T. Chau and J. Zhao (eds.), *Bifurcation and Degradation of Geomaterials in the New Millennium*, Springer Series in Geomechanics and Geoengineering,
DOI 10.1007/978-3-319-13506-9_21

the actual field conditions under which the soil deforms, doubt is cast upon the reliability of the test results and the selection of appropriate soil strength parameters used in the practical engineering analysis and design. Besides the specimen scale effects, other testing-related criticisms of the interface shear test include the non-uniformity of stress and strain fields, and the rotation of principal stresses. In the literature, the term "interface" often refers to a thin zone of soil grains where shear localization takes place. This zone acts as a transition region between the stiffer structural elements and the softer soil medium. For the following numerical simulations, however, it is convenient to distinguish between the skin friction in the interface and the inter-granular friction angle of the soil material. In particular, the term "interface" is defined as the surface between the soil body and the rough surface of the bounding structure. On the other hand, the term "shear band" is used for the zone of finite thickness where shear localization occurs within the soil body close to the bounding structure. The interface friction angle, which usually depends, in a complex manner, on the interaction between the surface roughness of the bounding structure and the soil morphology, is a key parameter not only for its role in indicating the maximum shear resistance along the interface, but it may also affect the shear deformation of the neighboring soil body. In particular, only small shear deformations occur in the soil at the beginning of shearing for an interface friction angle lower than the critical friction angle of the granular soil, while for continuous shearing the soil body slides against the interface like a rigid body. For high interface friction angles, however, the adjacent boundary grains can be captured by the rough surface of the wall, so that no relative displacements between the wall and the grains take place. Consequently, shear strain localization accompanied with dilatancy and grain crushing develops within the soil material. In the standard experiments with shear test apparatus, the evolution of shear stresses, volume strains and mobilized friction angles is measured in an average sense for the whole specimen while the local distribution of these state variables is difficult to quantify and usually remains unknown.

To gain further insight into the mechanics of interface shear box tests, numerical simulations are carried out based on a micro-polar continuum approach, where additional kinematic variables called micro-rotations and additional static quantities called couple stresses are introduced into the mathematical formulations (Mühlhaus 1986; Gudehus 1994; Unterreiner et al. 1994). Due to the presence of a characteristic length, the boundary value problems become mathematically well-posed at the onset of shear strain localization. Plane shearing along a rough wall was numerically investigated with a micro-polar hypoplastic model by several authors, e.g. Tejchman and Gudehus (2001), Huang and Bauer (2003). In these calculations, the assumption was made that no sliding of soil grains along the interface takes place, i.e. relative displacements between the wall and the adjacent boundary grains were excluded. Recently, the evolution of sliding along the interface was investigated by Ebrahimian and Bauer (2012) for the case of the infinite shear layer. In contrast to an infinite shear layer, the evolution of the state variables is no longer independent on the direction of shearing in the interface shear box apparatus. The evolution of inhomogeneous deformations within the granular soil specimen strongly depends on the size of the specimen and also influences the evolution of the shear resistance along the interface particularly at the beginning of shearing. In the following the case is discussed where the interface friction angle is greater than the inter-granular friction angle in the critical state, but smaller than the peak friction angle of the granular soil material. It is shown that for this case shear localization takes place partially at the interface and partially in the granular soil.

2 Finite Element Modeling of Direct Shear Interface Test

The numerical investigation of direct shear interface test is conducted under the conditions of monotonic plane shearing and free dilatancy. A constant vertical load, F, is applied to the rigid top platen, which leads to an average vertical pressure of 100 kPa at the top surface of specimen (Fig. 1). The soil specimen is located within the rigid shear box frame where the surfaces of the lateral walls are assumed to be frictionless. In order to prevent developing tensile stresses along the boundaries, the specimen is allowed to be detached from the shear box frame. This condition can occur for instance at the left boundary of the specimen when the shear box is moving to the left, i.e. $U_{1Box} > 0$.

The interface friction angle between the surface of the bottom plate and the specimen is prescribed, i.e. the limit shear stress is determined by the classical formulation of Coulomb's friction law in the form of $\tau_{\text{limit}} = \mu_F \sigma_{22}$, where μ_F is the coefficient of interface friction and σ_{22} denotes the local contact stress perpendicular to the interface. For the numerical simulations, the assumption is made that no relative displacement takes place between the soil and the bottom plate as long as the mobilized interface friction angle, $\varphi_{\text{mob}} = \arctan\left(\sigma_{12}/\sigma_{22}\right)$, is lower than the limit value, $\varphi_{\text{int}} = \arctan(\mu_F)$. When $\varphi_{\text{mob}} = \varphi_{\text{int}}$ meets, the method of contact surfaces and the finite sliding contact algorithm is used (ABAQUS 2002). For the present investigations, a value of $\varphi_{\text{int}} = 40°$ ($\mu_F = 0.84$) is assumed for the interface friction angle, which is higher than the critical friction angle, but lower than the peak friction angle of the granular soil material.

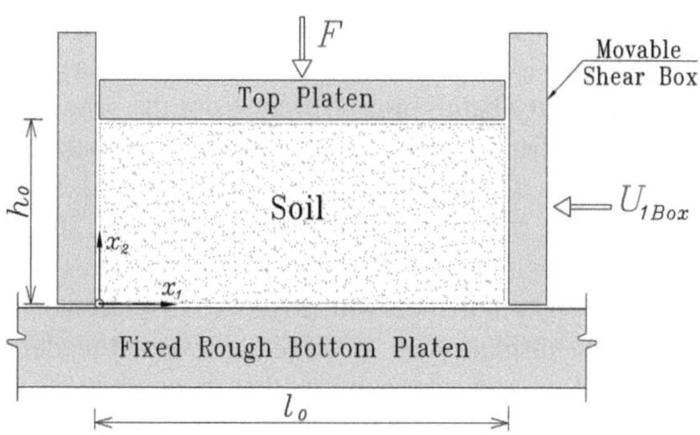

Fig. 1 Typical section of the interface shear box

The material properties of the soil specimen are described by the micro-polar hypoplastic model proposed by Huang and Bauer (2003). The evolution equations for the stresses and couple stresses are non-linear tensor valued functions depending on the pressure dependent relative density, non-symmetric Cauchy stress tensor, couple stress tensor, rate of deformation tensor and rate of micro-curvature. The model includes 10 material constants, which, for the medium quartz sand, are Huang and Bauer (2003): the critical friction angle $\varphi_c = 30°$, the solid hardness $h_s = 190$ MPa, the compression exponent $n = 0.4$, the maximum void ratio $e_{i0} = 1.2$, the minimum void ratio $e_{d0} = 0.51$, the critical void ratio $e_{c0} = 0.82$, the exponent $\alpha = 0.11$ of the density factor, the exponent $\beta = 1.05$ of the stiffness factor, the mean grain diameter $d_{50} = 1$ mm and the polar constant $a_c = 1.0$. For numerical simulations, the micro-polar hypoplastic model is implemented into the finite element program as outlined in more details by Huang and Bauer (2003). The soil specimen with an initial height of $h_0 = 40$ mm and initial lengths of $l_0 = 50$ and 100 mm for the small and large boxes, respectively, is discretized by four-node plane strain Cosserat elements with an initial size of 1.25 mm × 1.25 mm. In order to obtain mesh independent results for a mean grain diameter of $d_{50} = 1$ mm, the size of elements is chosen to be less than $5 \times d_{50}$ (Tejchman and Gudehus 2001). The micro-rotations of the element nodes along the top boundary are zero, while along the bottom interface no restriction is considered for the micro-rotations. Thus, no couple stresses can develop along the interface. The influence of gravity is not taken into account and for the initial state a homogeneous distribution of the initial void ratio of $e_0 = 0.6$ is assumed.

3 Discussion of the Numerical Results

In order to simulate the influence of the specimen size on the evolution of the soil deformation and the shear resistance finite element simulations are carried out with two different specimen lengths, i.e. $l_0 = 50$ and 100 mm. It is found that the distributions of stresses and strains are significantly non-uniform inside the specimen from the beginning of shearing. Two different stress regions can be distinguished in the specimen; a region with high stresses at the right side boundary (due to compression caused by this boundary being pushed into a sand specimen) and a region with low stresses at the left side boundary (due to extension caused by the side boundary being moved away from the sand specimen). For different applied horizontal displacements, U_{1Box}, of the shear box the distribution of the mobilized interface friction angle, φ_{mob}, is shown in Fig. 2. Sliding along the interface is only possible for $\varphi_{mob} = \varphi_{int}$. It is obvious that the shear resistance is fully mobilized from the beginning of shearing at the right hand side of the interface and grows from the right to the left with advanced displacements of the shear box. At a certain shear box displacement, however, a reversed tendency can be observed and φ_{mob} slightly decreases which means that there are local changes of the interface behavior from sliding to sticking. This is also visible in Fig. 4a where, for instance in

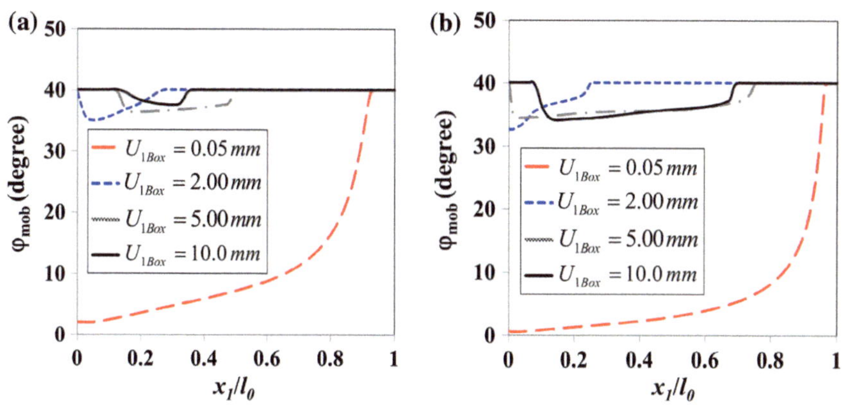

Fig. 2 Mobilized friction angle, φ_{mob}, against the normalized distance, x_1/l_0, at the interface for different applied U_{1Box}, and two different specimen lengths: **a** $l_0 = 50$ mm and **b** $l_0 = 100$ mm

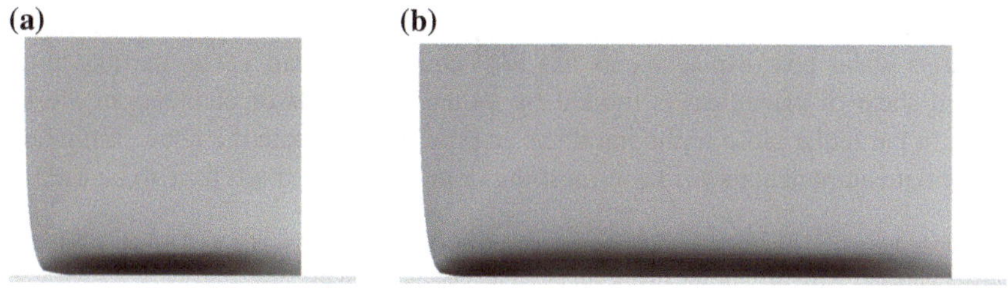

Fig. 3 Deformed sand specimen and contour plot of the void ratio at $U_{1Box} = 10$ mm for two different specimen lengths: **a** $l_0 = 50$ mm and **b** $l_0 = 100$ mm

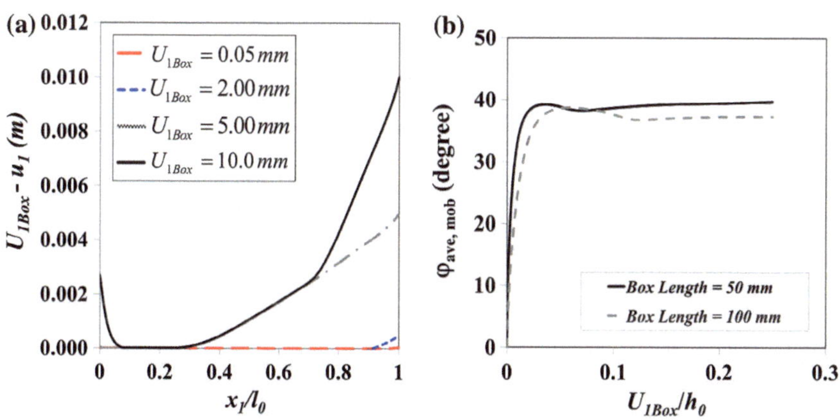

Fig. 4 Evolution of: **a** the relative displacement between shear box and the soil body along the interface for an initial specimen length of 100 mm; **b** the average mobilized interface friction angle, $\varphi_{\mathrm{ave,mob}}$, for the *small* and the *large* shear box lengths

the range between $U_{1Box} = 5$ and 10 mm, no relative displacement occurs in the middle of the interface. The inhomogeneous distribution of relative displacements along the interface is related to shear strain localization in the granular soil. The darker zone in Fig. 3 indicates a higher void ratio within the shear band, where the shear band thickness is higher in the middle of the specimen for both the smaller and the larger shear boxes. Figure 4b shows the scale effect on the evolution of the average mobilized interface friction angle, $\varphi_{ave,mob,}$ along the interface against the normalized shear box displacement.

4 Conclusions

The results obtained from numerical simulation of direct shear interface tests with two different shear box lengths show that the evolution of the shear resistance is affected by the specimen size at least at the beginning of shearing. For increasing the shear box displacement the average mobilized interface friction tends towards the limit value, but a certain scale effect still remains. Whether this scale effect vanishes for larger shear box displacements has not been investigated yet as the maximum applied shear displacement is limited by strong distortions of elements in the FE mesh, in particular close to the interface. In order to overcome the above limitation, appropriate numerical tools like re-meshing or mesh free methods have to be used.

Acknowledgment The first author wants to express his sincere gratitude to the Iran's National Elites Foundation (INEF) for the moral support and encouragement.

References

ABAQUS User's manual version 6.3. (2002) Hibbitt, Karlsson and Sorensen Inc.: Pawtucket, RI, USA

Boulon M (1989) Basic features of soil structure interface behavior. Comput Geotech 7(1–2):115–131

Ebrahimian B, Bauer E (2012) Numerical simulation of the effect of interface friction of a bounding structure on shear deformation in a granular soil. Int J Numer Anal Meth Geomech 36(12):1486–1506

Gudehus G (1994) Localization in granular bodies—position and objectives. In: Chambon R, Desrues J, Vardoulakis I (eds.) Proceedings of 3rd international workshop on localization and bifurcation theory for soils and rocks, Balkema, Rotterdam, pp 3–11

Huang W, Bauer E (2003) Numerical investigations of shear localization in a micro-polar hypoplastic material. Int J Numer Anal Meth Geomech 27(4):325–352

Mühlhaus HB (1986) Shear band analysis in granular materials by Cosserat theory. Ing Arch 56:389–399

Tejchman J, Gudehus G (2001) Shearing of a narrow granular layer with polar quantities. Int J Numer Anal Meth Geomech 76(2):513–536

Unterreiner P, Vardoulakis I, Boulon M, Sulem J (1994) Essential features of a Cosserat continuum in interfacial localisation. In: Chambon R, Desrues J, Vardoulakis I (eds.) Proceedings of 3rd international workshop on localization and bifurcation theory for soils and rocks, Balkema, Rotterdam, pp 141–155

Energy Criterion of In-plane Fracture Propagation in Geomaterials with Rotating Particles

Arcady Dyskin and Elena Pasternak

Abstract In-plane propagation of tensile fractures (Mode I cracks), shear fractures/bands (Mode II cracks) and compaction bands (Mode I anticracks) is routinely observed in geomaterials in the presence of high compressive stress. While the in-plane propagation of tensile cracks is expected, the mechanics of in-plane propagation of shear cracks is not clear. We propose a unified criterion of in-plane growth of these types of fractures based on the assumption that the grains are able to undergo independent relative rotations. The relative rotations break the binder between the grains even in the presence of high compressive stress. An asymptotic model is developed for long fractures showing that the energy release rate is controlled by the conventional Mode I and II stress intensity factors. The proposed unified criterion of fracture growth compares the energy release rate with the specific fracture energy consisting of three terms: the fracture energy of the bonds (present in all three types of fracture), specific energy of shear (for shear fractures/bands) and specific energy of compaction (for compaction bands). We developed estimates for all three components of the specific fracture energy.

1 Introduction

In-plane fracture propagation in geomaterials is often observed in different situations as long as an external compression is present. The obvious example is extensive hydraulic fracture propagation driven by internal pressure of the fracturing fluid (e.g. Valko and Economides 1995). Here the role of external compression is in directing the fracture parallel to the direction of the maximum compression and stabilizing its geometry (Valko and Economides 1995). Another example is the propagation of compactions bands in their plane, which is observed in loading of lab samples, both intact (Fortin et al. 2006) and in those that model boreholes (Katsman et al. 2009), and also in the

A. Dyskin (✉) · E. Pasternak
University of Western Australia, Perth, Australia
e-mail: arcady.dyskin@uwa.edu.au

K.-T. Chau and J. Zhao (eds.), *Bifurcation and Degradation of Geomaterials in the New Millennium*, Springer Series in Geomechanics and Geoengineering,
DOI 10.1007/978-3-319-13506-9_22

Earth's crust (Holcomb et al. 2007). In-plane shear crack propagation is observed in the Earth's crust (see review in Reches and Lockner 1994) and in the lab tests (e.g. Puzrin and Germanovich 2005), the latter being a part of shear failure mechanism of failure in compression of rocks and concrete. While the mechanism of in-plane hydraulic fracture propagation is based on mode I crack tending to grow in its own plane, the mechanics of in-plane propagation of compaction bands and shear fractures is less understood.

Compaction bands are considered as mode I anti-cracks and subsequently their growth is assumed to mirror that of a conventional tensile crack with failure in tension at the crack tip (or in the process zone) being replaced with failure in compression (e.g. Sternlof et al. 2005). There are however two arguments against the application of such 'naïve' symmetry. Firstly, even in uniaxial compression, rock strength is of an order of magnitude higher than in tension; the strength in triaxial compression being even (considerably) higher. Therefore very high magnitude of compression would be needed to effect the compaction band propagation. Secondly, the tensile crack is free to open, as much as needed to maintain the stress singularity at the crack tip (crack counter in 3D). Contrary to it in compaction bands when the rock at the crack tip is crushed under compression, it has nowhere to go. Therefore the compacted rock can resist further crack closure such that the crack profile can no longer be considered as elliptical, contrary to what was assumed in the literature (e.g. Sternlof et al. 2005; Rudnicki and Sternlof 2005).

Even more difficult is to understand the mechanics of in-plane growth of shear fractures. It is well known that cracks under applied shear load do not grow in their planes, but rather kink. This was observed in a considerable variety of materials including rocks, concrete and artificial brittle materials (see a review in Germanovich et al. 1994). Yet, in some cases in-plane propagation of what looks like shear fractures or bands is observed. We note however that individual in-plane shear fracture propagation observed in rock samples is a part of formation of shear failure and it happens after the peak load is reached (e.g. Lockner et al. 1992). Thus the shear bands start at a boundary of the sample and then propagate through it, rather than appear as a result of solution of a boundary-value problem in a material with instabilities.

In this paper we attempt to propose a general framework explaining all three phenomena from the same perspective. We note that there are two common features in these types of fracture propagation. Firstly, we are dealing with heterogeneous materials consisting of grains connected by some binder. Secondly, the presence of high magnitude compressive stress, either ambient (hydraulic fractures and shear bands) or a driving force of fracture propagation (compaction bands) produce considerable damage. This is the case for both compaction bands when high compression is required to overcome high compressive strength of the rock and for shear failure starting after the peak, when the rock acquires considerable internal damage (microcracks) sufficient to reduce the deformation modulus to zero. The damage is primarily accumulated in the binder thus leading to partial detachment of the grains. As a result, the grains assume the ability to rotate independently (at least infinitesimally). Independent rotation of the grains can lead to bending and eventual failure of remaining intact links (remaining parts of the binder). Thus, at the micro level, the failure starts with mutual rotation of the grains leading to their full detachment through the link (binder) failure and followed by either re-compaction in the case of compaction bands or sliding over the rolling grains in the case of shear fractures. The detailed explanation of this mechanism is presented in the following section. The next section discusses conventional cracks with rotational mechanism operating only at their process zones. The last section discusses the energy criterion and analyses the differences in fracture energy associated with all three abovementioned types of fracture propagation.

2 Micro-rotational Mechanism of In-plane Crack Growth

Consider a geomaterial consisting of grains joined together by a binder. If the loading induces independent grain rotation $\varphi = (\varphi_1, \varphi_2, \varphi_3)$, the difference in the rotational angles of neighbouring grains can produce bending in the link (binder) and finally its breakage through a propagation of a flexural microcrack, Fig. 1. Suppose the grain centres are connected by vector $\mathbf{l} = (l_1, l_2, l_3)$; its length $l_{micro} = \|\mathbf{l}\|$ is a characteristic length of microstructure or microscopic length. In the isotropic geomaterials we consider here the characteristic length is usually the average grain size. At a larger scale, say in a blocky rock mass the role of the characteristic scale will be played by the average block size.

In the Cosserat continuum the relative rotations of the particles are represented by components of the curvature-twist tensor, which at a point $\mathbf{x} = (x_1, x_2, x_3)$ reads

$$\kappa_{ji} = \varphi_{i,j} \approx [\varphi_i(\mathbf{x} + \mathbf{l}) - \varphi_i(\mathbf{x})] \big/ l_j, \quad i,j = 1,2,3. \tag{1}$$

Its counterpart—the moment stress tensor as well as the strain tensor, $\gamma_{ji} = u_{i,j} - \varepsilon_{kji}\varphi_k$, ($u_i$ is the displacement) and its counterpart—the non-symmetric stress tensor are related, in the case of isotropic Cosserat medium through the following constitutive equations (Nowacki 1970)

$$\begin{aligned} \sigma_{ji} &= (\mu + \alpha)\gamma_{ji} + (\mu - \alpha)\gamma_{ij} + \lambda\delta_{ij}\gamma_{kk} \\ \mu_{ji} &= (\gamma + \varepsilon)\kappa_{ji} + (\gamma - \varepsilon)\kappa_{ij} + \beta\delta_{ij}\kappa_{kk}, \end{aligned} \tag{2}$$

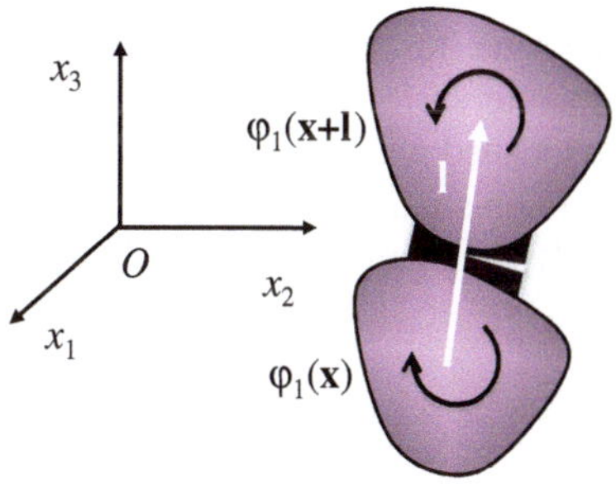

Fig. 1 A mechanism of local failure caused by relative particle rotations

where μ, λ are Lame constants and α, γ, ε, β are Cosserat moduli. The full set of governing equations is obtained by adding equations of equilibrium $\sigma_{ji,j} = 0$, $\mu_{ji,j} + \varepsilon_{ijk}\sigma_{jk} = 0$.

3 Asymptotic Model of Crack Growth in Geomaterials with Microrotations

Contrary to the classical continuum the Cosserat continuum possesses characteristic lengths that can be constructed as square roots of ratios of moduli from the second equation of (2) (units of Pa m^2) and moduli from the first equation (units of Pa). We showed previously (Dyskin and Pasternak 2008; Pasternak and Dyskin 2009; Dyskin and Pasternak 2010) that these lengths are of the order of the microstructural size l_{micro}. We consider a special case when the fracture length $L \gg l_{micro}$. This is often the case in the field, since grain sizes are in the millimeter or submillimeter range (Tembe et al. 2008), while the fracture lengths are from meters to hundreds meters (Valko and Economides 1995; Puzrin and Germanovich 2005; Tembe et al. 2008). It can also be shown that in the limit $l_{micro}/L \to 0$ the leading asymptotics terms of stress and moment stress concentrations at the crack tip are controlled by traditional stress intensity factors K_I and K_{II} (we restrict ourselves to the plane strain conditions). Subsequently, the main asymptotic term of the energy release rate is given by the conventional formula based on these stress intensity factors (see the next section).

4 Energy Criterion of Crack Growth in Geomaterials with Microrotations

We now formulate the fracture propagation criterion associated with particle rotation for all three abovementioned cases: Mode I crack (for hydraulic fractures), Mode I anti-crack (for growth of compaction bands) and Mode II shear crack (for growth of shear bands).

$$\left(K_{\mathrm{I}}^2 + K_{\mathrm{II}}^2\right)\left(1 - \nu^2\right)E^{-1} = \gamma_b + \gamma_s + \gamma_c. \tag{3}$$

Here γ_b is the specific fracture energy of fractured bonds, Fig. 1 (This term is the same for all three cases.), γ_s is the specific energy related to shear of the new fracture surfaces γ_c is the specific energy related to the geomaterial compaction associated with rearrangement of the grains, E and ν are the Young's modulus and Poisson's ratio of the geomaterial.

If we assume that the width of the bond is of the order of $l_{micro}/2$, then the fracture area is $\sim l_{micro}^2/4$ and $\gamma_b \sim v k_f \gamma_f / 4$, where v is volumetric fraction of the particles,

k is the coordination number and γ_f is the specific fracture energy of the material of the bonds. The specific energy related to shear of the new fracture surfaces γ_s could be estimated if one knows the magnitude of compressive stress p acting on the fracture plane and the friction coefficient μ associated with particle rolling. Then $\gamma_s = \mu p$ times unit length in the direction of crack contour. In the case of compaction bands the specific compaction energy, $\gamma_c \sim pV$, where p is the compressive stress causing compaction and V is the compaction displacement.

5 Conclusions

A unified in-plane fracture propagation mechanism in geomaterials is formulated based on the ability of grains to undergo independent relative rotations. In this case, the main microfailure mechanism is the breakage of bent binder between the neighbouring grains and their full detachment. For Mode I fractures the full grain detachment produces local tensile failure at the fracture tip resulting in a step of its propagation. For compaction bands, which are Mode I anticracks, the full grain detachment enables grain rolling with subsequent compaction leading to volume reduction and fracture closure. For Mode II fractures, which are shear fractures or bands the full grain detachment also enables grain rolling with subsequent sliding forming a step in shear fracture propagation.

Modelling of a geomaterial with independent grain rotations calls for the use of the Cosserat theory. The Cosserat continuum possesses characteristic lengths that in the case of geomaterials with bonds/links between the grains are of the order of the grain size. When the fracture length is considerably larger than the Cosserat characteristic lengths asymptotic of small grain size can be used. The leading term of the energy release rate is given by conventional Mode I and II stress intensity factors. The energy criterion of fracture propagation is formulated by comparing the energy release rate with the specific fracture energy. The latter generally consists of three terms: the fracture energy of the bonds (present in all three types of fracture), specific energy of shear (for shear fractures/bands) and specific energy of compaction (for compaction bands). The fracture energy of the bonds can be estimated from the specific fracture energy of the binder, volumetric fraction of grains and the coordination number. The specific energy of shear can be estimated from the magnitude of compressive stress p acting on the fracture plane and the friction coefficient μ associated with particle rolling. The specific energy of compaction can be estimated from the magnitude of compressive stress p causing compaction and the compaction displacement.

Acknowledgments We acknowledge financial support from ARC Linkage Grant LP120200797. The paper is a part of research under the initiative 'Engineering for Remote Operations of the Faculty of Engineering', mathematics and Computing of the University of Western Australia.

References

Dyskin AV, Pasternak E (2008) Rotational mechanism of in-plane shear crack growth in rocks under compression. In: Potvin Y, Carter J, Dyskin A, Jeffrey R (eds) Proceedings of 1st southern hemisphere international rock mechanics symposium SHIRMS 2008, vol 2, Australian Centre for Geomechanics, Australia, pp 111–120

Dyskin AV, Pasternak E (2010) Cracks in Cosserat continuum—Macroscopic modelling. In: Maugin GA, Metrikine AV (eds) Mechanics of generalized continua: one hundred years after the Cosserats. Advances in mechanics and mathematics, vol 21, Springer, New York, pp 35–42

Fortin J, Stanchits S, Dresen G, Guéguen Y (2006) Acoustic emission and velocities associated with the formation of compaction bands in sandstone. J Geophys Res 111(B10):B10203

Germanovich LN, Salganik RL, Dyskin AV, Lee KK (1994) Mechanisms of brittle fracture of rock with multiple pre-existing cracks in compression. Pure Appl Geophys 143(1–3):117–149

Holcomb D, Rudnicki J, Issen K, Sternlof KR (2007) Compaction localization in the Earth and the laboratory: state of the research and research directions. Acta Geotech 2(1):1–15

Katsman R, Aharonov E, Haimson B (2009) Compaction bands induced by borehole drilling. Acta Geotech 4(3):151–162

Lockner DA, Byerlee JD, Kuksenko V, Ponomarev A, Sidorin A(1992) Observations of quasi-static fault growth from acoustic emissions. In: Evans B, Wong T.-F (eds) Fault mechanics and transport properties of rocks, vol 51, Academic Press, London, pp 3–31

Nowacki W (1970) Theory of micropolar elasticity. Springer, Wien

Pasternak E, Dyskin AV (2009) Intermediate asymptotics for scaling of stresses at the tip of crack in Cosserat continuum. In: Proceedings of 12th international conference on fracture ICF12, Ottawa, paper T40.014

Puzrin AM, Germanovich LN (2005) The growth of shear bands in the catastrophic failure of soils. Proc Royal Soc: A 461(2056):1199–1228

Reches Z, Lockner DA (1994) Nucleation and growth of faults in brittle rocks. J Geophys Res 99(B9):18159–18173

Rudnicki JW, Sternlof KR (2005) Energy release model of compaction band propagation. Geophys Res Lett 32:L16303

Sternlof KR, Rudnicki JW, Pollard DD (2005) Anticrack inclusion model for compaction bands in sandstone. J Geophys Res 110(B11):B11403

Tembe S, Baud P, Wong TF (2008) Stress conditions for the propagation of discrete compaction bands in porous sandstone. J Geophys Res 113(B9):B09409

Valko P, Economides MJ (1995) Hydraulic Fracture Mechanics, John Wiley & Sons

Dynamic Instability in Geomaterials Associated with the Presence of Negative Stiffness Elements

Elena Pasternak and Arcady Dyskin

Abstract One of the mechanisms of instability in geomaterials is associated with the descending branch of stress-strain curve caused by internal fracturing. The descending stress-strain curve can also be produced by negative stiffness elements, one of the mechanisms of negative stiffness being the rotation of non-spherical grains. The presence of negative stiffness elements can cause dynamic instability in the geomaterials. We investigate a simple elastic model of dynamic instability consisting of chains of oscillators some with negative stiffness springs. We show that in a stable chain only one oscillator can have negative stiffness spring and formulate the criterion of stability.

1 Introduction

The conventional mechanisms of instability in geomaterials are based on reduction of strength after the strain, displacement or velocity exceeding a certain critical values. This is typically associated with post-peak softening and non-associated dilatancy law (Cook 1965; Salamon 1970; Rudnicki and Rice 1975; Tarasov and Dyskin 2005), or friction coefficient reduction with velocity (e.g., Turcotte 1992). It was recently found (Dyskin and Pasternak 2011, 2012a, b, c; Pasternak and Dyskin 2013) that another mechanism could be responsible for instability—the rotation of non-spherical (non-circular in 2D) particles, grains or blocks.

When non-spherical particles start rotating the moment equilibrium dictates that the shear force—displacement diagram is descending giving rise to the negative tangent stiffness. As the systems exhibiting negative stiffness are intrinsically unstable (the elastic energy is no longer positive definite) they can only exist as a part of an encompassing stable system provided that the system is sufficiently stiff

E. Pasternak (✉) · A. Dyskin
University of Western Australia, Perth, Australia
e-mail: elena.pasternak@uwa.edu.au

A. Dyskin
e-mail: arcady.dyskin@uwa.edu.au

K.-T. Chau and J. Zhao (eds.), *Bifurcation and Degradation of Geomaterials in the New Millennium*, Springer Series in Geomechanics and Geoengineering,
DOI 10.1007/978-3-319-13506-9_23

to effect the stabilization. A typical example is the post-peak softening curve in rocks, which can only be obtained if the loading devise is sufficiently stiff (Cook 1965; Salamon 1970). In some cases, the role of such a stabilizing system can be played by the remaining part of the geomaterial (Cook 1965). The stability conditions for materials with different types of negative stiffness inclusions were formulated in Dyskin and Pasternak (2011, 2012a, b, c), Pasternak and Dyskin (2013). It was shown that there exists a critical concentration of negative stiffness inclusions: when the critical concentration is reached the whole material becomes unstable (the effective stiffness abruptly becomes negative).

The stability analysis for the systems with negative stiffness elements (Dyskin and Pasternak 2011, 2012a, b, c; Pasternak and Dyskin 2013) was based on the concept of the effective characteristics and as such was essentially concerned static behaviour. Here we consider the conditions of dynamic instability. We model the geomaterial by linear chains of masses connected by normal springs, with a mixture of positive and negative stiffnesses. We derive the stability conditions and discuss the factors affecting the stability.

2 Negative Stiffness Created by Rolling of Non-spherical Particles

Consider a geomaterial under compression and suppose that due to defect accumulation under loading the bonds between the particles are broken and some particles are now partially detached from the rest of geomaterial, Fig. 1. For the following only the particle diameter (the largest size) d is important that is why the particle itself is greyed out.

Hereafter the model we consider will be based on force-displacement relationships. The transition to stress-strain relationship will involve the introduction of volume elements and relating the forces to elementary areas.We write the balance of moments of shear and normal forces about point 0

$$T \sin\varphi + P \cos\varphi = 0, \quad \pi/2 \le \varphi \le \pi. \tag{1}$$

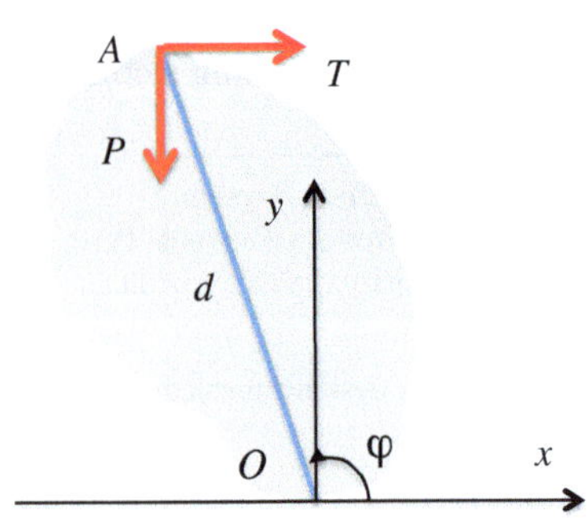

Fig. 1 A moment balance of rotating (rolling) partially detached particle, which temporarily admits only two contacts: forces P and T are exerted by the surrounding geomaterial

Here T and P are the magnitudes of the shear and normal forces and angle φ is related to the position of the particle at the moment of detachment. Obviously, the moment equilibrium is only maintained for the range of angles φ in (1) corresponding to the inclinations similar to shown in Fig. 1; when $\varphi < \pi/2$ the particle becomes unstable. We concentrate on the situations when $\varphi > \pi/2$.

An infinitesimal rotation $\delta\varphi$ will increment coordinates (x, y) of point A by $dx = -d\sin\varphi\, d\varphi$ and $dy = d\cos\varphi\, d\varphi$. Suppose the resistance of the geomaterial to dilation associated with the rotation increment increase the normal force P by

$$dP = k_m dy. \tag{2}$$

where k_m is the stiffness of the surrounding parts of the particulate material.

Taking into account (1) and the relation $dy = -dx\cos\varphi/\sin\varphi$ one obtains:

$$dT = kdx; \quad k = -\frac{P}{d\sin^3\varphi} + k_m\frac{\cos^2\varphi}{\sin^2\varphi}. \tag{3}$$

It is seen that the coefficient between incremental shear force dT and horizontal displacement dx can assume negative values, when

$$P > k_m d\sin\varphi\cos^2\varphi. \tag{4}$$

This is the condition of the emergence of *negative stiffness*. This condition is essentially an interplay between the magnitude of compressive force and the stiffness of the geomaterial. It is seen that no matter how stiff the geomaterial is there always exists a magnitude of compressive force when negative stiffness emerge.

3 Simple Model of Dynamic Instability Caused by Negative Stiffness

In order to investigate dynamic instability associated with the presence of negative stiffness elements we consider a simple model consisting of a chain of $n - 1$ equal masses, m connected by n springs of different stiffnesses, Fig. 2; some of them being negative. The ends of the chain are fixed.

The movements of the masses represented by vertical displacements v_1, v_2, …, $v_{n\text{-}1}$ are governed by the following system of equations

$$m\ddot{v}_i + (k_i + k_{i+1})v_i - k_i v_{i-1} - k_{i+1}v_{i+1} = f_i, \quad i = 1, \ldots n-1, \quad v_0 = v_n = 0. \tag{5}$$

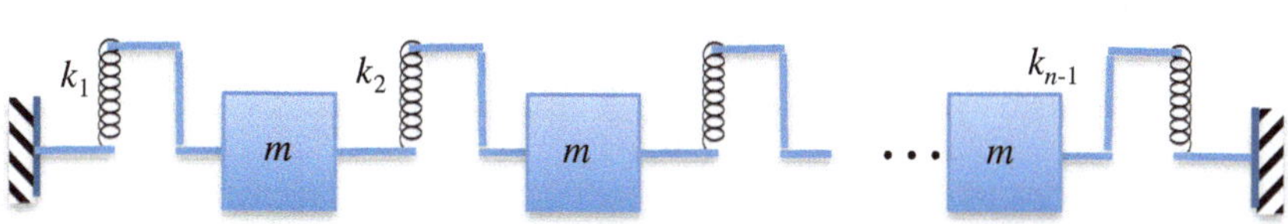

Fig. 2 Constrained chain of equal masses connected by springs of different stiffness

The solution of this system is dynamically stable if its potential energy is positive definite, which is always the case for positive stiffnesses. When some of the stiffnesses are negative, we need to check the positive definiteness of the energy, which is equivalent to positive definiteness of the matrix of the system:

$$A_n = \begin{bmatrix} k_1+k_2 & -k_2 & & & & & \\ -k_2 & k_1+k_2 & -k_3 & & & & \\ & \ddots & \ddots & \ddots & & & \\ & & -k_i & k_i+k_{i+1} & -k_{i+1} & & \\ & & & \ddots & \ddots & & \\ & & & & -k_{n-1} & k_{n-1}+k_n \end{bmatrix}. \tag{6}$$

where the empty spaces refer to zero entries. The condition of the positive definiteness of A_n is that all main diagonal minors Δ_i of A_n are positive. This can be expressed in the following form Dyskin et al. (2014):

$$\sum_{i=1}^{n}\prod_{\substack{j=1\\ j\neq i}}^{n} k_j > 0. \tag{7}$$

It can be shown that this condition is only satisfied if no more than one spring in the chain has negative stiffness.

As a simple model of dynamic instability of a geomaterial, consider the case when there is one negative stiffness spring with stiffness k given by the second equation of (3), while all other springs have the same stiffness k_m. Thus the springs with stiffness k_m represent the geomaterial, while the spring with negative stiffness k represents the effect of a rotating non-spherical particle.

Due to symmetry of (7) the spring stiffnesses can be renumbered, so without the loss of generality we can assume that

$$k_1 = k < 0, \quad k_2 = k_3 = \ldots k_n = k_m. \tag{8}$$

From here the criterion of dynamic stability (7) assumes the following form

$$k > -\frac{k_m}{n-1}. \tag{9}$$

Substituting here the second equation of (3) one obtains the critical magnitude of compressive force when the geomaterial is still dynamically stable. Subsequently, the stability criterion reads

$$P < P_{cr} = k_m d\left(\frac{\sin^2\varphi}{n-1} + \cos^2\varphi\right)\sin\varphi. \tag{10}$$

We note that when expressed in stress, the critical magnitude of compression is of the order of the deformation modulus of the geomaterial. While it might seem to be very high, it is to remember that the mechanism considered is only possible when the damage accumulated in the geomaterial in the process of loading is high enough to make the particle detachment and rotation possible. Therefore one can hypothesize that the accumulated damage can make the deformation modulus of the geomaterial sufficiently low such that the critical magnitude of compression is attainable.

4 Conclusions

We consider the negative shear stiffness effect of rotating non-spherical particles in geomaterials. We show that the emergence of negative stiffness depends on the relationship between the magnitude of compressive force and the stiffness of geomaterial. Furthermore, there exist the critical magnitude of compression starting from which negative stiffness always emerges.

Dynamic interaction between the negative stiffness elements and the rest of the geomaterial (essentially a system of positive stiffness elements) was modelled using a chain of masses connected by shear springs. The condition of dynamic stability dictates that the presence of two or more negative stiffness springs makes the chain unstable. Only one negative stiffness spring is allowed in a stable chain as long as the value of the negative stiffness does not exceed a certain critical value determined by the values of the positive stiffnesses and the number of springs in the chain; the larger the number of springs the smaller the critical value.

As the magnitude of compressive force increases the value of the negative spring increases such that after a certain value of compression the geomaterial becomes unstable.

Acknowledgments The authors acknowledge the financial support through ARC Discovery Grant DP120102434.

References

Cook NGW (1965) The failure of rock. Int J Rock Mech Min Sci 2(4):389–403

Dyskin AV, Pasternak E (2011) Rock mass instability caused by incipient block rotation. In: Qian Q, Zhou Y (eds) Harmonising rock engineering and the environment. Proceedings of 12th international congress on rock mechanics. CRC Press Balkema 4

Dyskin AV, Pasternak E (2012a) Elastic composite with negative stiffness inclusions in antiplane strain. Int J Eng Sci 58:45–56

Dyskin AV, Pasternak E (2012b) Mechanical effect of rotating non-spherical particles on failure in compression. Phil Mag 92(28–30):3451–3473

Dyskin AV, Pasternak E (2012c) Rock and rock mass instability caused by rotation of non-spherical grains or blocks. Rock engineering and technology for sustainable underground construction. Proceedings of Eurock paper 102P

Dyskin AV, Pasternak E, Sevel G (2014) Chains of oscillators with negative stiffness elements. J Sound Vibr 333(24):6676–6687

Pasternak E, Dyskin AV (2013) Instability and failure of particulate materials caused by rolling of non-spherical particles. In: Proceedings of 13th international conference on fracture, Beijing. Paper S09-002
Rudnicki JW, Rice JR (1975) Conditions for the localization of deformation in pressure-sensitive dilatant materials. J Mech Phys Solids 23(6):371–394
Salamon MDG (1970) Stability, instability and design of pillar workings. Int J Rock Mech Min Sci 7(6):613–631
Tarasov BG, Dyskin AV (2005) The phenomenon of anomalous rock embrittlement. In: Potvin Y, Hudyma M (eds) Sixth international symposium rockburst and seismicity in mines. ACG, Australia, pp 311–317
Turcotte DL (1992) Fractals and chaos in geology and geophysics. Cambridge University Press, Cambridge

Numerical Modeling of Particle Breaking Process in Granular Materials: Compaction and Evolution of Size Distribution

Duc-Hanh Nguyen, Emilien Azéma, Farhang Radjai and Philippe Sornay

Abstract The compaction of powders depends both on grain rearrangements and grain breakage. We introduce a grain fracture model prescribed in the framework of the contact dynamics method for the simulation of uniaxial compaction. We find that the grain size reduction is highly heterogeneous as a consequence of inhomogeneous stress transmission as observed in real grinding processes or in natural degradation of geomaterials. Even under high stresses, a significant fraction of grains survive whereas many grains are fully shattered. The grain size distribution tends to a power-law distribution with increasing size span. We analyze the progressive evolution of compressibility during compaction as well as the effect of grain shape and size distribution.

1 Introduction

Grain breakage occurs very commonly in natural granular flows and in industrial processes involving the transport, handling and compaction of granular materials. The grain size reduction is often undesired or uncontrolled, and it is referred to as attrition process. In contrast, the fragmentation of grains under controlled conditions is used in comminution processes such as the milling of vegetal products or grinding of mineral materials. The evolution of grain size distribution and energy dissipation

D.-H. Nguyen (✉) · E. Azéma · F. Radjai
Université Montpellier 2, CNRS, LMGC, Place Eugène Bataillon,
34095 Montpellier, France
e-mail: duc-hanh.nguyen@univ-montp2.fr

E. Azéma
e-mail: Emilien.azema@univ-montp2.fr

P. Sornay
Commissariat à l'énergie Atomique et aux énergies Alternatives, DEN, DEC, SPUA,
LCU, 13108 Saint Paul lez Durance, France
e-mail: philippe.sornay@cea.fr

K.-T. Chau and J. Zhao (eds.), *Bifurcation and Degradation of Geomaterials in the New Millennium*, Springer Series in Geomechanics and Geoengineering,
DOI 10.1007/978-3-319-13506-9_24

in all such processes depend on many factors such as the grain properties (shape, crushability), initial size distribution, loading history and mobility of the grains.

The manufacture of compact shapes by molding powdered materials is the archetypal example of a process in which the bulk crushing of the grains plays as much a role as grain rearrangements (Fuerstenau et al. 1996; Hosten and Cimilli 2009; Esnault and Roux 2013). However, despite its industrial importance, the compaction process and its underlying micromechanical mechanisms are still poorly understood partially due to short length and time scales governing grain breakage. The bulk comminution process and relocalization of the fragments during compaction are controlled by the structural disorder of the packing and highly inhomogeneous distribution of contact forces. Moreover, the process is both nonlinear and nonlocal as cascading events follow a local grain breakage and seem to play an important role in grain size reduction (Esnault and Roux 2013).

Numerical simulations by the molecular dynamics or discrete element method (DEM) have been increasingly employed in order to get a better understanding of the micromechanical mechanisms of the comminution process. Such methods combine the general framework of the DEM, based on rigid-body dynamics and frictional contact interactions, with a grain fracture model (Thornton et al. 1996; Liu et al. 2010; Bolton et al. 2008; Wang and Yan 2012; Timar et al. 2012; Esnault and Roux 2013).

In this paper, we introduce a new approach based on the contact dynamics (CD) method (Moreau 1994; Radjai and Richefeu 2009). The grains have polygonal shapes composed of primary polygonal cells generated by a Voronoi tessellation and interacting by cohesive frictional forces. In the following, we apply this approach to investigate the effect of grain shape and size distributions on the uniaxial compaction of granular materials. The evolution of solid fraction, grain size distribution, grain shape and damage ratios are studied as a function of the applied stress.

2 Methodology

The simulations were carried out by means of the contact dynamics (CD) method, which is suitable for simulating large assemblies of undeformable particles (Moreau 1994; Radjai and Richefeu 2009). In this method, the equations of motion are integrated by an implicit time-stepping scheme by taking into account the kinematic constraints resulting from frictional contact interactions. The implicit integration makes the method unconditionally stable. Moreover, since in this method no overlaps are introduced between particles, the time step can be larger than that in the MD or usual DEM method where the time step should be small enough to resolve elastic contact deflections at the contact points and ensure numerical stability. In the CD, an iterative algorithm based is used to determine the contact forces and grain velocities simultaneously at all potential contacts between the grains.

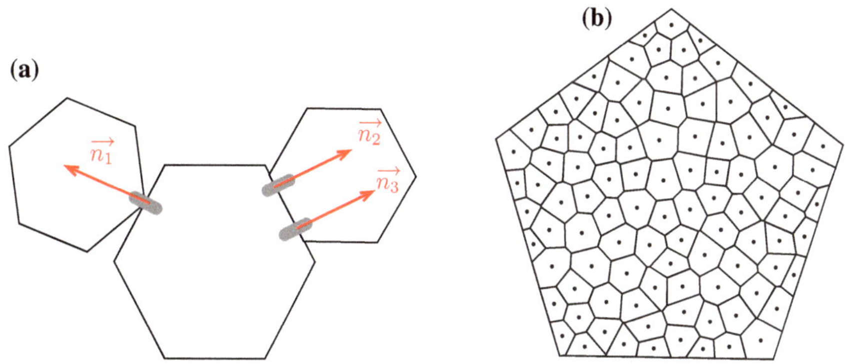

Fig. 1 **a** Different types of contact between two polygonal grains, and **b** discretization of a grain into Voronoi cells

The determination of the set of contacts for a packing made up of polygonal grains is important in this method. Two grains may touch either at a vertex/side contact or at a side/side contact. In both cases, the contact normal is imposed by the direction of the side(s) concerned, as illustrated in Fig. 1a. Note that contacts between two vertices are extremely rare, but when they occur, the normal direction is chosen according to the overlapping history from the path followed by the grains. The interaction between the grains is treated by applying the contact laws to a single point in the case of a side/vertex contact, and to a couple of points in the case of a side/side contact since a side/side contact represents two independent geometrical constraints. Thus, in practice, two forces are calculated at each side/side contact. However, only their resultant and point of application are physically meaningful and are independent of the choice of the two representative points. In this model, each grain is divided into cells by a Voronoi tessellation; see Fig. 1b. The cells represent primary rigid particles that interact by frictional cohesion. Hence, the breaking threshold is simply the Mohr-Coulomb plastic criterion enhanced by cohesion. Initially, all cells interact by side/side contacts. The cohesion threshold fc between two cells depends linearly on the contact surface S so that $f_c = 0.5S\sigma_c$, where σ_c is the strength of material. The coefficient 0.5 in this equation is introduced to account for the two contact points used to represent a side/side contact. The fracture is irreversible and a side/side contact transforming into a side/vertex contact looses its cohesion. Each contact point carries a normal force f_n and a tangent force f_t. The normal force f_n satisfies the unilateral contact condition and the tangential force f_t obeys Coulomb's law of friction, as shown in Fig. 2.

In the simulations, we use irregular pentagonal grain shapes. Let us consider a regular pentagon. The angular position of a vertex i is given by $\theta_i = \theta_0 + i2\pi/5$, where θ_0 is the position of the first vertex. This regular pentagon can be transformed into an irregular pentagon by perturbing the position of each vertex i: $\theta_i = \theta_0 + i2\pi/5 \pm \text{rand}(\delta)\pi/5$, where δ is a parameter varying in the range [0, 1]

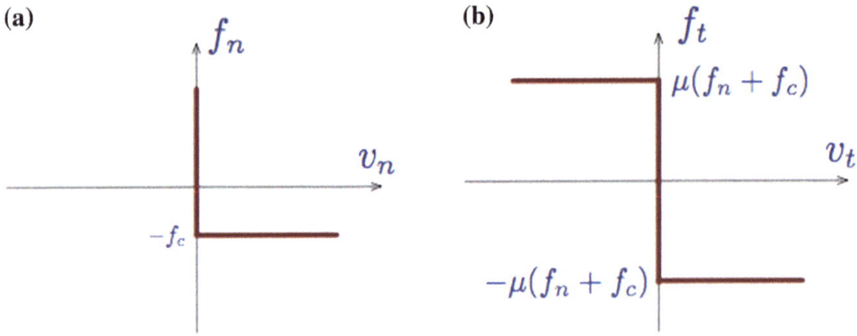

Fig. 2 Unilateral contact condition relating the normal relative velocity and the normal force at a contact point (**a**) and Coulomb's friction law for an adhesive contact threshold f_c relating the tangential relative velocity and the friction force (**b**)

and reflects the degree of irregularity of the shape, the function rand(δ) is a random number in the range [0, δ]. Note that the resulting grains have an irregular shape as compared to regular pentagons. The grains can be made much more irregular if the restriction of the angular deviation to $\pi/5$ is removed.

The size of a pentagonal grain is defined by the diameter d of the circumscribed disk. The diameter of the grains are varied in the range of [d_{min}, d_{max}] with a uniform distribution by grain volume fractions. We define the size span s of the distribution by $s = (d_{max} - d_{min})/(d_{max} + d_{min})$. Four samples are considered with two values of the size span ($s = 0$ and $s = 0.5$) and two values of δ ($\delta = 0$ and $\delta = 1$). Each sample consists of 200 grains which are deposited by gravity into a rectangular box. Then, each grain is discretized into 100 Voronoi cells, and a cohesive stress σ_c is imposed at the contacts between cells. Next, the sample is compressed by stress increments applied on the top wall from 0 to $10\sigma_c$. We note that the number of cells may influence the breaking threshold of the grains. However, we checked that the breaking stress declines rapidly as the number of cells is increased. Beyond nearly 100 cells per grain, used in our simulations, the breaking stress is nearly independent of the number of cell.

3 Uniaxial Compression

The displacement of the top wall leads to displacement of the grains that rearrange and/or break. Figure 3 shows an image of a sample at the beginning and the end of a compression test. The color level represents the degree of degradation of the particles. We see that at the end of the test, a significant fraction of grains survive without fracture whereas many particles are fully shattered.

Figure 4a shows a typical example of the evolution of the void ratio e and proportion P_b of broken grains as a function of the vertical stress normalized by

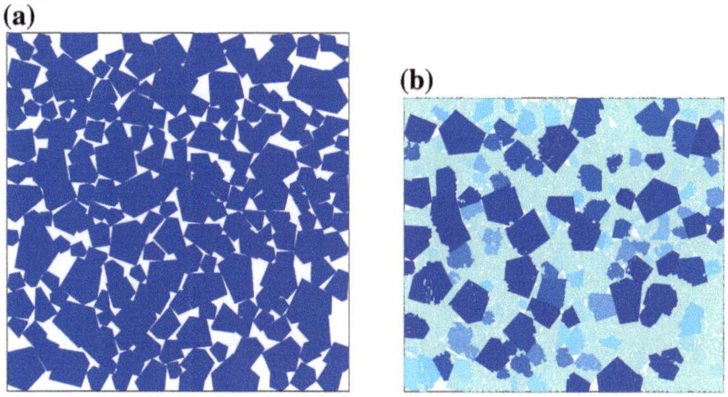

Fig. 3 Snapshot of a sample under uniaxial compaction at the beginning (**a**) and the end (**b**) of the test. The color level represents the degree of degradation (*blue* least damaged, *green* most damaged)

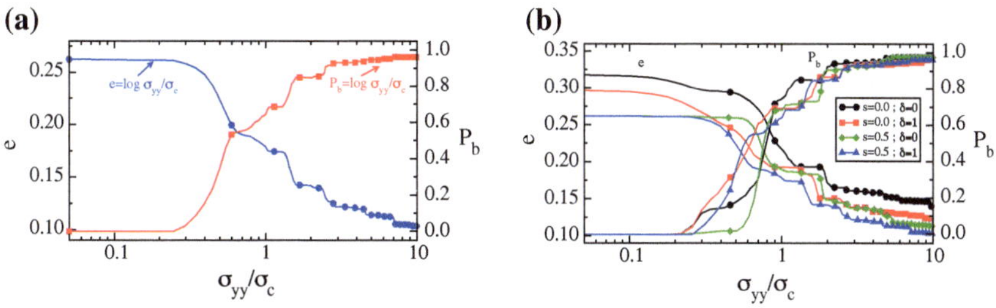

Fig. 4 Evolution of the void ratio e and proportion P_b of broken grains as a function of the vertical stress normalized by the cohesive stress: **a** typical example, and **b** for four different values of size span and shape irregularity

σ_c. We distinguish three stages of evolution. In the first stage, at low stresses from 0 to $0.4\sigma_c$, the grains rearrange and do not break, or they break only at the contact points, and the void ratio declines slightly. In the second stage, for stresses ranging from $0.4\sigma_c$ to $5\sigma_c$, many grains break and rearrange and the void ratio decreases rapidly. In the third stage, for $5\sigma c$–$10\sigma c$, the grains stop breaking and the void ratio decreases slightly and then saturates. Figure 4b shows the evolution of e and P_b for four values of s and δ. We see that the trends are very similar but the value of the break-point stress increases with size span, and the samples of irregular-shape grains undergo more fragmentation than regular-shaped grains. These results are consistent with experimental data (Nakata et al. 2001; Ueda et al. 2013; Hagerty et al. 1993).

Our simulations show that grain breakage begins at $0.3\sigma_c$ in the class of smallest grains, at $0.5\sigma_c$ in the class of average grain size and at $0.6\sigma_c$ in the class of largest grain. In other words, the breaking stress increases with grain size. At the end of compression, the proportion of fully broken grains decreases as the size

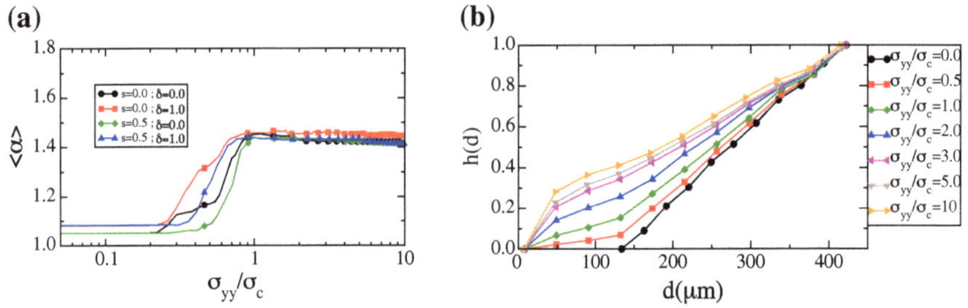

Fig. 5 **a** Evolution of the mean aspect ratio of the grains, and **b** evolution of grain size distribution

of grains increases. At very high stress ($10\sigma_c$) there are still 20 % of intact grains and 50 % of only partially broken grains in the class of the largest grains. This effect can be explained by the fact that larger grains bear low shear stresses when surrounded by a population of smaller grains.

Figure 5a shows the evolution of the mean aspect ratio of the grains defined as the ratio between the long axis and short axis of the fragments. The aspect ratio increases sharply and it tends to the silver ratio ($\sqrt{2}$) as recently also found out for lunar samples (Ueda et al. 2013). From $\sigma_{yy} = \sigma_c$, the aspect ratio keeps a constant value of $\sqrt{2}$ even if the grains continue to break. This clearly suggest that the effects of the initial size and shape distributions have been erased and the fragmentation proceeds through a self-similar process. Figure 5b shows the cumulative volume fraction h(d) of the grain diameters d for increasing level of the applied stress. At very low stress, the grain size distribution is uniform by volume fractions (h(d) is a straight line). During compression, the grain size distribution keeps a power-law shape with a cut-off at the size of Voronoi cells, which can not break.

4 Conclusion

In this work, we introduced a numerical approach based on the contact dynamics method and cell-discretization of the grains for the simulation of granular materials with crushable grains. This approach conserves grain shapes (polygonal) and total solid volume during the communication process. It was applied to the uniaxial compaction of packing of crushable grains with different size spans and shape irregularities. We showed that the degree of grain damage decreases as grain size increases. During compaction, the grain size distribution keeps a power-law form. The aspect ratio of the fragments tends to the silver ratio ($\sqrt{2}$) as recently also found out for lunar samples and argued to be a signature of self-similarity. We showed that the value of break-point stress increases with size span, and the samples of irregular-shaped particles undergo more fragmentation than regular-shaped particles.

References

Bolton MD, Nakata Y, Cheng YP (2008) Micro- and macro-mechanical behaviour of DEM crushable materials. Geotechnique 58:471–480

Esnault VPB, Roux JN (2013) 3D numerical simulation study of quasistatic grinding process on a model granular material. Mech Mater 66:88–109

Fuerstenau DW, Gutsche O, Kapur PC (1996) Confined particle bed comminution under compressive loads. Elsevier, Amsterdam, pp 521–537

Hagerty MM, Hite DR, Ullrich CR, Hagerty DJ (1993) One-dimensional high-pressure compression of granular media. J Geotech Eng 119(1):1–18

Hosten C, Cimilli H (2009) The effects of feed size distribution on confined-bed comminution of quartz and calcite in piston-die press. Int J Mineral Process 91:81–87

Liu L, Kafui KD, Thornton C (2010) Impact breakage of spherical, cuboidal and cylindrical agglomerates. Powder Technol 199(2):189–196

Moreau JJ (1994) Some numerical methods in multibody dynamics. Application to granular materials. Eur J Mech A/Solids 4:93–114

Nakata Y, Hyodo M, Hyde A, Kato Y, Murata H (2001) Microscopic particle crushing of sand subjected to high pressure one-dimensional compression. Soils Found 41(1):69–82

Radjai F, Richefeu V (2009) Contact dynamics as a nonsmooth discrete element method. Mech Mater 41(6):715–728

Thornton C, Yin KK, Adams MJ (1996) Numerical simulation of the impact fracture and fragmentation of agglomerates. J Phys D Appl Phys 29(2):424

Timar G, Kun F, Carmona HA, Herrmann HJ (2012) Scaling laws for impact fragmentation of spherical solids. Phys Rev E 86:016113

Ueda T, Matsushima T, Yamada Y (2013) DEM simulation on the one-dimensional compression behavior of various shaped crushable granular materials. Granular Matter 1–10. doi:10.1016/j.sandf.2012.07.006

Wang J, Yan H (2012) Dem analysis of energy dissipation in crushable soils. Soils Found 52(4):644–657

Onset of Strain Localization in Unsaturated Soils Subjected to Constant Water Content Loading

Dunja Perić, Gaofeng Zhao and Nasser Khalili

Abstract Analytical solutions for the inception of strain localization in unsaturated soils were implemented into the constitutive driver for a bounding surface plasticity model. Effects of the initial net mean stress and initial suction on the inception of strain localization in a porous material subjected to constant water content plane strain compression (PSC) were investigated. It was found that decreases in both, the initial suction and the initial net mean stress decreased the axial strain at onset, thus effectively increasing the susceptibility to strain localization. The corresponding deformation bands were largely contractant shear bands. Dilatant shear bands occurred only for the initial over consolidation ratios (OCR) larger than 3.25.

1 Introduction

Although the inception of strain localization is an important failure precursor, the majority of research related to this topic has covered monophasic materials. Some studies have been carried out in two phase materials while the amount of research related to three phase porous materials remains scarce. Thus, the main objective of this study was to investigate the inception of strain localization in unsaturated soils subjected to constant water content PSC. To this end, analytical solutions for onset of strain localization were found and implemented into the bounding surface plasticity model developed by Khalili et al. (2008).

D. Perić (✉)
Kansas State University, Manhattan, KS 66506-5000, USA
e-mail: peric@ksu.edu

G. Zhao · N. Khalili
The University of New South Wales, Sydney, NSW 2052, Australia

K.-T. Chau and J. Zhao (eds.), *Bifurcation and Degradation of Geomaterials in the New Millennium*, Springer Series in Geomechanics and Geoengineering,
DOI 10.1007/978-3-319-13506-9_25

2 Preliminaries

An unsaturated porous material is experiencing an infinitesimal strain and obeying a general non-associative flow rule. The material can be viewed as a mixture of three independent overlapping continua including solid phase s, water phase w, and air phase a. Each constituent has its mass M_ξ and volume V_ξ. A volume content of a phase per unit reference volume v_ξ, and a fluid mass content of each fluid phase m_ξ per unit reference volume can defined as:

$$v^\xi = \frac{V_\xi}{V_0} \quad \text{and} \quad m^\xi = \frac{M_\xi}{V_0} \tag{1}$$

Bishop's definition of an effective stress is adopted. It is given by:

$$\sigma'_{ij} = \sigma_{ij} + p_a - \chi s \delta_{ij} \tag{2}$$

where p_a is an air pressure, s is a suction, δ_{ij} is Kronecker delta, and χ is the effective stress parameter given by Khalili et al. (2008) as:

$$\chi = \left\{ \begin{array}{ll} 1, & \frac{s}{s_{ae}} \leq 1 \\ \left(\frac{s}{s_{ae}}\right)^{-0.55}, & else \end{array} \right\} \tag{3}$$

Normal components of stress σ_{ij} and solid strain tensors ε^s_{ij} are negative in compression, while mean stress p, is positive in compression. A suction value at air entry is denoted by s_{ae}, A time rate of effective stress tensor σ'_{ij} is obtained from Eq. (2) as:

$$\dot{\sigma}'_{ij} = \dot{\sigma}_{ij} + \dot{p}_a \delta_{ij} - \psi \dot{s} \delta_{ij} \tag{4}$$

The elastic-plastic stress strain relationship includes a suction hardening and it is given by:

$$\dot{\sigma}'_{ij} = D_{ijkl} \dot{\varepsilon}^s_{kl} - \frac{D^e_{ijkl} g_{kl}}{A} \left(\frac{\partial F}{\partial s} \dot{s} \right) \tag{5}$$

where F denotes a yield surface, D^e_{ijkl} denotes elastic stiffness moduli tensor, and A is given by:

$$A = H + f_{ij} D^e_{ijkl} g_{kl} \tag{6}$$

where H is plastic hardening modulus while f_{ij} and g_{ij} denote gradients of yield and plastic potential functions respectively.

3 Inception of Strain Localization

The pore fluids are assumed to be immiscible. Furthermore, water and solid phases are each assumed to be incompressible. It is assumed that a weak discontinuity emerges with continuing loading resulting in a discontinuous strain rate across an associated singular surface C. Pursuing a standard approach the jump in the solid strain rate is given by:

$$\left[\dot{\varepsilon}_{ij}^{s}\right] = \frac{1}{2}\left(c_i^{cw} N_j + c_j^{cw} N_i\right) \tag{7}$$

where N_i denotes the unit normal to the surface C, and c_i^{cw} is the associated eigenvector. The equilibrium requires no jump in the total traction rate across C, thus implying the following:

$$\left[\dot{\sigma}_{ij}\right] N_j = 0 \tag{8}$$

Constant water content loading implies no change in the mass of water while the mass of air can change, from where it follows that:

$$\dot{s} = \bar{K}_s \dot{\varepsilon}_v^s \quad \text{where } \bar{K}_s = -\frac{\psi}{n(\partial S_r/\partial s)} \tag{9}$$

Equation (9) relates the rate of volumetric solid strain to the rate of suction. A degree of water saturation is denoted by S_r while n denotes the porosity. Furthermore, the jump in suction across C can be obtained from Eq. (9) as:

$$[\dot{s}] = \bar{K}_s \left[\dot{\varepsilon}_v^s\right] \tag{10}$$

Assuming that the material on both sides of C responds plastically and combining Eqs. (4), (5), (7) and (10) results in:

$$Q_{ik}^{cw} c_k^{cw} = 0 \quad \text{and} \quad Q_{ik}^{cw} = N_j D_{ijkl}^{cw} N_l \tag{11}$$

where

$$D_{ijkl}^{cw} = D_{ijkl} - \bar{K}_s \frac{D_{ijst}^{e} g_{st}}{A}\left(\frac{\partial F}{\partial s}\right)\delta_{kl} + \psi \bar{K}_s \delta_{ij}\delta_{kl} \tag{12}$$

And Eq. (11) defines the condition for onset of strain localization. The corresponding critical hardening modulus can be obtained by considering the following eigenvalue problem (EVP):

$$Q_{ik}^{cw} c_k^{cw(i)} = \lambda^{cw(i)} Q_{ik}^{e,cw} c_k^{cw(i)} \quad i = 1, 2 \tag{13}$$

where Q_{ik}^{cw} and $Q_{ik}^{e,cw}$ are so called total and total elastic acoustic tensors that correspond to constant water content loading, while λ^{cw} are the corresponding eigenvalues. Pursuing the standard approach the EVP can be solved to yield the following solution for the plastic eigenvector as:

$$c_j^{cw(P)} = \zeta P_{ji}^{e,cw} b_i \tag{14}$$

where ζ is an arbitrary scalar and $P_{ij}^{e,cw}$ is the inverse of the undrained elastic acoustic tensor $Q_{ij}^{e,cw}$.

The critical hardening modulus is obtained by setting the plastic eigenvalue $\lambda^{cw(P)}$ equal to zero, thus giving the following:

$$H^{cw} = -f_{ij} D_{ijkl}^{e} g_{kl} + a_i P^{e,cw} b_j + \bar{K}_s \frac{\partial F}{\partial s} N_i P_{ij}^{e,cw} b_j \tag{15}$$

where vectors b_i and a_i are given by:

$$b_i = N_j D_{ijst}^{e} g_{st} \quad \text{and} \quad a_i = f_{mn} D_{mnij}^{e} N_j \tag{16}$$

The orientation of surface C is found by setting the first derivative of expression in Eq. (15) with respect to N_i equal to zero. The corresponding maximum critical hardening modulus is then obtained by substituting the critical direction N_i into Eq. (15).

4 Application to Bounding Surface Plasticity Model

The above solution was implemented in the constitutive driver for bounding surface plasticity model developed by Khalili et al. (2008). The model was calibrated against the experimental results obtained on Bourke silt from Bourke region in New South Wales, Australia. The resulting model parameters are: $M = 1.17$, $\lambda = 0.09$, $\nu = 0.25$, $A = 2$, $N = 3$, $R = 2$, $k_{\mathrm{m}} = 200$, $\lambda_{\mathrm{p}} = 0.41$ and $s_{\mathrm{ae}} = 18$ kPa. In this case only the intercept of isotropic consolidation line depends on suction and it is equal to 2.049, 2.058, and 2.068 at suctions equal to s_{ae}, 100 and 300 kPa respectively.

Figure 1 depicts a deviatoric stress versus axial strain response obtained for PSC at initial net mean stress of 30 kPa. It is seen that increase in initial suction delays the onset of strain localization. Square symbols indicate the inception of strain localization.

Figure 2 shows the angle between the eigenvector and unit normal N_i. The angles larger than 90° indicate contractant shear bands, those smaller than 90° correspond to dilatant shear bands, while 90° corresponds to a pure shear band. Dilatant shear bands occurred only for the initial OCR larger than 3.25. The past maximum effective stress is equal to 200 kPa.

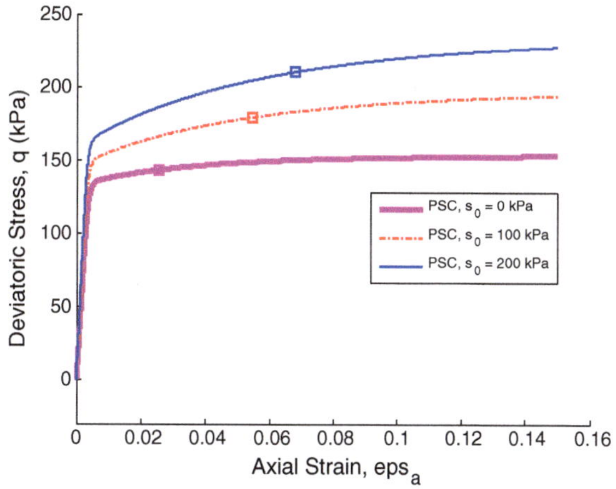

Fig. 1 Deviatoric stress versus axial strain for PSC tests at initial net stress of 30 kPa

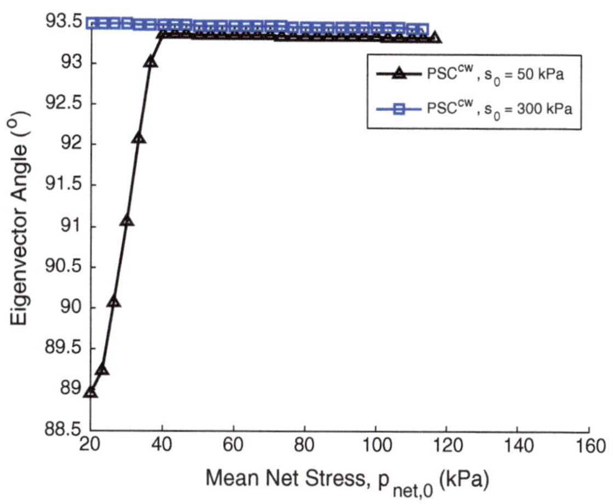

Fig. 2 The angle between eigenvector and unit normal N_i for variety of PSC tests

5 Conclusions

The example of Bourke silt subjected to constant water content PSC tests illustrates that a decrease in the initial OCR generated by an increasing suction delays the onset of strain localization.

Reference

Khalili N, Habte M, Zagarbashi S (2008) A fully coupled flow deformation model for cyclic analysis of unsaturated soils including hydraulic and mechanical hystereses. Comput Geotech 35:872–889

Double-Scale Assessment of Micro-mechanics Based Constitutive Models for Granular Materials Undergoing Mechanical Degradation

Alessandro Tengattini, Edward Andò, Giang D. Nguyen, Gioacchino Viggiani and Itai Einav

Abstract The richness across the scales of geomaterials has long been known. Yet only recently, thanks to the development of new experimental techniques, it has been possible to study the micro (grain scale) origin of some of the phenomena observed at the macro (specimen) scale. This unprecedented insight calls for new models able to build rational links between these two scales. Some recently proposed models for cemented and uncemented granular materials take advantage of this understanding to conjugate the macroscopic irreversible strains with internal variables representing a statistically averaged evolution of the micro-structure. While these models have shown their capability to reproduce the macroscopic behavior of the geomaterials they were designed for, to fully assess them and to prioritize possible enhancements, a comparison between the predicted evolution of the micro-structure and appropriate experimental data is desirable. In this contribution we study the possibility of extracting robust and statistically meaningful measurements of microstructural evolution from X-ray computed tomography images which are then compared with the micro-scale predictions of the existing micro-mechanics based models.

A. Tengattini (✉) · E. Andò · G. Viggiani
Grenoble-INP/UJF-Grenoble 1/CNRS UMR 5521, Laboratoire 3SR,
38041 Grenoble, France
e-mail: alessandro.tengattini@sydney.edu.au

A. Tengattini · I. Einav
School of Civil Engineering, University of Sydney, Sydney, Australia

G.D. Nguyen
The School of Civil, Environmental and Mining Engineering,
The University of Adelaide, Adelaide, Australia

K.-T. Chau and J. Zhao (eds.), *Bifurcation and Degradation of Geomaterials in the New Millennium*, Springer Series in Geomechanics and Geoengineering,
DOI 10.1007/978-3-319-13506-9_26

1 Introduction

The behaviour of cemented and uncemented granular materials is controlled by inelastic phenomena occurring at the grain scale. Grain crushing, cement degradation and fragment rearrangement are the key mechanical degradation processes (Menendez 1996). Thanks to the ever-increasing detail of full field experimental techniques as X-ray CT and ultrasonic tomography, it has been possible to broaden our understanding of the micro-scale and introduce it into macro-scale constitutive models. Models such as Breakage mechanics (Einav 2007a, b) explicitly connect the micro (grain) and macro (specimen) scales. These models have already shown their capability to reproduce the experimentally observed behaviour at the specimen scale (Einav 2007b; Das et al. 2011), and have shown advantages with respect to classic elasto-plastic models. However, the possibility of an experimental assessment of double scale models at both scales will be explored for its notable practical and theoretical implications.

In this contribution we briefly introduce micro-mechanics based models and some recent advancements in the experimental quantification of the micro-scale variables inherent in these models.

2 Constitutive Models for (Cemented) Granular Materials with Micro-mechanics Based Internal Variables

Recently, new constitutive models have been proposed, which describe the inelastic mechanisms occurring at the micro-scale that govern the macroscopic behaviour. In some of these models the internal variables are, at least in principle, directly identifiable.

The model described in Einav (2007a) and Tengattini et al. (2013a) for granular materials, explicitly introduces the role of the grain size distribution (GSD) and its evolution as grain crushing develops. To describe the state of this process, the internal variable Breakage (B), has been proposed (defined as the area ratio in Fig. 1a). Acknowledging the scaling of the stored energy with grain size, it is possible, through statistical homogenisation, to describe dependence of the internal elastic energy on B. Assuming then that the energy dissipated through Breakage coincides with the loss in residual Breakage Energy, the dependence of the dissipation on B is established. Casting these considerations into a thermomechanical framework allows for the constitutive description of Granular Materials the performances of which have been established at the macroscopic scale (Einav 2007b; Das et al. 2011).

For sake of brevity Breakage mechanics will not be further detailed herein, and only the end product in the form of a set of master equations is listed in Table 1.

In Tengattini et al. (2013a) and Das et al. (2013), an extension of this model has been proposed, to include the effect of cement in granular materials.

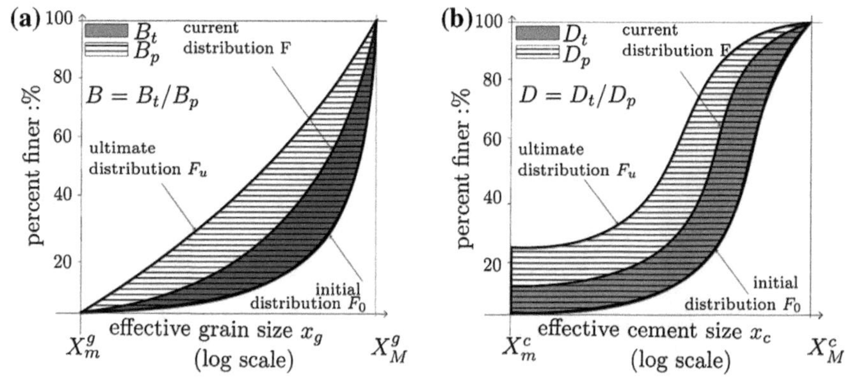

Fig. 1 Scalar internal variables. **a** Breakage (B) (Einav 2007a). **b** Damage (D) (Einav 2007b)

Table 1 Essential equations of the breakage model, using the same notation as in Alikarami et al. (2014)

Element of the model	Equation	No.
Stresses	$q = 3G\varepsilon_S^E(1-\theta B)$	(1)
	$p = K\varepsilon_V^E(1-\theta B)$	(2)
Breakage energy	$E_B = \frac{1}{2}\theta\left(K\varepsilon_V^{E^2} + 3G\,\varepsilon_S^{E^2}\right)$	(3)
Yield criterion	$y = \frac{E_B(1-B)^2}{E_\Gamma} + \left(\frac{q}{Mp}\right)^2 - 1 \leq 0$	(4)
Flow rules	$B = 2\delta\lambda\cos^2(\omega)\frac{E_B(1-B)^2}{E_R E_\Gamma}$	(5)
	$\varepsilon_P^V = 2\delta\lambda\frac{E_B}{E_\Gamma}\frac{(1-B)^2}{p}\sin^2(\omega)$	(6)
	$\varepsilon_P^s = 2\delta\lambda\frac{q}{(Mp)^2}$	(7)

To describe the role of cement, an effective cement size distribution has been envisaged. The mechanical degradation of the cement can be then described through the scalar internal variable Damage (*D*), defined as the area ratio in Fig. 1b. With energy considerations analogous to the ones described for the grain phase, it has been possible to develop a constitutive model for cemented granular materials with fewer mechanical parameters than the reference ones available in the literature (Lagioia and Nova 1995). More importantly, the internal variables are once again experimentally identifiable, at least in principle. The feasibility of practical observations of this internal variable has been discussed in Tengattini et al. (2013b). The following section will tackle, on the other hand, the measurability of the Breakage variable B.

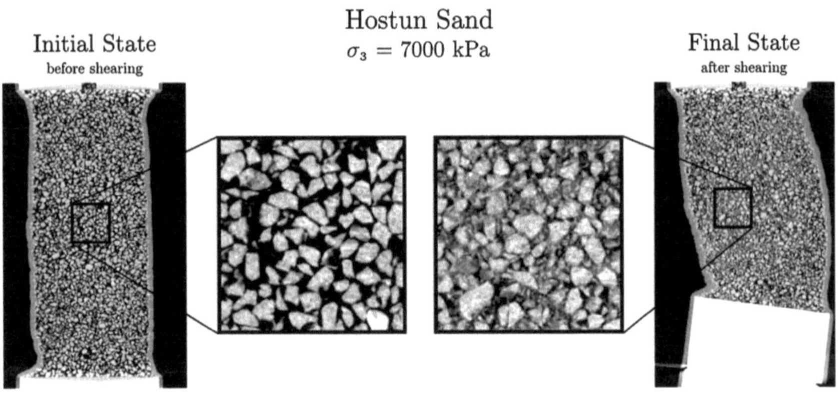

Fig. 2 Typical X-ray CT scans obtained in Alikarami et al. (2014)

3 Experimental Quantification of Grain Crushing

The evolution of the grain size distribution in classical geotechnical tests on sand has long been studied by means of *post-mortem* sieving (Nakata et al. 1999). It is only recently, though, that tools such as X-ray Computed Tomography have given us access to in situ, full field observations of the phenomena occurring at the grain scale. A recent set of experiments (Alikarami et al. 2014) focuses on high confinement triaxial tests on Hostun sand undergoing severe grain crushing. In Fig. 2, typical observations obtained from the X-ray scans of a triaxial test run at 7 MPa confinement are presented.

In Gkiousas-Kapnisis et al. (2014) a novel technique is proposed to quantify the observed grain crushing by reconstructing the evolution of the grain size distribution of the grains above a threshold determined by the image acquisition (5–7 voxels, ~15 μm size). Despite the current impossibility to directly measure the full range of the grain size distribution, it is the authors' contention that this level of detail is sufficient to attempt a double scale assessment of the model, inferring the granulometry (where not accessible) under the assumption of the size independence of Breakage.

4 Double Scale Experimental Assessment of Breakage Mechanics

The implementation of the model as described in Eqs. (1–7) with the parameters reported in Table 2, allows the reproduction of the macroscopic experimental behaviour, as reported in Fig. 3a. The novelty of this contribution lies in the assessment of the model's capability of predicting the evolution of the microstructure,

Table 2 Model parameters used in the current simulations

K (GPa)	G (GPa)	M	E_C (kPa)	ω (○)	θ
0.38	0.13	1.35	80	50	0.35

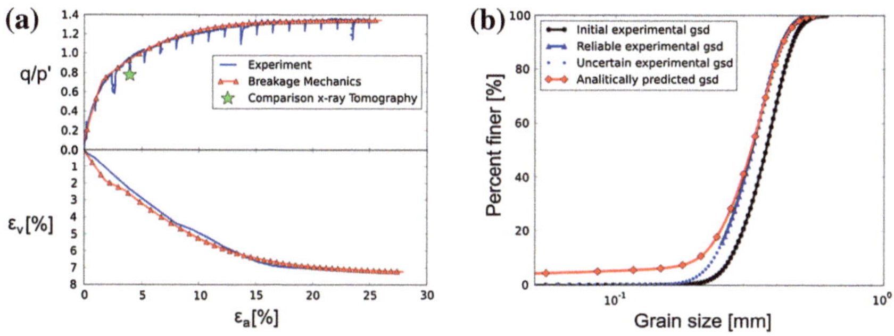

Fig. 3 Model assessment. **a** Macroscopic scale. **b** Micro (grain) scale

described here through the grading of the sand. Figure 3b shows the experimentally measured granulometry at the beginning of the test (black line) and at a point in the test where crushing is occurring but most of the fragments are still above the resolution of the technique (blue line, the solid line represents the reliable part of the data, while the dotted section is uncertain). The predicted curve (red line) seems to be close to the measured one, particularly over the reliable part of the measurement.

Where K, G are the bulk and shear Stiffness, M the critical state stress ratio and E_C, ω, θ are respectively the Critical Breakage Energy, the coupling angle and the grading index as in Einav (2007a).

Although these should be regarded as preliminary results, the encouraging match highlights the potential of a synergic effort: on one hand, the experiments allow a direct calibration of the model; on the other, the model provides access to the scales below the experimental resolution.

5 Conclusions and Perspectives

In this contribution we have presented some preliminary results exploring the possibility of assessing micro-mechanics based constitutive models both at the macro (specimen) and micro (grain) scale. There are evident practical implications of models that reliably predict the evolving microstructure of geomaterials (e.g., for the evolution of permeability in hydrocarbon extraction), allowing us to go beyond phenomenological models. As remarked in Krajcinovic (1998), "models based on phenomenological evidence and selected to fit a particular stress strain-curve may provide a result that pleases the eye but seldom contributes to the understanding of the processes represented by the fitted curve". A more systematic application of the tool, possibly taking into account the spatial heterogeneity of the specimen as well as the exploration of a wider range of boundary conditions and materials, presents itself as a natural extension of the current work.

References

Alikarami R, Andò E, Gkiousas-Kapnisis M, Torabi A, Viggiani G (2014) Strain localisation and grain breakage in sand under shearing at high mean stress: insights from in-situ X-ray tomography (submitted)

Das A, Nguyen GD, Einav I (2011) Compaction bands due to grain crushing in porous rocks: a theoretical approach based on breakage mechanics. J Geophys Res 116:B08203

Das A, Tengattini A, Nguyen GD, Viggiani G, Hall S, Einav I (2013) A thermomechanical constitutive model for cemented granular materials with quantifiable internal variables. Part II—validation and localization analysis (submitted)

Einav I (2007a) Breakage mechanics—Part I: theory. J Mech Phys Solids 55(6):1274–1297

Einav I (2007b) Breakage mechanics—Part II: modelling granular materials. J Mech Phys Solids 55:1298–1320

Gkiousas-Kapnisis M, Andò E, Tengattini A, Einav I, Viggiani G (2014) Development of image analysis tools to evaluate in-situ evolution of the grain size distribution in sand subjected to breakage (submitted, this volume)

Krajcinovic D (1998) Selection of damage parameter art or science. Mech Mater 28:165–179

Lagioia R, Nova R (1995) An experimental and theoretical study of the behaviour of a calcarenite in triaxial compression. Geotechnique 45(4):633–648

Menendez B, Zhu W, Wong T-F (1996) Micromechanics of brittle faulting and cataclastic flow in Berea sandstone. J Struct Geol 18(1):1–16

Nakata Y, Hyde AFL, Hyodo M, Murata H (1999) A probabilistic approach to sand crushing in the triaxial test. Geotechnique 49(5):567–583

Tengattini A, Das A, Nguyen GD, Viggiani G, Hall S, Einav I (2013a) A thermomechanical constitutive model for cemented granular materials with quantifiable internal variables. Part I—theory (submitted)

Tengattini A, Einav I, Viggiani G (2013b) Experimental study of inelastic deformation at the micro scale in cemented granular materials: some recent results. In: Proceedings of the III international conference on particle-based methods: fundamentals and applications. International Center for Numerical Methods in Engineering (CIMNE) Barcelona, Spain

Hypoplastic Constitutive Modelling of Grain Damage Under Plane Shearing

Erich Bauer, Linke Li and Wenxiong Huang

Abstract In this paper a new concept is proposed for the constitutive modeling of grain damage of a cohesionless and unweathered granular material within the framework of micro-polar hypoplasticity. The effect of the change of the grain size distribution as a result of grain abrasion and grain rupture is taken into account in a simplified manner by reducing the mean grain diameter and modifying the constitutive relation for the incremental stiffness. For constitutive modeling it is convenient to distinguish different causes of particle damage such as the increase of the isotropic and deviatoric stress, the increase of the rotation resistance of particles and the abrasion caused by large particle rotation. The focus of this paper is mainly on modeling particle damage as a result of an increase of the curvature and particle rotation, which is investigated for monotonic plane shearing of an infinite granular layer under a constant normal stress.

1 Introduction

Grain damage in cohesionless granular materials like sand, gravel and rockfills under different loading conditions has been observed in field tests and laboratory experiments and can have a significant influence on the mechanical behavior of the crushed granular material (Sadrekarimi and Olson 2011; Daouadji et al. 2001). The amount of grain crushing is not only a question of the stress level, but it is also influenced by different factors related to the morphology of the granular material and the loading history. Under shearing, for instance, abrasion and pronounced damage of grains can already occur under small mean stresses caused by the shear induced torque and the rotation resistance of individual particles within

E. Bauer (✉) · L. Li
Institute of Applied Mechanics, Graz University of Technology, 8020 Graz, Austria
e-mail: erich.bauer@tugraz.at

W. Huang
College of Mechanics and Materials, Hohai University, 210098 Nanjing, China

K.-T. Chau and J. Zhao (eds.), *Bifurcation and Degradation of Geomaterials in the New Millennium*, Springer Series in Geomechanics and Geoengineering,
DOI 10.1007/978-3-319-13506-9_27

the grain skeleton (Luzzani and Coop 2002). Large shearing of sand specimens in a ring shear device shows that grain breakage is concentrated in the shear band while almost no grain damage can be observed outside the shear band. Particle damage may change the internal structure and is consequently accompanied by a reduction of the shear resistance, changes in particle size distribution and additional compaction. The appropriate constitutive modeling of these different effects is a difficult task and needs experimental data usually not available from standard laboratory tests. New technologies used for experimental investigations like X-ray micro-tomography allow a deeper insight into the interaction between loading history and grain breakage on the micro-level and also provide new opportunities to measure the evolution of particle rotation and grain breakage (Ando et al. 2012).

In the following it is convenient to distinguish different causes of particle damage like for instance the increase of the isotropic and deviatoric stress, the increase of the rotation resistance of particles related to an increase of the curvature and the abrasion caused by large particle rotation. It is the aim of the present paper to propose a new concept for modeling the effects of these causes of particle damage using the framework of hypoplasticity. With respect to the so-called "solid hardness" the effect of pressure dependent grain crushing on the stress strain behavior is already included in the non-polar hypoplastic model proposed by Gudehus (1996) and Bauer (1996). Recent investigation of compression tests shows that for lower stress ranges the adaptation to the experimental data leads to much higher values for the parameter of the solid hardness than for larger stress ranges. In order to model the influence of the evolution of grain breakage the constant solid hardness is replaced by an evolution equation. The change of the grain size distribution is linked to the corresponding reduction of the mean grain diameter, while for the sake of simplicity a change of the curvature of the particle size distribution is neglected. Herein different causes of particle damage are separated, which allows an easier calibration of the constants involved. As relevant experimental data has not been fully quantified yet, the proposed constitutive relation and the calibration is hypothetical in character. The focus of this paper is mainly on studying grain crushing under monotonic plane shearing. In particular a reduction of the mean grain diameter is related to the change of the curvature and to the evolution of the difference between micro-rotation and macro-rotation in a micro-polar continuum description.

2 Micro-polar Material Description of Grain Damage

Grain damage is described based on a micro-polar hypoplastic continuum with the following extensions: the constant mean grain diameter d_{50} and constant solid hardness h_s in the original version (Huang 2002) are replaced by history dependent functions with the following hypothetical evolution equations:

$$\dot{d}_{50} = -\left[b_\kappa \|\dot{\boldsymbol{\kappa}}\| + b_\omega \left\|\dot{\omega}_3^c - \dot{\omega}_3\right\| + b_p f_p(\dot{p}) + b_\eta f_\eta(\dot{\eta}, \ldots)\right] d_{50}, \tag{1}$$

and

$$\dot{h}_s^* = b_s h_s^* \dot{d}_{50}/d_{50}. \tag{2}$$

Herein d_{50} and h_s^* are the current values of these state quantities and their rates are $\dot{d}_{50}$ and $\dot{h}_s^*$ · $\|\dot{\kappa}\|$ denotes the norm of the rate of the curvature, $\left\|\dot{\omega}_3^c - \dot{\omega}_3\right\|$ denotes the norm of the difference of the rate of the micro- and macro-rotations, $f_p(\dot{p})$ is a function depending on the rate of the mean stress, $f_\eta(\dot{\eta},\ldots)$ is a function depending on other state quantities $(\dot{\eta},\ldots)$ not specified yet. Factors $b_\kappa, b_\omega, b_p, b_\eta$ and b_s can also depend on the current state variables, but are assumed to be constant for the present investigations. Equations (1) and (2) describe a rate independent behavior. For the extended constitutive model the evolution equations for the objective stress rate, the objective couple stress and of the rate of the void ratio, $\dot{e}$, read:

$$\overset{\circ}{\boldsymbol{\sigma}} = f_{\mathrm{s}}\left[\hat{a}^2\dot{\boldsymbol{\varepsilon}}^{\mathrm{c}} + \left(\hat{\boldsymbol{\sigma}} : \dot{\boldsymbol{\varepsilon}}^{\mathrm{c}} + \hat{\boldsymbol{\mu}} : \dot{\bar{\boldsymbol{\kappa}}}\right)\hat{\boldsymbol{\sigma}} + f_{\mathrm{d}}\left(\hat{\boldsymbol{\sigma}} + \hat{\boldsymbol{\sigma}}^{\mathrm{d}}\right)\sqrt{\hat{a}^2\left\|\dot{\boldsymbol{\varepsilon}}^{\mathrm{c}}\right\|^2 + a_{\mathrm{m}}^2\left\|\dot{\bar{\boldsymbol{\kappa}}}\right\|^2}\right], \tag{3}$$

$$\overset{\circ}{\boldsymbol{\mu}} = d_{50} f_{\mathrm{s}}\left[a_{\mathrm{m}}^2\dot{\bar{\boldsymbol{\kappa}}} + \left(\hat{\boldsymbol{\sigma}} : \dot{\boldsymbol{\varepsilon}}^{\mathrm{c}} + \hat{\boldsymbol{\mu}} : \dot{\bar{\boldsymbol{\kappa}}} + 2f_{\mathrm{d}}\sqrt{\hat{a}^2\left\|\dot{\boldsymbol{\varepsilon}}^{\mathrm{c}}\right\|^2 + a_{\mathrm{m}}^2\left\|\dot{\bar{\boldsymbol{\kappa}}}\right\|^2}\right)\hat{\boldsymbol{\mu}}\right], \tag{4}$$

$$\dot{e} = (1+e)\dot{\boldsymbol{\varepsilon}}^c : \dot{\boldsymbol{\varepsilon}}^c. \tag{5}$$

In these equations $\hat{\boldsymbol{\sigma}} = \boldsymbol{\sigma}/tr\boldsymbol{\sigma}$ is the normalized non-symmetric Cauchy stress tensor, $\hat{\boldsymbol{\sigma}}^d = \hat{\boldsymbol{\sigma}} - \boldsymbol{I}/3$ its deviator, $\hat{\boldsymbol{\mu}} = \boldsymbol{\mu}/(d_{50}tr\boldsymbol{\sigma})$ is the normalized couple stress tensor, $\dot{\boldsymbol{\varepsilon}}^c$ is the rate of deformation tensor, and $\dot{\bar{\boldsymbol{\kappa}}} = d_{50}\dot{\boldsymbol{\kappa}}$ denotes the normalized rate of curvature tensor. The stiffness factor f_s and density factor f_d are functions of the current void ratio e, the pressure dependent limit void ratios, e_i and e_d, and the critical void ratio e_c (Gudehus 1996; Bauer 1996). Function $\hat{a}$ is related to critical stress states, i.e. to the critical friction angle φ_c, and a_m is related to the rotation resistance of particles as outlined in more detail in Huang et al. (2002). It should be noted that for a refined modeling of the effect of grain crushing factors f_s, f_d, $\hat{a}$ and a_m are also dependent in a more complex manner on the change of d_{50}, which, however, is not further discussed in the present paper.

3 Shearing of an Infinite Granular Strip

For plane shearing the relevant kinematic and static quantities for the 2-D micropolar continuum are shown in Fig. 1.

The evolution of grain crushing is investigated for the case of monotonic plane shearing of a lateral infinite granular layer under a constant vertical stress of $\sigma_{22} = -1{,}000$ kPa, an initial void ratio of $e = 0.62$, an initial mean grain diameter of $d_{50} = 0.5$ mm and an initial value of $h_{s0}^* = 10{,}000$ MPa. The granular layer is located between two parallel very rough walls, so that no sliding and no particle rotation can occur along the bottom surface and the top surface. Depending on dilatant

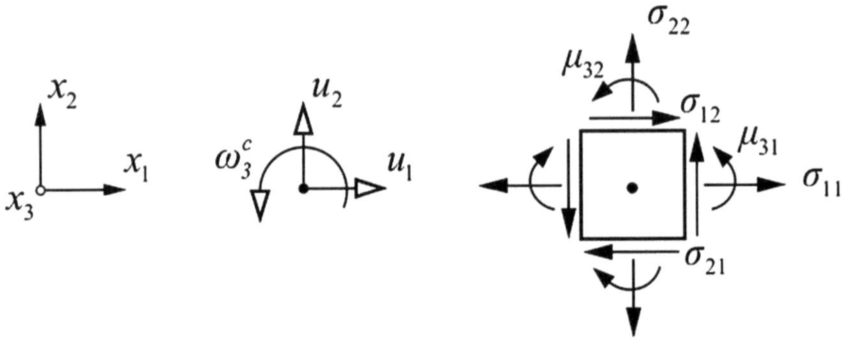

Fig. 1 Illustration of kinematic and static quantities of the 2-D micro-polar continuum

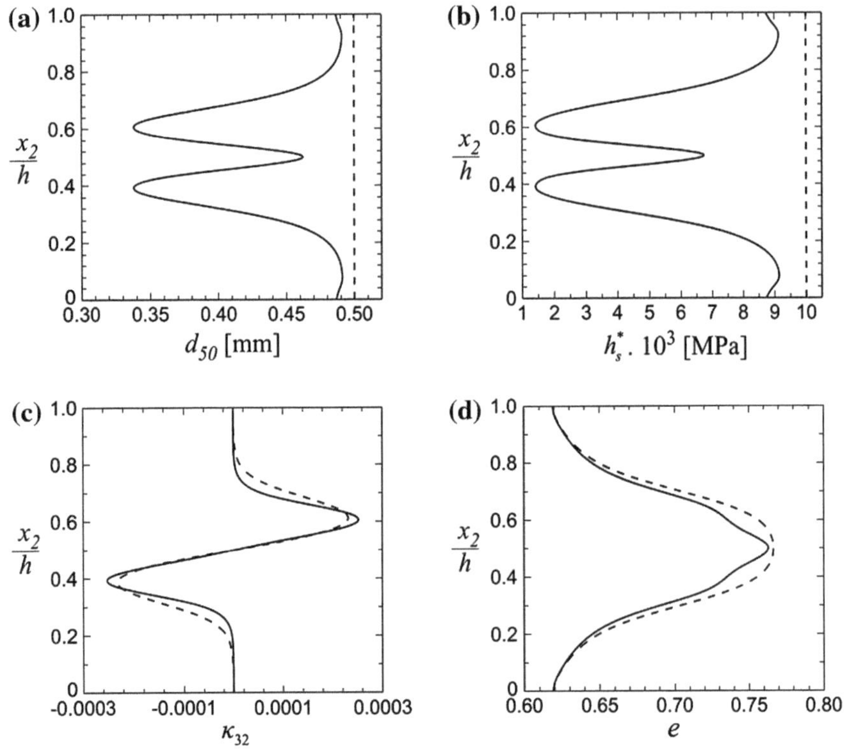

Fig. 2 Case 1: influence of the evolution of the amount of the rate of curvature on grain damage

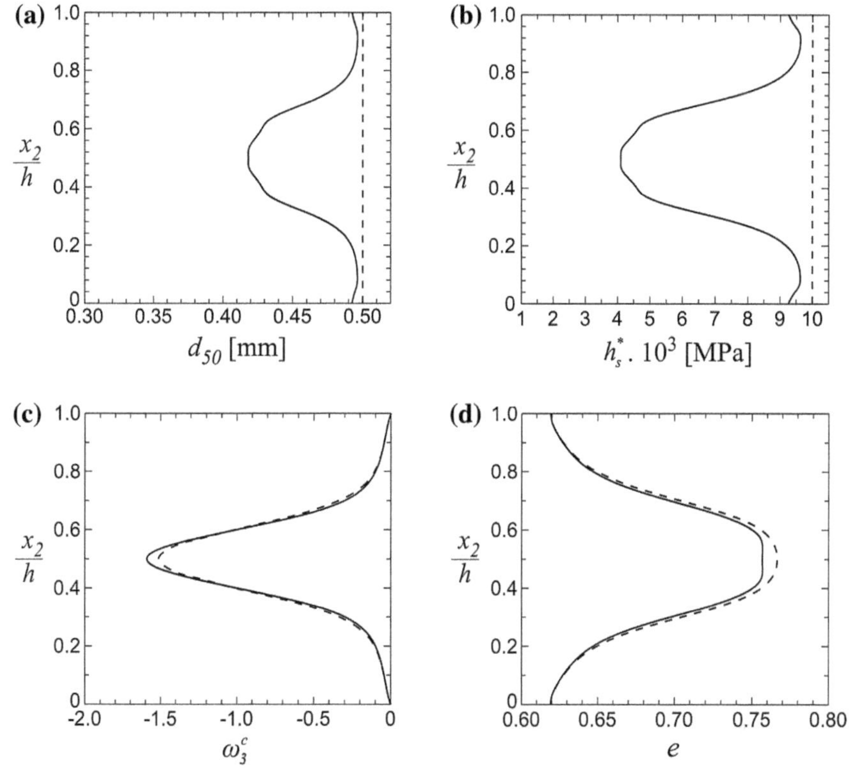

Fig. 3 Case 2: influence of the evolution of the rate of micro-rotation on grain damage

or contracting behavior of the material the top surface can move up and down in the vertical direction. Because of the constant vertical stress the change of the mean pressure in the specimen during shearing is rather small, i.e. $\dot{p} \ll$, so that the influence of the third term in Eq. (1) is neglected. For the present investigations the following values for the material parameters in the constitutive relations (1) and (2) are assumed: $b_\kappa = 4.0, b_\omega = 2.0, b_p = b_\eta = 0$ and $b_s = 5.0$. For the material parameters of Eqs. (3) and (4) similar values are taken as proposed in Huang et al. (2002).

For numerical simulations with the finite element program ABAQUS the extension of the constitutive model is implemented into the version originally developed by Huang (2000). In order to separate the different influences on the reduction of the mean grain diameter numerical simulations are carried for two different cases. In Case 1 only the first term in Eq. (1) is active, i.e. $\dot{d}_{50} = -b_\kappa \|\dot{\kappa}\| d_{50}$, while in Case 2 only the second term in Eq. (1) is active, i.e. $\dot{d}_{50} = -b_\omega \left\| \dot{\omega}_3^c - \dot{\omega}_3 \right\|$. For a horizontal displacement of the top surface the results obtained for Case 1 and Case 2 are shown in Figs. 2 and 3, respectively. Herein the dashed curves represent the result obtained for a constant mean grain diameter, i.e. for $b_\kappa = b_\omega = 0$. Because of the symmetric boundary conditions for the rough surfaces at the top and at the bottom shear strain localization occurs combined with a pronounced reduction of d_{50} in the middle of the shear layer. The numerical results show that outside the shear band only at the beginning of shearing small grain damage takes place. Within the shear band a higher void ratio is visible due to pronounced dilatancy for the initially medium dense specimen as shown in Figs. 2d and 3d. As a consequence of grain damage the void ratio is somewhat smaller compared to the undamaged material. In Case 1 (Fig. 2) a significant reduction of d_{50} and h_s^* takes place close to the boundaries of the shear band where the curvature κ_{32} are extremal. These zones are characterized by extreme values for the rate of the curvatures and for couple stresses. By contrast in Case 2 (Fig. 3) the minimum value of d_{50} and h_s^* occurs in the middle of the shear layer, where the maximum micro-rotation ω_3^c takes place.

4 Conclusions

In this paper the effect of particle rotation and curvature on particle damage is modelled with an extended micro-polar hypoplastic description. It is demonstrated that for monotonic plane shearing under constant vertical stress grain damage becomes dominant within the localized zone. In particular grain damage in the middle of the localized zone is mainly related to the particle rotation while the effect of large curvature on grain damage is pronounced close to the boundaries of the localized zone. Grain damage also leads to an additional compaction of the material. For a more realistic quantification of the constitutive parameters involved appropriate experimental methods are needed to analyze the distribution of grain crushing across the shear band.

References

ABAQUS User's manual version 6.3. Hibbitt. Karlsson and Sorensen Inc., Pawtucket

Ando E, Hall SA, Viggiani G, Desrues J, Besuelle P (2012) Grain-scale experimental investigation of localised deformation in sand: a discrete particle tracking approach. Acta Geotech 7:1–13

Bauer E (1996) Calibration of a comprehensive hypoplastic model for granular materials. Soils Found 36(1):13–26

Daouadji A, Hicher PY, Rahma A (2001) An elastoplastic model for granular materials taking into account grain breakage. Eur J Mech-A Solids 20:113–137

Gudehus G (1996) A comprehensive constitutive equation for granular materials. Soils Found 36(1):1–12

Huang W (2000) Hypoplastic modelling of shear localisation in granular materials. PhD thesis, Graz University of Technology, Austria

Huang W, Nübel K, Bauer E (2002) Polar extension of a hypoplastic model for granular materials with shear localization. Mech Mater 34:563–576
Luzzani L, Coop MR (2002) On the relationship between particle breakage and the critical state of sand. Soils Found 42(2):71–82
Sadrekarimi A, Olson SM (2011) Critical state friction angle of sands. Géotechnique 61(9):771–783

Hierarchy of Failure Indicators in the Failure Analysis of Geomaterials

Richard Wan and Xu Gong

Abstract The failure of geomaterials presents distinctive features that arise from material response being dependent on pressure, density and fabric. In conventional plasticity, these dependencies are described by a non-associated plastic flow rule which provides mathematical sources for material or constitutive instability, thereby admitting a multiplicity of material responses for the same initial loading history. The non-symmetry of the tangent constitutive matrix due to non-associated plasticity triggers different failure indicators during material response history, leading to various failure modes such as diffuse or localized. The genesis of failure is analysed in a finite element computation of a drained compression test on sand in plane strain as a boundary value problem.

1 Introduction

The failure of geomaterials can be seen as an instability of an otherwise uniform state such as density variations in the form of localized dilation, leading to patterned bands of localized shear deformations, so-called shear bands (Desrues 1990; Rudnicki and Rice 1975). On the other hand, the instability can also emerge in the absence of any localization of deformations giving way to a so-called diffuse deformation mode (Darve 1994; Hill 1958; Nicot and Darve 2006) with a sudden loss in bearing capacity depending on loading mode. In contrast to shear localized failure which occurs around plastic limit or peak state, diffuse failure usually appears well before the Mohr-Coulomb plastic limit surface is reached, and hence introduces an instability surface as a lower limit. Both numerical and experimental studies of

R. Wan (✉) · X. Gong
University of Calgary, Calgary, Canada
e-mail: wan@ucalgary.ca

X. Gong
e-mail: xgon@ucalgary.ca

K.-T. Chau and J. Zhao (eds.), *Bifurcation and Degradation of Geomaterials in the New Millennium*, Springer Series in Geomechanics and Geoengineering,
DOI 10.1007/978-3-319-13506-9_28

this phenomenon have been conducted, see Desrues (1990) and Pinheiro (2009) among others. In fact, the circumstances under which bifurcation of shear response occurs are directly linked to the nature of the underlying constitutive relation, especially its non-symmetry. Numerically, the variation of the non-symmetric constitutive tensor during loading history activates various failure indicators based on the characteristics of the matrix following a certain hierarchy, see Pinheiro (2009) and Wan et al. (2013) for more details. In general, diffuse failure precedes localization for non-associated elasto-plasticity, but arises concurrently with localization in the associated plasticity case (Bigoni and Hueckel 1991; Wan et al. 2013). However, Desrues (1990) shows that strain localization can also occur before the peak deviatoric stress condition in drained compression of sand.

In this paper, the hierarchy of the various failure indicators and their evolution with loading history are examined numerically at material points in a finite element analysis of plane strain drained test on sand as a boundary value problem. An advanced elasto-plastic model (Wan and Guo 1998) encompassing the key features of geomaterials is adopted in this study. Various failure indicators at material points are examined and compared through their evolution in both time and space.

2 Diffuse and Localization Failure Indicators

The mathematical formulation of diffuse and localization failure (Pinheiro 2009; Wan et al. 2013) is herein briefly summarized. Recall the basic stress strain relationship:

$$d\sigma = \mathbf{D} \cdot d\boldsymbol{\varepsilon}, \tag{1}$$

where $d\sigma$ and $d\varepsilon$ are incremental stress and strain, and $\mathbf{D}$ the tangent constitutive matrix. Loss of uniqueness in the stress-strain response is signaled by $\det(\mathbf{D}) = 0$ at the plastic limit.

Diffuse failure is closely associated with the stability of equilibrium states of deforming solids to vanishingly small disturbances examined in terms of the second order work (Hill 1958). Locally, the second order work criterion is the product of incremental stress and strain during a loading increment which has to remain positive for stability to prevail. Thus, noting Eq. (1),

$$W_2 = d\boldsymbol{\varepsilon} \cdot \mathbf{D} \cdot d\boldsymbol{\varepsilon}, \;\; d\boldsymbol{\varepsilon} \neq 0, \;\; \det(\mathbf{D}_{sym}) > 0, \tag{2}$$

such that a positive second order work signals the positive definiteness of the constitutive matrix, i.e. the determinant of its symmetric part must be positive. According to Bromwich Bounds in linear algebra, the smallest eigenvalue of $\mathbf{D}_{sym}$ presents a lower bound to the smallest eigenvalue of $\mathbf{D}$, and thus the loss of stability in the form of $W_2 = 0$ precedes the plastic limit condition.

This condition of sufficient stability can be enlarged and it turns out that for a given stress strain there are certain directions of loading increments that violate the second order work, giving rise to the notion of directional stability or bifurcation cone (Darve 1994; Wan et al. 2013). In terms of kinematics, diffuse failure maintains continuity in both the displacement and displacement gradient fields. In contrast to diffuse failure mode, localization failure involves weak discontinuities in displacement gradient while the displacement field still remains continuous. Loss of ellipticity of the static equilibrium equations admitting weak discontinuities in the strain field across surfaces with local normal **n** are sought, leading to the well-known localization criterion (Rudnicki and Rice 1975) written as:

$$\det\{\mathbf{A}(\mathbf{n})\} = 0; \;\; \mathbf{A}(\mathbf{n}) = \mathbf{n} \cdot \mathbf{D} \cdot \mathbf{n}, \tag{3}$$

where **n** is the unit vector normal to the localization surface and **A** is the so-called acoustic tensor. This local criterion eventually leads to global loss of ellipticity or ill-posedness of the static boundary value problem, hence highlighting the fact that the problem is dynamic.

3 Finite Element Numerical Analysis

In this section, the hierarchy of the above introduced failure indicators and their evolution with loading history are examined numerically at material (Gauss integration) points in the finite element analysis of a plane strain drained test on sand as a boundary value problem. An enriched elasto-plastic (WG) model (Wan and Guo 1998) is herein used due with constitutive attributes that provide mathematical sources for material instability. These are non-associativity of plastic yield and flow, as well as non-linearities arising from density, stress level and fabric dependencies on loading history, see Pinheiro (2009).

Loss of uniqueness in the solution of the boundary value problem essentially relies on the characteristics of the constitutive law, and hence the instability is material based. To evaluate the sensitivity of results to (displacement controlled) loading step size and the loss of uniqueness in the post peak response of the global structure, three displacement increment values are examined: 0.01, 0.005 and 0.001 m. Figure 1

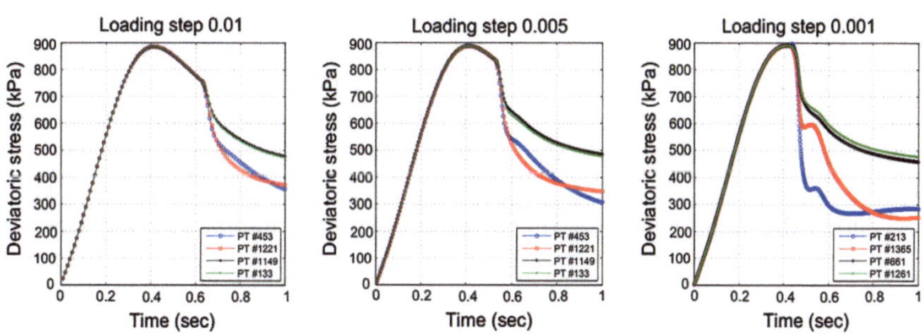

Fig. 1 Deviatoric stress versus time for various Gauss points at different displacement loading steps

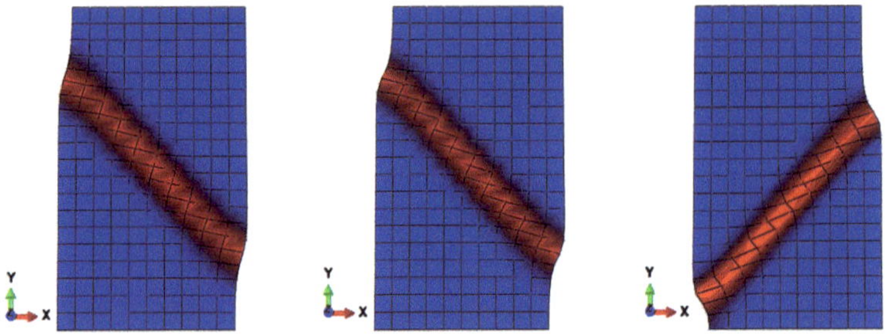

Fig. 2 Shear band patterns at various displacement loading steps

shows the evolution of deviatoric stress at four selected Gauss points for each loading step size. Numerical results discussed pertain to points located inside (blue and red) and outside (black and green) the shear band. Clearly, the deviatoric stresses for all above-mentioned points follow a unique curve until a so-called bifurcation point signaling a loss of homogeneity in the initial deformation field is reached where the solution branches into different paths. For displacement step sizes 0.01 and 0.005 m, the point of bifurcation occurs in the softening regime, while for the finest displacement step size of 0.001 m, it appears at peak. Also, a different shear band location is found for step size 0.001 m compared with the two others, see Fig. 2. This is typical of a bifurcation problem in the post peak regime where different failure modes are obtained depending on the path followed due to ill-posedness of the boundary value problem. While shear band width can be easily reproduced by introducing a characteristic length, the position of the shear band cannot be predicted in advance, which is in line with experiments where a large variability of failure patterns is normally observed.

For illustration purposes, the case of displacement step size 0.005 m is discussed with regard to the evolution of all failure indicators at selected four Gauss points 453 and 1,221 (inside the shear band) and 1,149 and 133 (outside the shear band); see Fig. 3. The initial evolution of the second order work is the same for all points until the peak effective mean stress ($p \approx 640$ kPa) where W_2 first vanishes. Then, second order work becomes negative for Gauss points 453 and 1,221 inside the shear band and positive for 1,149 and 133 outside the shear band. The determinant of the acoustic tensor, det($\mathbf{A}$), increases slightly initially and monotonically decreases to zero ($p \approx 410$ kPa) which is much earlier than at peak condition ($p \approx 640$ kPa) for all Gauss points. Therefore, localization can be met during the strain hardening regime (prepeak) for a drained test, see Desrues

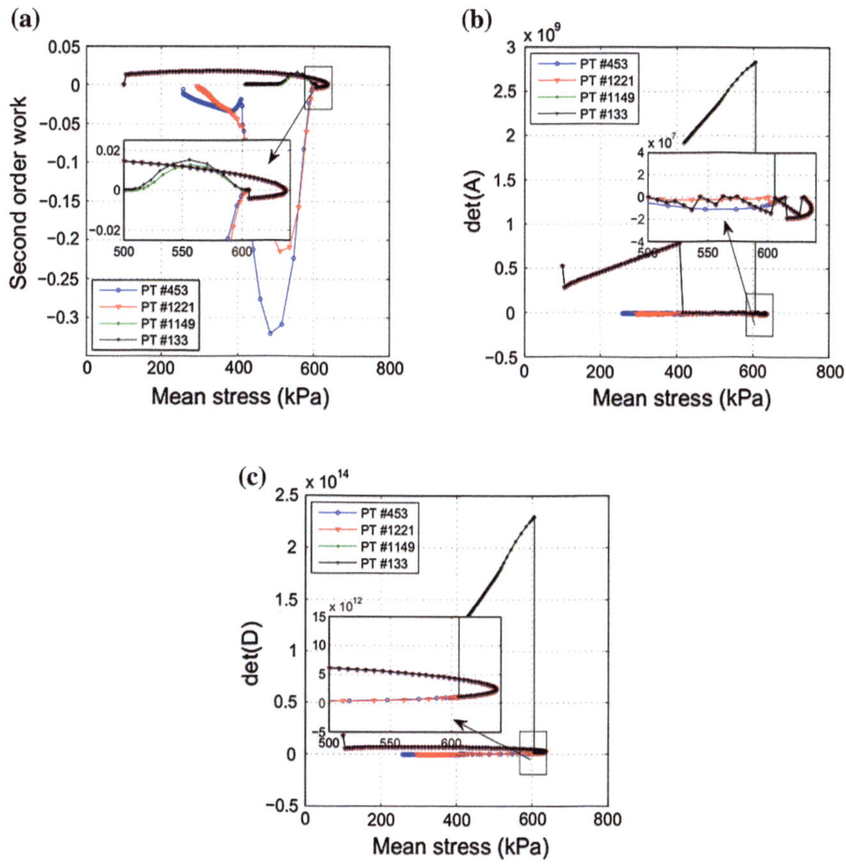

Fig. 3 Failure behaviours of Gauss points inside shear band with respect to stress path, second order work, localization and plastic limit: **a** Second order work, **b** det (**A**), **c** det (**D**)

(1990). As anticipated, looking at points 453 and 1,221, the plastic limit failure (det(**D**) ≤ 0) is ultimately reached at critical state long after both diffuse and localization failures have occurred.

Figures 4, 5 and 6 illustrate the genesis of failure throughout the plane strain sample (FEM mesh) through the evolution of the deviatoric (shear) plastic strain field over which are superimposed Gauss points that activate the various failure indicators at selected representative (pseudo) times during loading history. The loss of homogeneity in the plastic strain field occurs at peak deviatoric stress conditions where patterned shear localization zones appear with a two-phase structure. At the

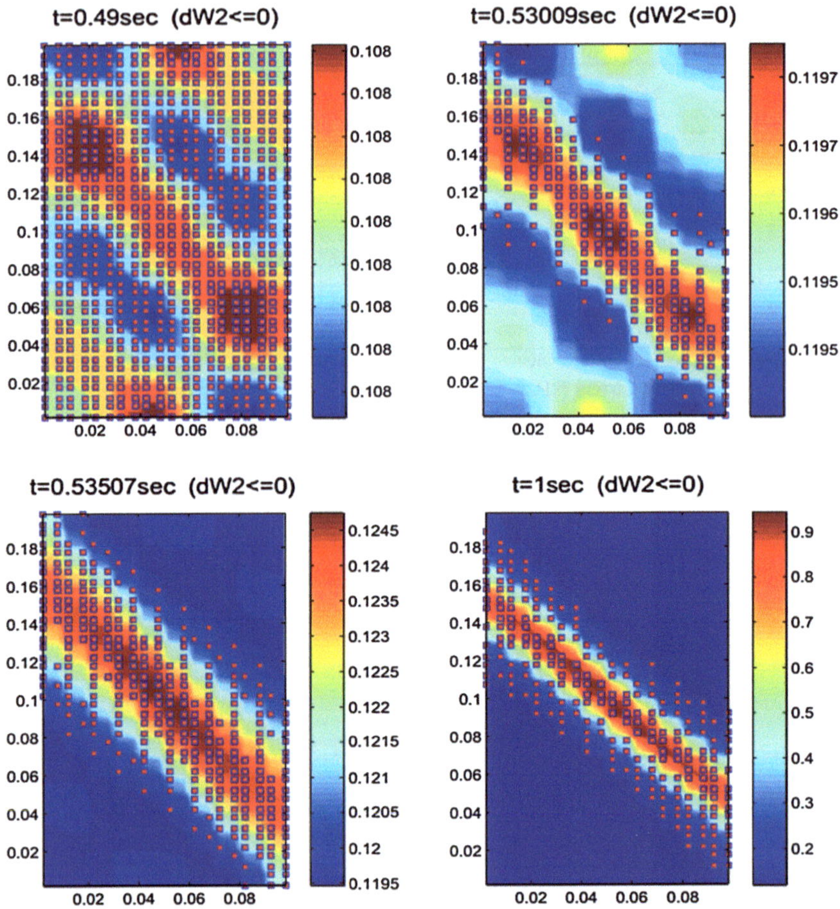

Fig. 4 Plastic strain field plot at various time steps and negative second order work data points

same time, all Gauss points simultaneously violate the second order work criterion together with the satisfaction of the localization criterion which was already reached before peak (Fig. 3). Going further along loading history, both violated second order work and localized Gauss points cluster within an emerging central localization zone which ultimately concentrates into a clear single shear band. At this point, all failure indicators (second order work, localization and plastic limit) coincide.

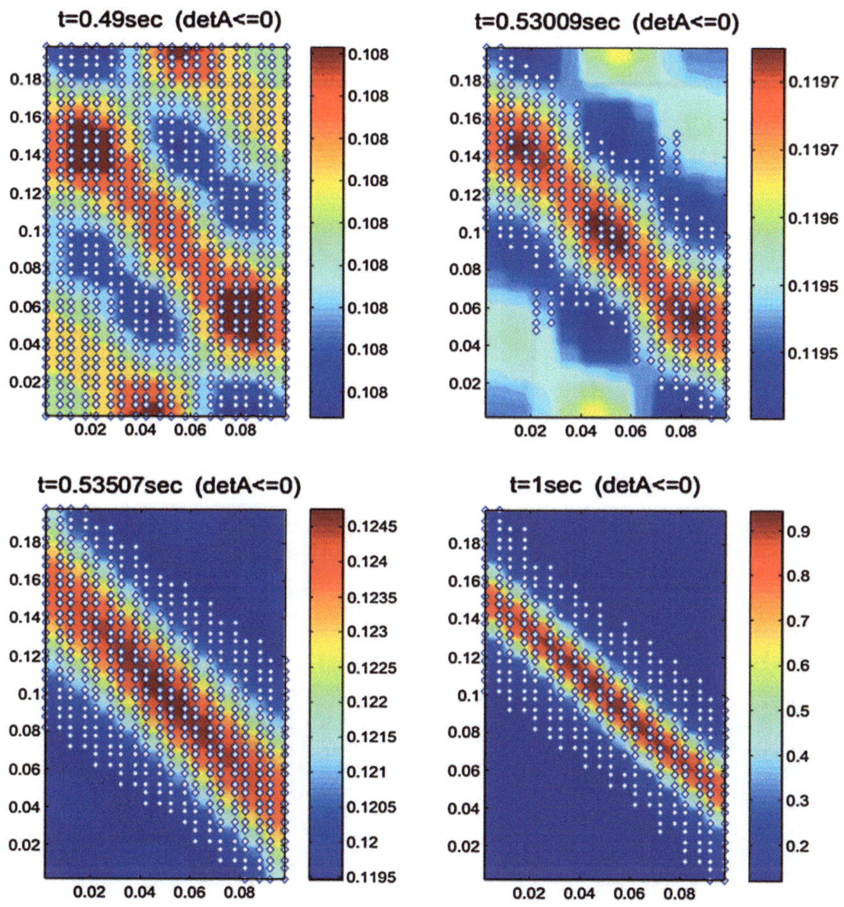

Fig. 5 Plastic strain field plot at various time steps and negative det(**A**) data points

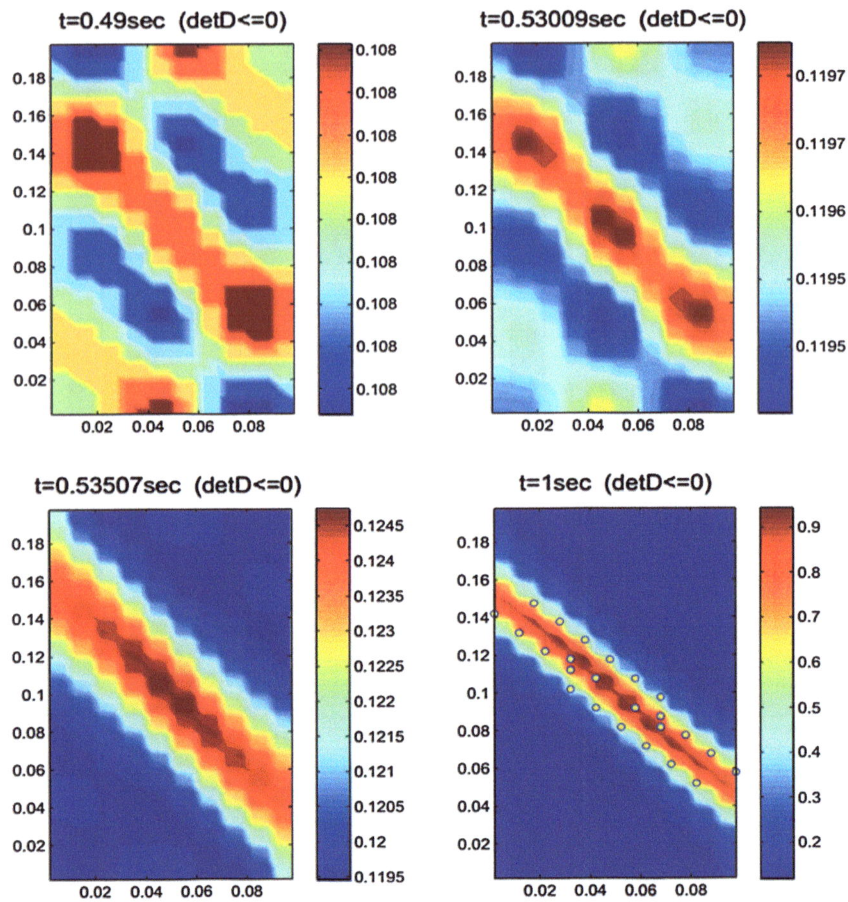

Fig. 6 Plastic strain field plot at various time steps and negative det(**D**) data points

4 Conclusions

A drained plane strain compression test was modelled as a boundary value problem together with the analysis of localization, second order work and plastic limit failure indicators. The localization indicator is first signaled at all Gauss points before peak deviatoric stress followed by the violation of the second order work throughout the entire specimen at peak conditions. These are only indicative of potential failure throughout the specimen. The loss of homogeneity in the strain field follows a clustering of activated failure indicator points that gradually concentrate within the shear band further along during loading. Ultimately, all failure indicators coincide with the plastic limit appearing the last. The finite element analysis shows that a continuum-level description of strain localization is possible by capturing the instability of an otherwise uniform state through a gradual degradation of kinematics with localization of density variations to eventually give rise to clear failure pattern.

Acknowledgments This work is jointly funded by the Computer Modelling Group (CMG) and the Natural Science and Engineering Council of Canada through a CRD grant.

References

Bigoni D, Hueckel T (1991) Uniqueness and localization—I. Associative and non-associative elastoplasticity. Int J Solids Struct 28(2):197–213

Darve F (1994) Stability and uniqueness in geomaterials constitutive modelling. In: Chambon et al (eds) Localisation and bifurcation theory for soils and rocks. Balkema, Rotterdam, pp 73–88

Desrues J (1990) Shear band initiation in granular materials: experimentation and theory. In: Darve F (ed) Geomaterials constitutive equations and modelling. Elsevier, Amsterdam, pp 283–310

Hill R (1958) A general theory of uniqueness and stability in elastic-plastic solids. J Mech Phys Solids 6(3):236–249

Nicot F, Darve F (2006) Micro-mechanical investigation of material instability in granular assemblies. Int J Solids Struct 43(11–12):3569–3595

Pinheiro M (2009) Constitutive and computation aspects of localized and diffuse instabilities in geomaterials. PhD thesis, Dept Civil Eng, University of Calgary, Canada

Rudnicki JW, Rice JR (1975) Condition for the localization of deformation in pressure sensitive dilatant material. J Mech Phys Solids 23(6):371–394

Wan RG, Guo PJ (1998) A simple constitutive model for granular soils: modified stress-dilatancy approach. Comput Geotech 22(2):109–133

Wan R, Pinheiro M, Daouadji A, Jrad M, Darve F (2013) Diffuse instabilities with transition to localization in loose granular materials. Int J Numer Anal Meth Geomech 37(10):1292–1311

Static Liquefaction of Granular Soils: The Role of Grain Shape and Size

J. Yang and L.M. Wei

Abstract This paper presents some interesting results from a specifically-designed experimental study that aimed to explore the role of particle shape and size in static liquefaction of sandy soils. The experimental program included a series of laboratory tests on a range of sand-fines mixtures in terms of grain shape and size. The stress-strain behavior of the mixed soils, including such characteristic states as instability state and critical state, is studied in detail. The effect of fines is shown to vary with the shape and size of both coarse and fine particles. The critical state friction angle of a sand-fines mixture is dominated by the roundness of sand grains, but the roundness of the fines also plays a role. Evidence is also given for the impact of grain shape and size on the onset of instability of granular soils.

1 Introduction

A large number of failures involving earth structures such as hydraulic-placed fills and slopes were caused by instability or, more specifically, static liquefaction of granular soils. This type of failure, characterized by a sudden loss in soil strength and a rapid development of large deformation under monotonic loading, usually occurs in saturated loose sandy soils that contain some amount of fines. Several laboratory studies have produced valuable data showing the effect of fines on the undrained behavior of sandy soils (e.g., Pitman et al. 1994; Lade and Yamamuro 1997; Thevanayagam et al. 2002; Chu and Leong 2002; Georgiannou 2006). Nevertheless, as noted by Yang and Wei (2012), the existing views on the effect of fines are diverse or even contradictory, due in part to the oversight of the role of grain characteristics. Based on a series of laboratory experiments on sand-fines mixtures, formed by adding glass beads and crushed silica at silt size into two

J. Yang (✉) · L.M. Wei
Department of Civil Engineering, The University of Hong Kong, Pokfulam, Hong Kong
e-mail: junyang@hku.hk

K.-T. Chau and J. Zhao (eds.), *Bifurcation and Degradation of Geomaterials in the New Millennium*, Springer Series in Geomechanics and Geoengineering,
DOI 10.1007/978-3-319-13506-9_29

clean quartz sands, Yang and Wei (2012) obtained several novel findings on the role of grain shape, and proposed the concept of combined roundness to characterize the coupled effects of grain roundness and fines content on the friction angle at critical state. This paper reports a further study built upon complementary experiments. Particular attention is paid to the onset of liquefaction or instability in the framework of critical state soil mechanics.

2 Test Program

The test program included three host sands and three non-plastic fines, giving nine types of sand-fines mixtures. The SEM images of the test materials are shown in Fig. 1. In addition to Toyoura sand (TS) and Fujian sand (FS) that were tested in the previous study, Leighton Buzzard sand (BS) was also used in the further study as the host sand. The grading curves of the three sands are shown in Fig. 2. Apart from mean grain size, the three host sands also differ from each other in grain roundness. BS and FS have more rounded grains with roundness $R = 0.602$ and $R = 0.548$ while TS has a more angular grain shape with $R = 0.486$. Among the three fines, glass beads are most rounded, with the roundness of approximately unity, and crushed silica fines are most angular, with $R = 0.377$. In the further study, crushed glass beads ($R = 0.431$), obtained by crushing coarse glass beads, were also used as fines to mix with host sands. More details on the quantification of grain shape were given in Yang and Wei (2012).

Crushed silica fines, glass beads and crushed glass beads were added into TS to form mixed soils of TSS, TG and TCG, respectively. Similarly, the three fines were added into FS to form FSS, FG and FCG, respectively. Crushed silica fines were

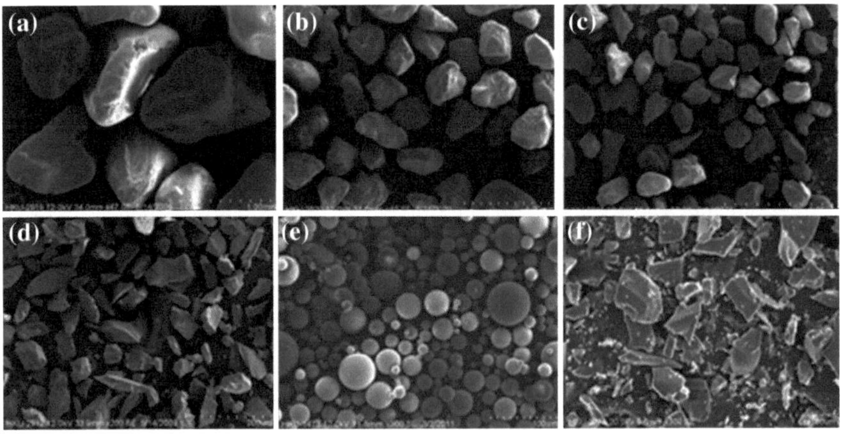

Fig. 1 Microscopic images of test materials. **a** Leighton Buzzard sand. **b** Fujian sand. **c** Toyoura sand. **d** Crushed silica. **e** Glass beads. **f** Crushed glass beads

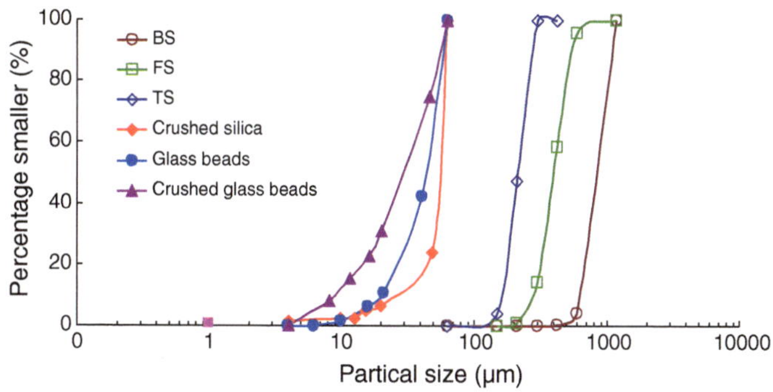

Fig. 2 Particle size distribution curves of test materials

also added into BS to form BSS, giving a binary mixture of larger size disparity in comparison with FSS and TSS. All mixed soil specimens were prepared by under-compaction moist-tamping method, brought to saturation, and then subjected to undrained, strain-controlled triaxial compression. The strain-controlled tests allowed reliable collection of data for the post-peak response, thus giving a well-defined stress path and stress-strain curve for a wide range of strains.

3 Results and Discussions

The test results for a pair of specimens, BS and BSS(5), are compared in q–p' plane and q–ε_a plane in Fig. 3. The number in the parentheses indicates the fines content being at 5 %. While both were sheared from similar initial states, the BS specimen reached an extremely high strength with no strain softening, but the BSS(5) specimen went through a sudden loss of strength at small strains and then reached a low residual strength. Compared with the pair of specimens TS and

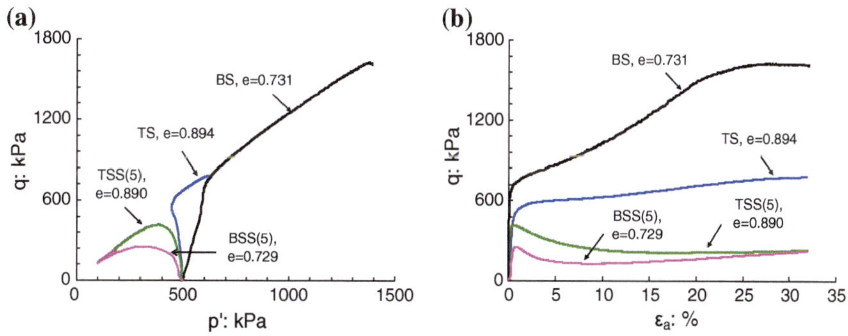

Fig. 3 Effect of fines on undrained shear: behavior (**a**) stress path, (**b**) stress-strain curve

TSS(5) in Fig. 3, a markedly large strength reduction was observed when crushed silica fines of the same quantity were added into BS, indicating a more detrimental impact on BS. This observation is thought to be due to the coarser size of BS in comparison with TS; that is, when the size difference between sand grains and fines increases, a larger fraction of fines may roll into the void space and make no contribution to resist external force. On the other hand, the more rounded grain shape of BS could also contribute to the strength reduction because the intergranular contacts formed between the more rounded BS and crushed silica fines tended to be less stable than that formed between the more angular TS and crushed silica grains, as discussed in detail in Yang and Wei (2012).

Figure 4 shows the undrained responses of specimens of FS and FS modified by 5 % of the three fines, all sheared from a similar void ratio. Evidently, the three fines all introduced a less stable soil structure to FS. In particular, the presence of glass beads led to an extremely unstable intergranular contacts in FG(5), resulting in saw-tooth fluctuations in the stress-strain curve. Interestingly, when rounded glass beads were crushed and added into FS, the fluctuations vanished [see the stress-strain curve of FCG(5) in Fig. 4]. This set of tests provides clear evidence showing the impact of grain shape: it is the rounded shape of glass beads that led to the unstable binary structure and the unusual fluctuations in the stress-strain curve.

In the previous study of Yang and Wei (2012), the effect of grain shape was found to couple with fines content on the friction angle at critical state. To account for this coupling, an index, termed combined roundness (R_{com}), was proposed as

$$R_{com} = R_{HS}(1 - F_c) + R_F F_c \tag{1}$$

where R_{HS} is the roundness of the host sand, F_c is the fines content and R_F is the roundness of fines. Based on test results for mixed soils containing crushed silica and glass beads, Yang and Wei (2012) proposed a linear relationship between the critical state friction angle and the combined roundness, as shown in Fig. 5. When the combined roundness was calculated for TCG(5) and FCG(5) and plotted against the critical state friction angle determined, it is encouraging that the new

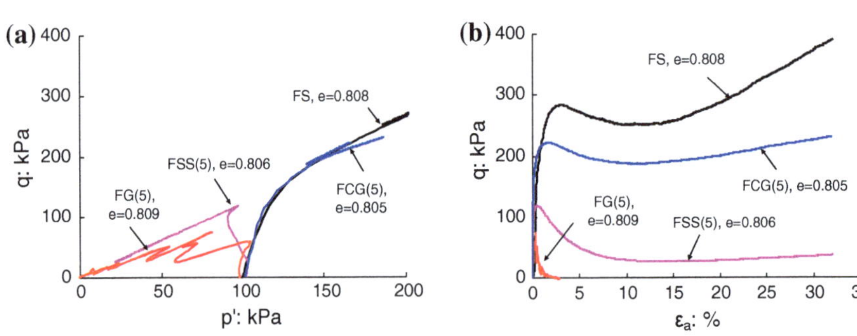

Fig. 4 Shear behavior of sand with fines of distinct shape: **a** stress path, **b** stress-strain curve

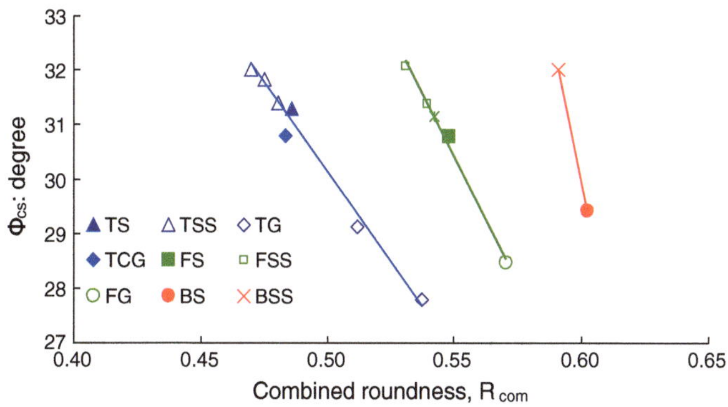

Fig. 5 Variation of critical state friction angle with combined roundness

data well fit the linear relations. Although data on BS mixed with fines is limited, it also shows a trend consistent with the other two types of mixtures.

Now, let's examine the undrained instability state (UIS), defined as the local peak deviatoric stress at which the strain-softening response occurs. The locus of UIS in the q–p' plane was also referred to as flow liquefaction line indicating the onset of liquefaction (Yang 2002). By examining several data sets for clean sands in the literature, Yang (2002) proposed an exponential relationship linking the stress ratio at the UIS (η_{UIS}) with the state parameter of Been and Jefferies (1985) that collectively accounts for the initial confining stress and void ratio:

$$\eta_{UIS} = \frac{M_{cs}}{B} \exp(A\psi) \tag{2}$$

where M_{cs} is the slope of the critical state line in q–p' plane; A and B are two calibrating parameters.

The state parameter was calculated for the specimens with strain-softening behaviour and the relationship between ($\eta_{\mathrm{UIS}}/M_{\mathrm{cs}}$) and Ψ was examined, as shown in Fig. 6. In Fig. 6a, the three host sands follow three distinct relationships whereas

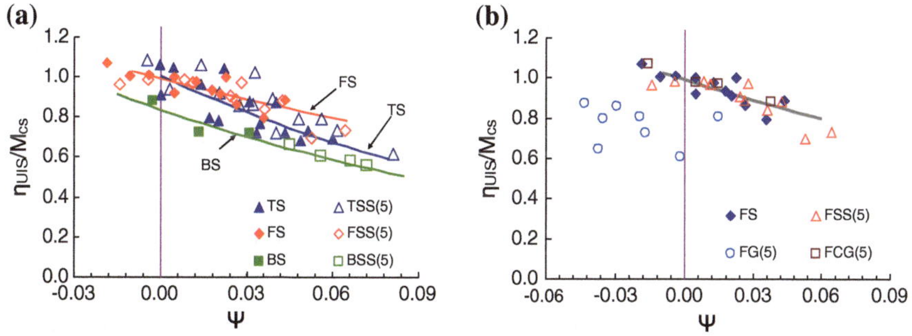

Fig. 6 Variation of stress ratio at undrained instability state with state parameter

the data for each of the sands with the addition of 5 % crushed silica fall in the vicinity of the relationship for the host sand, implying that the onset of liquefaction or instability of the mixed soil can be approximately predicted by the state parameter in the same way as that for the host sand. When fines of distinctively different shape are involved, however, this observation does not appear to hold true, as shown in Fig. 6b: while data for FSS(5) and FCG(5) generally follow the trend of FS, data points for FG(5) obviously deviate from the trend. The scatter may be attributed to the difficulty in determining the UIS in the fluctuated stress-strain curves which, as discussed before, resulted mainly from the extremely unstable intergranular contacts.

4 Conclusions

This paper presents several experimental findings on the role of grain shape and size in undrained shear response of sand-fines mixtures and, particularly, in strain softening behavior. The main points are summarized as follows:

- The presence of crushed silica fines brought about a more detrimental impact on Leighton Buzzard sand than on Toyoura sand and Fujian sand. This is attributed mainly to the coarser size and the more rounded shape of Leighton Buzzard sand.
- Adding rounded glass beads to Fujian sand led to a metastable soil structure exhibiting fluctuated stress-strain response, whereas adding crushed glass beads of more angular shape at the same quantity to the same sand gave a more stable soil structure with much higher shear strength.
- The combined roundness proves a useful index for characterizing the coupled effects of grain shape and fines content on the friction angle of mixed soils at the critical state.
- The stress ratio at undrained instability state tends to correlate with the state parameter in an exponential form for both clean sands and sand-fines mixtures tested, except for the mixture with a metastable structure FG(5). This suggests that the onset of instability of sandy soils may be predicted using the state parameter.

Acknowledgments The authors wish to acknowledge the financial support provided by the University of Hong Kong through the Seed Funding for Basic Research scheme and the Outstanding Young Researcher Award scheme.

References

Been K, Jefferies MG (1985) A state parameter for sand. Géotechnique 35(2):99–122
Chu J, Leong WK (2002) Effect of fines on instability behaviour of loose sand. Géotechnique 52(10):751–755
Georgiannou VN (2006) The undrained response of sands with additions of particles of various shapes and sizes. Géotechnique 56(9):639–649

Lade PV, Yamamuro JA (1997) Effect of non-plastic fines on static liquefaction of sands. Can Geotech J 34(6):918–928

Pitman TD, Robertson PK, Sego DC (1994) Influence of fines on the collapse of loose sands. Can Geotech J 31(5):728–739

Thevanayagam S, Shenthan T, Mohan S, Liang J (2002) Undrained Fragility of clean sands, silty sands, and sandy silts. J Geotech Geoenvironmental Eng ASCE 128(10):849–859

Yang J (2002) Non-uniqueness of flow liquefaction line for loose sand. Géotechnique 52(10):757–760

Yang J, Wei LM (2012) Collapse of loose sand with the addition of fines: the role of particle shape. Géotechnique 62(12):1111–1125

Static Liquefaction and Instability in Granular Media Subjected to Monotonic Loading—A Micromechanical Investigation

J.C. Lopera Perez, C.Y. Kwok, C. O'Sullivan, K.J. Hanley and X. Huang

Abstract Static liquefaction has caused a number of failures involving dam tailings, hydraulically placed fills and slopes. In order to understand the failure mechanisms that induce static liquefaction, the discrete element method (DEM) was used to study the behavior of a representative model of a granular sample at the micro level. Samples with different void ratios at same confining pressures were sheared under constant-volume conditions and the changes in the macro and micromechanical responses of the granular media were quantified. Characteristic states such as the instability state, quasi-steady state, phase transformation and critical state were identified in the simulations. The transitions between different characteristic states were related to micromechanical characteristics such as coordination number. Finally, the orientation of the weak contacts was seen to be dependent on the characteristic state, while the orientation of the strong contacts coincided with the major principal stress direction.

1 Introduction

Static liquefaction has caused a number of tragic events in recent history involving the failure of slopes, tailings dams, and hydraulic placed fills, etc. Static liquefaction is a characteristic behavior of granular materials, and the problems associated with static liquefaction have motivated a number of prior studies (Ishihara 1996; Murthy et al. 2007). Most of this prior research involved laboratory testing where

J.C. Lopera Perez · C.Y. Kwok (✉) · X. Huang
Department of Civil Engineering, The University of Hong Kong,
Haking Wong Building, Pokfulam Road, Pokfulam, Hong Kong
e-mail: fkwok8@hku.hk

C. O'Sullivan · K.J. Hanley · X. Huang
Department of Civil and Environmental Engineering, Imperial College London,
Skempton Building, London SW7 2AZ, UK

K.-T. Chau and J. Zhao (eds.), *Bifurcation and Degradation of Geomaterials in the New Millennium*, Springer Series in Geomechanics and Geoengineering,
DOI 10.1007/978-3-319-13506-9_30

it is possible only to observe the overall material behavior, with the micromechanical response remaining unclear and so only hypotheses can be made regarding the micromechanical evolution which lack of solid physical bases.

In this work, the discrete element method (DEM) is employed to investigate both the macro and micro responses of the granular media. This can give insights of the evolution of the micromechanical variables that influence the general macro-mechanical response of the sample.

Depending on the initial packing density, specimen granular material may experience any of the following characteristic states: instability state, quasi-steady state, phase transformation and critical state as explained by Murthy et al. (2007). In the present work, samples are sheared under constant volume conditions with a range of initial densities, in order to capture all possible characteristic states. Both the macro- and micromechanical responses are studied, including contact force magnitudes and orientations.

2 Numerical Simulations

The numerical method DEM introduced by Cundall and Strack (1979) was employed for the simulations shown in this paper. Simulations were carried out using a modified version of the open-source code LAMMPS (Plimpton 1995). The particle size distribution (PSD) is shown in Fig. 1, and it is a realistic PSD that largely follows a Toyoura sand PSD (Huang et al. 2014). The input parameters applied in all simulations are summarized in Table 1. A simplified Hertz-Mindlin contact model was used with a shear modulus (G) of 29 GPa and a Poisson's ratio (ν) of 0.12. Periodic boundary conditions with 20,164 particles were used for all the simulations. The samples initial density was controlled by the application of different friction coefficient during isotropic compression. Samples were sheared monotonically under constant volume conditions applying a friction coefficient (μ) of 0.25. This low value of μ is chosen as if values of $\mu \geq 0.5$, the critical state void ratio initially would increase with increasing mean effective stress (p') which is

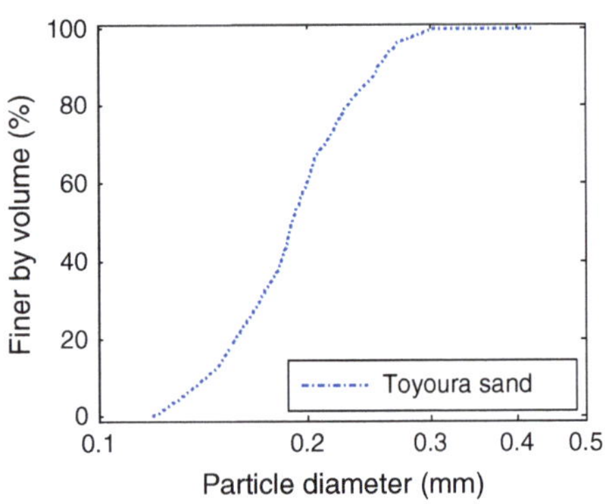

Fig. 1 PSD used in the simulations

Table 1 Input parameters used during the shearing stage

Particle density	2,650 kg/m^3
Contact model	Hertz-Mindlin
Particle shear modulus G	29 GPa
Particle Poisson's ratio ν	0.12
Coefficient of friction	0.25
Local damping ratio	0.10
Strain rate	1 s^{-1}
Number of particles	20,164

unreal under the critical state soil mechanics framework (Huang et al. 2013). A strain rate of 1/s was set during shearing stage, which can be considered still in the quasi-steady regime, i.e., the initial inertial number was below 2×10^{-3}.

3 Results and Discussion

3.1 Macro-mechanical Response

Figure 2a shows the variation of deviator stress (q) with p′, while Fig. 2b shows the evolution of q with axial strain (ϵ_a).The sample denoted as CVT1 is the densest tested, with samples CVT2 to CVT5 becoming progressively looser. The dependency of the macro response on the initial density is clear from Fig. 2a, b; loose samples showed an unstable behavior while dense samples showed a dilative stable response.

The densest sample CVT1 remained dilative throughout the entire shearing process. Sample CVT2 initially contracted, with a corresponding reduction in mean effective stress until about 2 % of axial strain; subsequent to this phase transformation, the sample dilated until the critical state. Samples CVT3 and CVT4 had initial peaks in their deviator stresses which characterize the instability state.

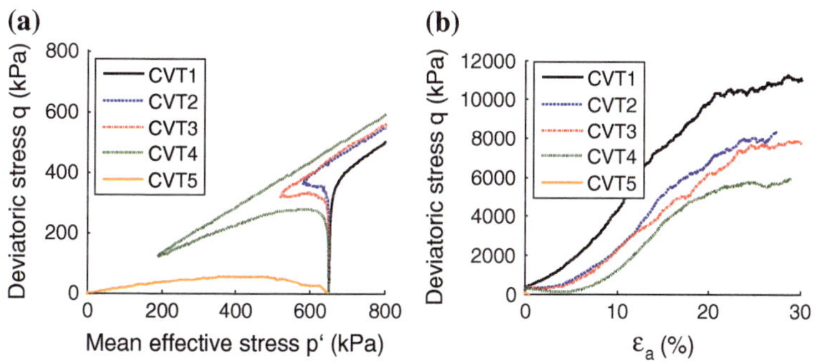

Fig. 2 **a** Variation of deviatoric stress against mean effective stress. **b** Evolution of deviatoric stress with axial strain

CVT3 and CVT4 achieved a local minimum in q before 5 %, which is associated with the quasi-steady state. The loosest sample (CVT5) reached a maximum deviator stress and then contracted to zero deviator and mean effective stresses, i.e., attaining a state that could be described as total liquefaction.

3.2 Micromechanical Response

The coordination number (*Z*) reflects the average contact intensity in the system and the evolution of coordination number with axial strain is presented in Fig. 3. The initial coordination numbers for all samples were between 4.8 and 5. In all cases the coordination number attained a local minimum during the first 5 % of axial strain; subsequently the coordination number increased so that the value at the critical state was similar to the initial value. As the void ratio increased, the local minimum value attained decreased. This appearance of a local minimum coordination number does not imply a point of instability as can be observed from sample CVT1 which is dilative throughout shearing and also presents a minimum in coordination number. Sample CVT5, which liquefied, exhibited an abrupt reduction in the coordination number to be less than one. This kind of behavior is a good indicator that this sample did indeed present total liquefaction.

Figure 4 shows rose diagrams of the contact directions at each characteristic state for the sample CVT4, which experienced all characteristic states. Each bin in the rose diagram is colored according to the total contact force magnitude lying within that bin. The contact forces and directions are divided into strong and weak contacts. Strong contacts are those which transmit forces exceeding the mean contact force, while weak contacts transmit forces less than the mean. At the initial state, both strong and weak contact directions and forces are isotropic as is expected from the sample preparation approach and the isotropic stress state. Once shearing commences, the strong contacts align parallel to the shearing direction and this trend remains unchanged throughout all states. However the weak contacts remain in an isotropic state in both direction and forces particularly at

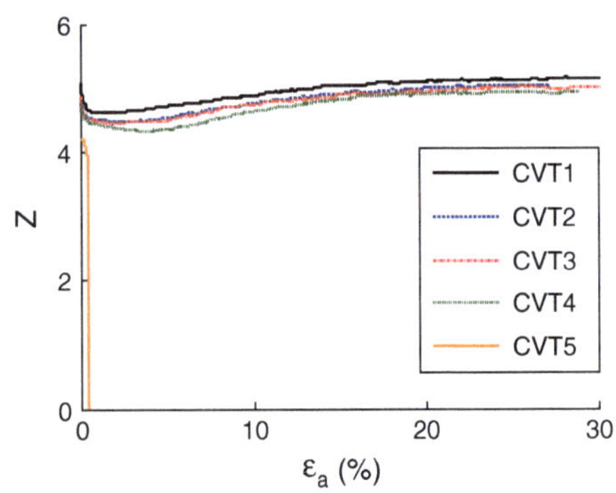

Fig. 3 Variation of coordination number with axial strain

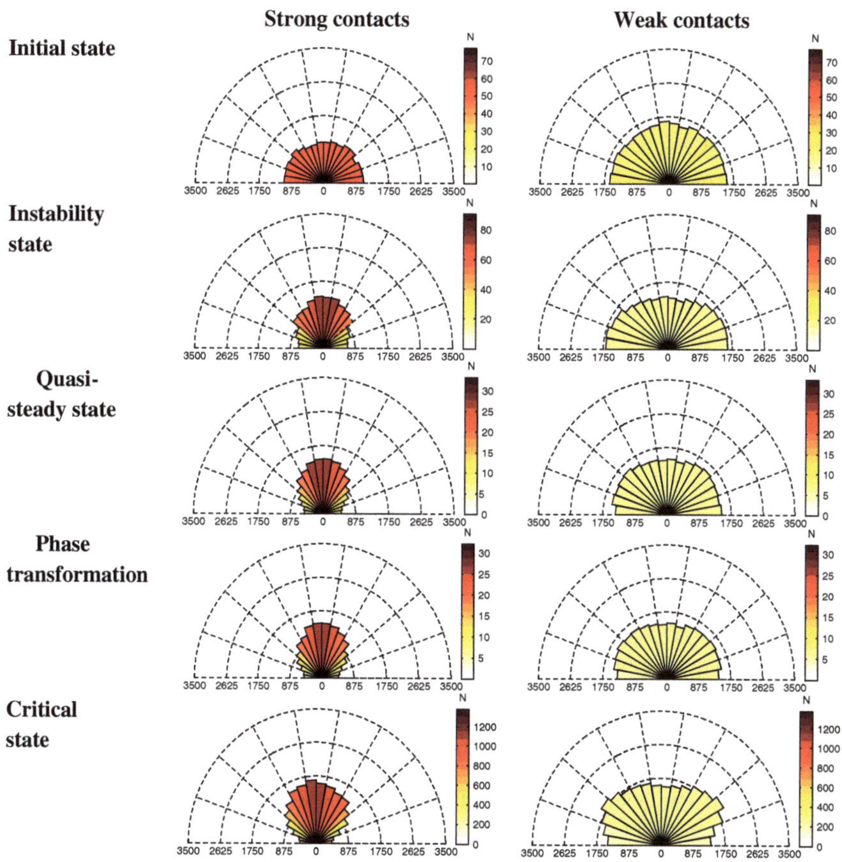

Fig. 4 Contact forces and directions for test CVT4 for which the *color bars* show the magnitudes of the contact forces in N

the quasi-steady state and phase transformation. For the instability state and critical state, it is possible to observe a small bias in the weak contact directions and forces in the radial direction. This particular tendency can also be observed in the simulation of tests under drained conditions (Huang et al. 2012).

4 Conclusions

Constant volume simulations were carried out considering numerical samples with a realistic particle size distribution. All characteristic states for granular materials were observed in the simulations, where the initial density is the main influence for both the macro and particle-scale response of each sample.

The magnitude of the peak deviator stress in the instability state depends on the initial density: denser samples present higher initial peaks. Very dense samples show only a dilative tendency during all shearing stages. The macro responses on the deviator stress-axial strain and deviator stress-mean effective stress planes are in good correspondence with laboratory tests.

Considering the particle scale response, a minimum value of the coordination number is not related to an unstable state (instability state, quasi-steady state); however, the magnitude of the minimum coordination number is related to the initial density.

Irrespective of which characteristic state the sample is facing during the shearing stage, strong contacts are always aligned parallel to the shearing direction. Weak contacts present an interesting tendency between the onset of instability to stable states where the directions and forces remain isotropic.

Acknowledgments Mr Lopera would like to thank the IT Department at The University of Hong Kong for generously providing access to the high performance computer HPCPOWER2 in order to conduct the simulations presented in this work.

References

Cundall PA, Strack ODL (1979) A discrete numerical model for granular assemblies. Geotechnique 29(1): 47–65

Huang X, Hanley KJ, O'Sullivan C, Kwok CY (2014) Effect of sample size on the response of DEM samples with a realistic grading. Particuology 15: 107–115

Huang X, Hanley KJ, O'Sullivan C, Kwok CY, Tham LG (2013) Effects of inter-particle friction on the critical state behavior of granular materials: a numerical study. Particle-Based Methods III Fundam Appl pp 165–175

Huang X, Kwok CY, O'Sullivan C, Tham LG (2012) A micro mechanical study on the critical behavior of a granular assemblage with a realistic grading. International conference on geomechanics and engineering, Seoul, Korea

Ishihara K (1996) Soil behavior in earthquake geotechnics. Oxford University Press, New York

Murthy TG, Loukidis JA, Carraro JAH, Prezzi M, Salgado R (2007) Undrained monotonic response of clean and silty sands. Geotechnique 57(3):273–288

Plimpton S (1995) Fast parallel algorithms for short-range molecular dynamics. J Comp Phys 117:1–19

Capillary Effect on Critical Length of Force Chains in Partially Saturated Granular Materials Within the Pendular Regime

Peijun Guo

Abstract This paper investigates the influence of capillary interaction on critical length of grain columns in partially saturated granular materials at pendular state. When relative rotation between a pair of elliptical particles (2D disks) takes place, the liquid meniscus between the particles is distorted and generates unbalanced capillary forces around the contact, which further induces capillary torque. Both the capillary force and capillary torque have some influence on the characteristic length of grain column or the force chain. The findings imply that the capillary interaction may have some influence on shear band thickness. For engineering soils, however, such effect may be undermined owing to the relative low level of capillary stress in partially saturated granular soils within the pendular regime.

1 Introduction

The importance of capillary effect has been recognized in wet discrete granular systems from granular soil to nanoparticles. In a moist granular assembly, the direction of the capillary bridge force (i.e., the capillary attraction) between two nearby solid surfaces is oriented normal to the plane of the three-phase (solid, liquid and air) contact line. The capillary attraction, which generally consists of contributions from the capillary pressure and surface tension, depends on the surface-to-surface separation and the characteristic curvature of the solid at the contact point. In most cases, the toroid (circle) approximation is applied to quantify the shape of capillary bridges and the capillary-bridge force (Scholtès et al. 2009; Richefeu et al. 2008; Soulié et al. 2006). However, models for the capillary attraction forces between two elliptical particles or uneven-sized spheres have also been proposed (Soulié et al. 2006; Lechman and Lu 2008; Richefeu et al. 2008).

P. Guo (✉)
Department of Civil Engineering, McMaster University, Hamilton, ON L8S 4L7, Canada
e-mail: guop@mcmaster.ca

K.-T. Chau and J. Zhao (eds.), *Bifurcation and Degradation of Geomaterials in the New Millennium*, Springer Series in Geomechanics and Geoengineering,
DOI 10.1007/978-3-319-13506-9_31

Regarding the behaviour of partially saturated granular materials within the pendular regime, the capillary effect can be simulated by applying discrete capillary forces F^{bridge} on particles between which liquid bridges form. The macroscopic stress (or the capillary stress σ^{cap}) associated with F^{bridge} is obtained via appropriate micromechanics analyses. A second-rank capillary stress tensor can be defined according to the distribution of capillary forces at particle contacts and a fabric tensor that characterizes the distribution of contact normals or branch vectors describing the connectivity between particles affected by liquid bridges (see, e.g., Richefeu et al. 2008; Scholtès et al. 2009; Guo 2012). All these studies have shown that the shear strength, cohesion and tensile strength of wet sand can be related to the capillary stress.

Recent progresses in micro-assembly technology and colloidal materials reveal that for non-spherical particles, the fluid meniscus at contacts is generally asymmetric about the contact normal, which results in asymmetric distribution of capillary forces (Lehle et al. 2008; Chau et al. 2010; He et al. 2013). As a result, a capillary torque develops around the contact point. Such a torque may cause rotation of particles floating on fluid surface.

The objective of this paper is to explore the existence and effect of capillary torque in partially saturated granular materials of non-spherical particles within the pendular regime. By investigating 2D capillary interaction between mono-sized elliptical disks, the capillary force and torque are found to vary with particle geometry, the volume of water menisci, and the relative rotation between particles. The effect of capillary torque is equivalent to an increase in the inter-particle rolling resistance. The influence of capillary effect on the characteristic length of particle chains within the pendular regime is examined.

2 Capillary Interaction Between Two Equal-Sized Ellipses

Without loss of generality, the discussion is limited to parallel elliptical cylinders (or equivalently 2D elliptical disks). As shown in Fig. 1a, two equal sized-ellipses, initially with parallel major axes, are in contact at vertex P. The major and minor radii of the ellipses are a and b respectively. The fluid meniscus is bounded by the two ellipses and the air-water interface with radius r_1. A contact angle α_w (the wetting angle) can be specified where the air-liquid interface is in contact with the ellipses. When the fluid is water, one has $\alpha_w = 0$. At the initial state, the capillary force per unit length of the cylinder on each side of the contact point P is determined as $F_A^{bridge} = F_B^{bridge} = \gamma_s(1 + h/r_1)$, in which h is the thickness of the liquid bridge at its neck, $s = u_a - u_w = \gamma_s/r_1$, and γ_s is the surface tension of water. Owing to the symmetry of the water meniscus, the capillary interaction does not induce any torque about P.

When the two ellipses roll against each other by angle θ without sliding (as shown in Fig. 1b), the symmetry of wet peripheries and the shape of water menisci are lost, which causes different capillary attraction forces F_A^b and F_B^b on both sides of the present contact point P', with the total capillary force being

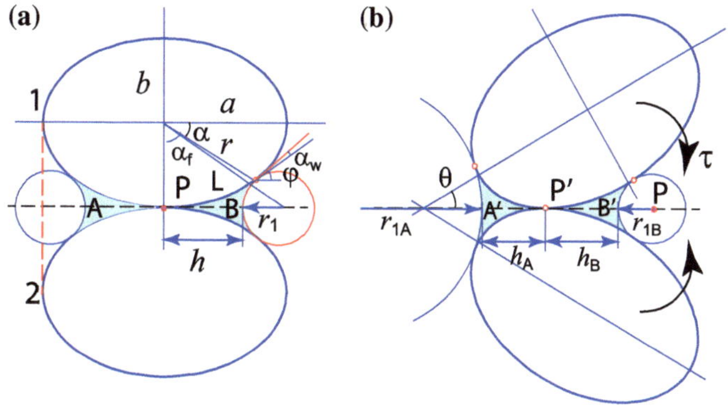

Fig. 1 Water meniscus between two ellipses in contact: **a** initial position with parallel major principal axes, **b** rotated ellipses

$$F^{bridge} = F_A^b + F_B^b = \gamma_s(2 + h_A/r_{1A} + h_B/r_{1B}) \quad (1)$$

The asymmetric capillary attraction forces further induce unbalanced net torque on the ellipses about the contact point P' as

$$\tau = \gamma_s\left[h_B - h_A + \left(h_B^2/r_{1B} - h_A^2/r_{1A}\right)/2\right] \quad (2)$$

To determine the net torque τ at a rotation angle θ, it is assumed that the volume of meniscus water on each side of the contact point P′ does not change when the ellipses are rolling. This assumption is reasonable for elliptical cylinders in contact, but it needs to be modified for two ellipsoids in contact.

3 Results and Discussions

In this section, the capillary attraction between two ellipses in contact are presented as a normalized capillary stress $\bar{\sigma}_{cap} = \sigma_{cap}/(\gamma_s \cdot 1)$ with σ_{cap} being the capillary stress defined as $\sigma_{cap} = F^{bridge}/(2a)$ in which a is the major principal radius and γ_s the surface tension of water. The capillary torque can be normalized either by the surface tension as $\tau/[(\gamma_s \cdot 1)a]$ or by the total capillary force as $\tau/(aF^{bridge})$.

Figure 2 presents the forces and torques induced by capillary interaction between two ellipses with aspect ratio $a/b = 1.5$ when they rotate against each other (as illustrated in Fig. 1b) at different degrees of saturation S_r. The torque is defined as positive when it resists rotation of particles. When there is no loss of meniscus fluid during rotation, an increase in the rotation angle θ reduces the total capillary force F^{bridge} monotonically, with $F^{bridge}_{\max}$ and $F^{bridge}_{\min}$ at $\theta = 0°$

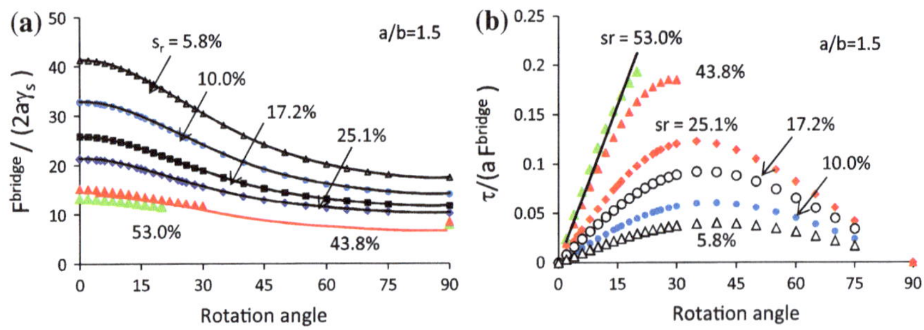

Fig. 2 Variation of **a** $F^{bridge}/(2a\gamma_s)$, **b** $\tau/(aF^{bridge})$ with rotation angle θ

(parallel major axes) and 90° (parallel minor axes) respectively. The torque induced by asymmetric capillary forces is positive at all rotation angles. Different from F^{bridge}, the capillary torque initially increases with θ at small rotation angles, and its maximum $\tau_{\max}$ appears at approximately $\theta = 28$–$30°$. Further increase of θ causes reduction of τ that becomes zero at $\theta = 90°$ when the symmetry of meniscus water is resumed about the contact point. The particle shape (herein the aspect ratio a/b for elliptical cylinders) also has a certain influence on F^{bridge} and the capillary torque.

For sufficiently small rotation angle ($\theta < \pi/9$), the dependency of normalized capillary torque $\tau/(aF^{bridge})$ on rotation angle can be approximately expressed as

$$\tau/(aF^{bridge}) = A(a/b - 1)S_r\theta \tag{3}$$

in which A is found to be $A = 0.0237$.

4 Effect of Capillary Force and Torque on Particle Chains

Let us consider an initially straight grain column with the presence of water menisci around particle contacts, as illustrated in Fig. 3. In the deflected particle column, the capillary force between elliptical particles is attractive and varies with the separation between particle surfaces as well as their local geometries. The capillary torque owing to the asymmetry of the capillary traction with respect to the point of contact increases the rolling resistance of particles against each other, which in turn increases the stability of the grain column against buckling.

The effect of capillary bridge on buckling of particle columns is analyzed following Guo (2012). The rolling resistance torque M_r at a contact point is related to the relative rotation $\Delta\theta_r$ between two particles in contact via $M_r = -k_r\Delta\theta_r$ with $M_r \le \mu_r F_n R_r$ in which $k_r = 2J_n R_r F_n$ is the rolling stiffness, R_r the rolling radius at contact, μ_r the dimensionless rolling resistance coefficient or the coefficient of rolling friction (CRF), J_n a dimensionless coefficient that varies theoretically between 0.25 and 0.5. Following Guo (2012), the moment equilibrium of the deflected grain column segment is

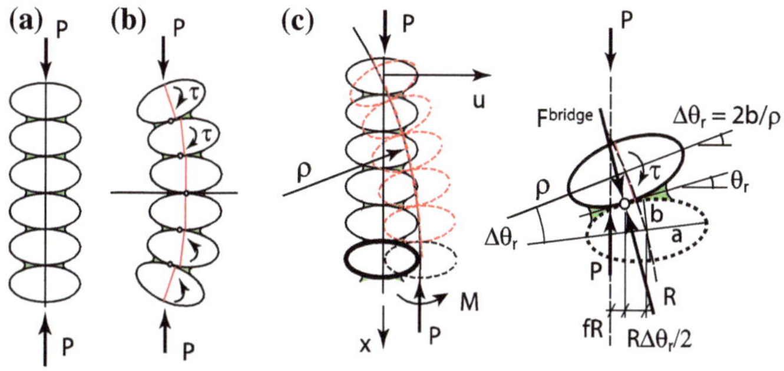

Fig. 3 **a** A straight, **b** a deflected grain column with water menisci, and **c** interaction between two particles in a deflected grain column

$$M'' + (Pu')' = 0 \tag{4}$$

in which the internal moment resisting bending consists of the torque M_{cap} from capillary interaction (including both the capillary force and the capillary torque) and M_r from the rolling resistance such that $M = M_{cap} + M_r$ with

$$M_{cap} = a\left(k_c\gamma_s b - 2F^{bridge}a\right)u''; \quad M_r = -k_r 2bu'' = -4a^2 J_n P u'' \tag{5}$$

As a result, when $P =$ constant, Eq. (4) becomes

$$\left[Pa^2\left(1 + 4J_n - 2F^{bridge}/P\right) + ak_c\gamma_s b\right]u^{(4)} + Pu'' = 0 \tag{6}$$

For a grain column between two rigid parallel plates, the particles at both ends of the column can rotate on the plates, corresponding to the boundary conditions $\Delta\theta(x = 0) = \Delta\theta(x = l) = bu''$ and $u(x = 0) = a^2u''$. By enforcing these constraints, the characteristic equation of the differential equation in Eq. (6) is given as

$$\omega l \sin \omega l + 2\cos \omega l - 2 = \omega^2 l^2 (2a^2/l^2 + b/l)(1 - \cos \omega l) \tag{7}$$

The above equation has two solutions:

$$\begin{aligned} &\text{Solution 1: } \omega l = 2\pi \\ &\text{Solution 2: } \tan(\omega l/2) - \omega l/2 + \omega^2 l^2/2\left(2a^2/l^2 + b/l\right)\tan(\omega l/2) = 0 \end{aligned} \tag{8}$$

Solution 2 in Eq. (8) can be obtained numerically and expressed generally as $\omega l = f(l/b, a/b, \xi)$ or $l/b = g(a/b, \xi)$. The critical length of the grain column expressed as the number of particles is finally determined as

$$N_{cr} = 1 + l/(2b) = 1 + 0.5g(a/b, \xi) \tag{9}$$

in which $\xi = [1 + 4J_n - (2 - AS_r + AS_r b/a)F^{bridge}/P]^{1/2}$. Figure 4 shows the variation of N_{cr} with the aspect ratio a/b and ξ. The numerical solutions of Eq. (9) can be satisfactorily reproduced by a simple expression $N_{cr} \simeq K\xi(a/b)$ with $K = 4.468$.

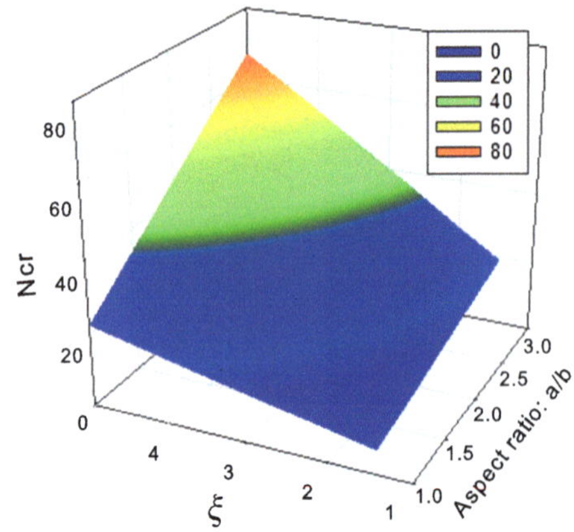

Fig. 4 Characteristic length of grain columns at different aspect ratios and ξ

Regarding the critical length of grain columns, based on the above analysis and the results shown in Fig. 4, the following conclusions can be withdrawn:

- Given the theoretical range of $J_n = 0.25$–0.5, for particles of $a/b = 1.5$–2, the effect of capillary torque is equivalent to as high as 30 % increase of the rolling resistance when σ^{cap} is in the same order of the net stress σ^{net};
- Since $(2 - AS_r + AS_r b/a)$ is generally positive, the capillary force tends to reduce the characteristic length of elliptical particle column;
- With a decrease of the $\sigma^{cap}/\sigma^{net}$ ratio, the overall capillary effect on the length of particle columns decreases and may even be neglected.

Since the critical length of the grain column can be considered as the thickness of shear band (Guo 2012), the above analysis implies that the shear band thickness in partially saturated sand may be smaller than that in dry sand. This conclusion is generally in agreement with the findings from DEM simulation by Schwarze et al. (2013), who observed that the shear band in wet sand gets a little narrower for small liquid content, then become wider for very high liquid content. Experimental observations on shear band in partially saturated sand can be found in Higo et al. (2011). It should be noted, however, the characteristic length of grain columns in a real granular material specimen may be different from that in Eq. (9) owing to differences of particle sizes and shapes.

5 Conclusions

The capillary interaction between mono-sized elliptical cylinders (or disks) within the pendular regime is investigated. By taking into account the relative rotation between a pair of elliptical particles (2D disks) and the resulting distortion of liquid menisci, the capillary torque induced by unbalanced capillary attractions around the contact is determined. The maximum capillary torque develops when the rotation angle reaches a critical value. Further relative rotation between the particles results in graduate decrease of the torque that eventually diminishes as the symmetry of capillary meniscus around the contact point is recovered. The net capillary attraction between particles, however, monotonically decreases as particles rotate. The capillary torque increases the rolling resistance between particles, particularly at low levels of net stress. For small rotation between particles, an approximate linear relation between the capillary torque and particle rotation angle is demonstrated. While increased rolling resistance from the capillary torque may increase the shear strength owing to capillary stresses, the characteristic length of grain columns of mono-sized elliptical particles does not change significantly, particularly when the capillary stresses are much lower than net stresses. The shear band thickness in partially saturated sand may be smaller than that in dry sand. Continued investigation on how the "brittle" or "softening" responses of capillary torque towards relative particle rotations may affect the mechanical behaviour of partially saturated granular materials within the pendular regime is still required.

Acknowledgments Funding provided by the Natural Sciences and Engineering Research Council of Canada is gratefully acknowledged.

References

Chau A, Régnier S, Delchambre A, Lambert P (2010) Theoretical and experimental study of the influence of AFM tip geometry and orientation on capillary force. J Adhes Sci Technol 24:2499–2510

Guo P (2012) Critical length of force chains and shear band thickness in dense granular materials. Acta Geotech 7(1):41–55

He A, Nguyen K, Mandre S (2013) Capillary interactions between nearby interfacial objects. Lett J Exploring Front Phys 102(38001):1–6

Higo Y, Oka F, Kimoto S, Sanagawa T, Matsushima Y (2011) Study of strain localization and microstructure changes in partially saturated sand during triaxial tests using microfocus X-ray CT. Soils Found 51(1):95–111

Lechman JB, Lu N (2008) Capillary force and water retention between two uneven-sized particles. J Eng Mech 135(5):374–384 (for two spheres of different sizes)

Lehle H, Noruzifar E, Oettel M (2008) Ellipsoidal particles at fluid interfaces. Eur Phys J E 26:151–160

Richefeu V, El Youssoufi MS, Peyroux R, Radjaï F (2008) A model of capillary cohesion for numerical simulations of 3D polydisperse granular media. Int J Numer Anal Meth Geomech 32(11):1365–1383

Scholtès L, Chareyre B, Nicot F, Darve F (2009) Micromechanics of granular materials with capillary effects. Int J Eng Sci 47(2009):64–75

Schwarze R, Gladkyy A, Uhlig F, Luding S (2013) Rheology of weakly wetted granular materials: a comparison of experimental and numerical data. Granular Matter 15(4):455–465

Soulié F, Cherblanc F, El Youssoufi MS, Saix C (2006) Influence of liquid bridges on the mechanical behaviour of polydisperse granular materials. Int J Numer Anal Meth Geomech 30:213–228

Study on the Initiation of Strain Localization in Soils by 3D Non-coaxial Plasticity

Xilin Lu, Jianju Qian and Maosong Huang

Abstract The strain localization in soil under plane strain and true triaxial conditions were predicted by bifurcation analysis on the non-coaxial plasticity. By using an elliptical shape function to modify the Mohr-Coulomb failure criterion, a 3D non-coaxial non-associated elasto-plasticity model was proposed by adding a non-coaxial plasticity flow rule. Comparing to the experiments, bifurcation analysis on the non-coaxial model gives out better prediction than the coaxial models. Strain localization occurs at the hardening regime of the soil under plane strain condition, the influence of confining stress on the strain localization could be described by non-coaxial model rightly. The prediction of true triaxial tests showed the non-coaxial model reflects the influence of the intermediate stress ratio on strain localization well.

1 Introduction

Strain localization (or shear band) is a well-known phenomenon in soils and rocks, this phenomenon has been widely studied experimentally, theoretically and numerically (Gutierrez et al. 1991; Huang et al. 2010; Miura et al. 1986). The main point on this topic is the prediction of its initiation. After the bifurcation theory has been proposed by Rudnicki and Rice (1975) and Rice (1976), it was widely used in the prediction of the strain localization.

X. Lu (✉) · J. Qian · M. Huang
Key Laboratory of Geotechnical and Underground Engineering of Ministry of Education, Tongji University, 200092 Shanghai, China
e-mail: xilinlu@tongji.edu.cn

M. Huang
e-mail: mshuang@tongji.edu.cn

X. Lu · J. Qian · M. Huang
Department of Geotechnical Engineering, Tongji University, 200092 Shanghai, China

K.-T. Chau and J. Zhao (eds.), *Bifurcation and Degradation of Geomaterials in the New Millennium*, Springer Series in Geomechanics and Geoengineering,
DOI 10.1007/978-3-319-13506-9_32

The results of bifurcation analysis strongly rely on the constitutive model used. For the coaxial model often give out unreasonable prediction, non-coaxial model, which is originated from the vortex theory of Rudnicki and Rice (1975), was used by Papamichos and Vardoulakis (1995) to improve the prediction. Qian et al. (2008) extended the model into 3D stress space, and reanalyzed the plane strain tests as a 3D problem. Huang et al. (2010) using the 3D non-coaxial plasticity to study true triaxial tests and obtain good predictions. Lu et al. (2011) studied the onset of strain localization on cross-anisotropic sand under true triaxial conditions. Recently, the non-coaxial plasticity model has been further formulated by Yang et al. (2011) and Cai et al. (2013), and these models are ready to be used in the bifurcation analysis.

This paper studied the initiation of strain localization by a simple 3D non-coaxial Mohr-Coulomb model. The influence of the non-coaxial hardening modulus on the strain at bifurcation point under plane strain drained condition were investigated. Then the method was used to study the onset of strain localization in plane strain undrained test and also in true triaxial test.

2 The 3D Non-coaxial Elasto-Plasticity

In order to consider the influence of the intermediate stress on soil strength, the Mohr-Coulomb criterion is modified by adding an elliptical shape function

$$\begin{aligned} F = {} & q - Mp \\ & \times \frac{2(1-\beta^2)\cos(\frac{\pi}{6}+\theta_\sigma) + (2\beta-1)\sqrt{4(1-\beta^2)\cos^2(\frac{\pi}{6}+\theta_\sigma)+\beta(5\beta-4)}}{4(1-\beta^2)\cos^2(\frac{\pi}{6}+\theta_\sigma)+(2\beta-1)^2} = 0 \end{aligned} \tag{1}$$

where $p = I_1/3 = \sigma_{kk}/3, q = \sqrt{3J_2} = \sqrt{3s_{ij}s_{ij}/2}, \theta_\sigma = \sin^{-1}(3\sqrt{3}J_3/2J_2^3/3$; I_1 is the first stress invariant; J_2, J_3 are the second and third invariant of the deviatoric stress; s_{ij} is the deviator stress tensor, $s_{ij} = \sigma_{ij} - \delta_{ij}p$, and δ_{ij} is Kronecker's delta; $J_3 = s_{ij}s_{jk}s_{kl}/3$; $\beta = sin\varphi_E(3-\sin\varphi_C)/[\sin\varphi_C(3+\sin\varphi_E)]$.

According to the non-coaxial flow theory of plasticity, the strain rate is

$$\dot{\boldsymbol{\varepsilon}} = \dot{\boldsymbol{\varepsilon}}^e + \dot{\boldsymbol{\varepsilon}}^p = \dot{\boldsymbol{\varepsilon}}^e + \dot{\boldsymbol{\varepsilon}}^{cp} + \dot{\boldsymbol{\varepsilon}}^{np} \tag{2}$$

The coaxial strain rate $\dot{\boldsymbol{\varepsilon}}^{cp}$ could be determined by

$$\dot{\boldsymbol{\varepsilon}}^{cp} = \dot{\lambda}\frac{\partial Q}{\partial \boldsymbol{\sigma}} \tag{3}$$

where plastic multiplier $\dot{\lambda}$ could be derived from the consistency condition.

The non-coaxial strain $\dot{\boldsymbol{\varepsilon}}^{np}$ could be obtained by

$$\dot{\boldsymbol{\varepsilon}}^{np} = \frac{1}{H_t}\hat{\mathbf{s}}^n \tag{4}$$

where H_t is non-coaxial hardening modulus, for simplicity, it is often assumed as a constant; the non-coaxial stress rate $\hat{\mathbf{s}}^n$ is

$$\hat{\mathbf{s}}^n = \hat{\mathbf{s}} - \frac{\hat{\mathbf{s}}:\mathbf{s}}{\mathbf{s}:\mathbf{s}}\mathbf{s} - \frac{\hat{\mathbf{s}}:\mathbf{S}}{\mathbf{S}:\mathbf{S}}\mathbf{S} \tag{5}$$

where $\mathbf{S} = \mathbf{s}{:}\mathbf{s} - 2J_2\boldsymbol{\delta}/3 - 3J_3\mathbf{s}/2J_2$.

The rate form of the stress-strain relationship is

$$\dot{\boldsymbol{\sigma}} = \mathbf{D}^{\mathrm{e}} - \mathbf{D}^{\mathrm{e}}\left(\frac{\frac{\partial Q}{\partial \boldsymbol{\sigma}}\left(\frac{\partial F}{\partial \boldsymbol{\sigma}}\right)^{\mathrm{T}}}{\left(\frac{\partial F}{\partial \boldsymbol{\sigma}}\right)^{\mathrm{T}}\mathbf{D}^{\mathrm{e}}\frac{\partial Q}{\partial \boldsymbol{\sigma}} + H_p} + \frac{H_t}{H_t + 2G}\mathbf{C}\right)\mathbf{D}^{\mathrm{e}}\dot{\boldsymbol{\varepsilon}} \tag{6}$$

where $\mathbf{D}^{\mathrm{e}}$ is elastic modulus tensor, H_{p} is the hardening modulus, $\mathbf{C}$ is the non-coaxial flexibility tensor (Qian et al. 2008), Q is the potential function.

$$Q = \begin{cases} q + \dfrac{A_d}{1-A_d}M_d p' g(\theta_\sigma)\left[1 + \left(\dfrac{p'}{p_{at}}\right)^{-(1-A_d)}\right] = 0 & A_d \neq 1 \\ q + M_d p' g(\theta_\sigma) = 0 & A_d = 1 \end{cases} \tag{7}$$

3 Criteria for the Onset Strain Localization

According the bifurcation theory of Rudnicki and Rice (1975), the condition for the onset of strain localization is

$$\det(A_{jk}) = \det\left[n_i D^{ep}_{ijkl} n_l\right] = 0 \tag{8}$$

where A_{jk} is the acoustic tenser; D^{ep}_{ijkl} is elasto-plastic modulus tenser and n_i is the unit vector which is normal to the localized band.

By introducing the angles α and δ which are defined in Fig. 1a, the components of the unit vector n_i become

$$\begin{cases} n_1 = \sin\alpha \\ n_2 = \cos\alpha\cos\delta \\ n_3 = \cos\alpha\sin\delta \end{cases} \tag{9}$$

In case of plane strain, for the normal vector of the shear band is perpendicular to the y-direction (as shown in Fig. 1b), Eq. (8) becomes

$$\det(A_{jk}) = C_1\tan^4\theta + C_2\tan^2\theta + C_3 = 0 \tag{10}$$

where $n_i = \{\cos\theta, 0, \sin\theta\}$ and θ is the inclination of shear band; $C_1 = D^{ep}_{1313}D^{ep}_{1111}$, $C_2 = D^{ep}_{1111}D^{ep}_{3333} - D^{ep}_{1133}D^{ep}_{1313} - D^{ep}_{3311}D^{ep}_{1313} - D^{ep}_{1133}D^{ep}_{3311}$, $C_3 = D^{ep}_{1313}D^{ep}_{3333}$.

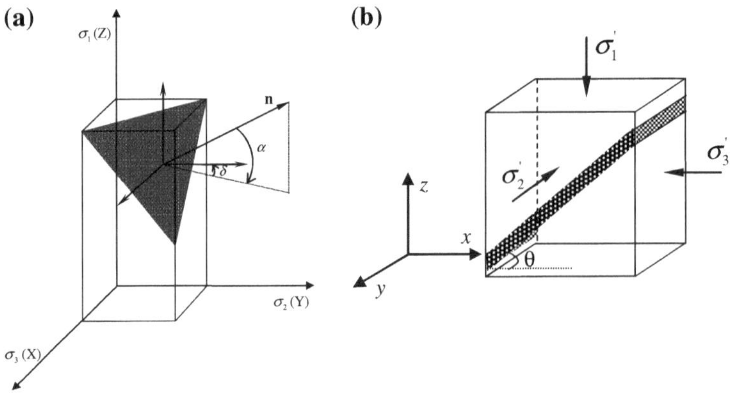

Fig. 1 Strain localization. **a** True triaxial condition, **b** plane condition

4 Experimental Validation

4.1 Plane Strain Drained Tests

In order to validated the proposed method in the prediction of strain localization, the plane strain tests of Han and Drescher (1993) was used, the test specimen was made of Ottawa sands with $D_{50} = 0.72$ mm, initial porosity $n_0 = 0.32$–0.33. The material parameters are summarized in Table 1.

The relationship between H_t/G and the major principle strain at bifurcation point is shown Fig. 2, the strain at bifurcation point decreases with H_t/G, the predicted value is close to experiment when $H_t/G = 0.04$. As shown in Fig. 3, both coaxial model and non-coaxial one give out bigger strain at bifurcation point with the confining pressure increases. While the predicted results by non-coaxial model are closer to the experiments than coaxial model.

Table 1 Material parameters for the simulation

	Ottawa sands	St Peter sandstone sand	Santa Monica beach sand
Elastic modulus E (Mpa)	175	175	175
Poisson ratio ν	0.1667	0.05	0.1667
Peak stress ratio M_f	1.70	1.70	1.75
Dilatancy stress ratio M_d	0.965	1.38	0.965
Strength difference β	0.705	0.66	0.705
Dilatancy parameter A_d	1	1.4	1
Fitting constant A	0.001	0.001	0.001
Non-coaxial modulus H_t (kPa)	1,500	2,900	290

Fig. 2 The bifurcation point

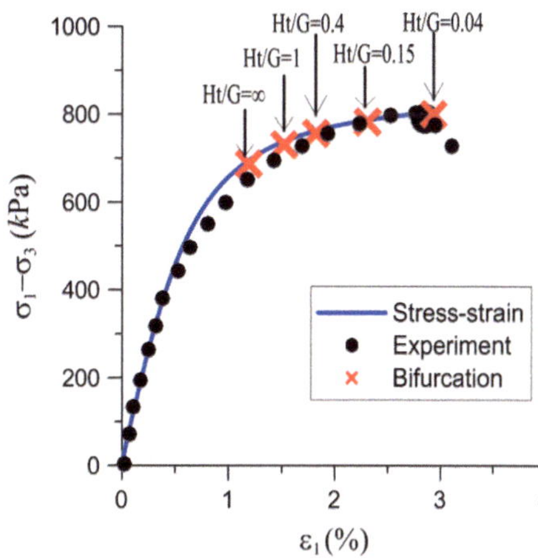

Fig. 3 Shear strain versus confining pressure ($H_t/G = 0.04$)

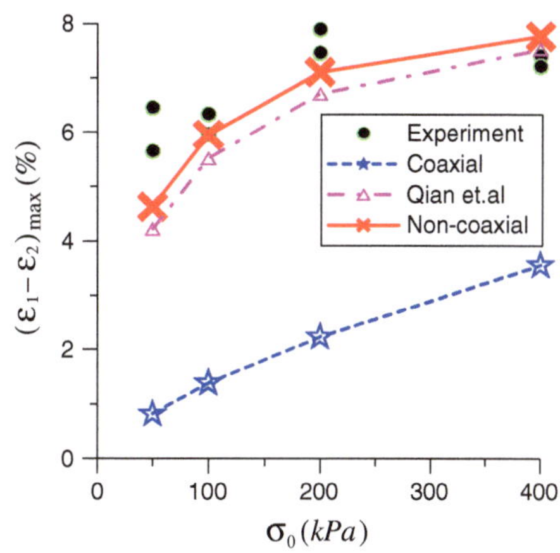

4.2 *Plane Strain Undrained Tests*

The proposed method was also used to predict the undrained plane strain test by Han and Vardoulakis (1991). The specimen in the test was made of St Peter Sandstone sand with specific gravity $G_s = 2.65$, mean size $D_{50} = 0.165$ mm, coefficient of uniformity $C_u = 1.4$, maximum void ratio $e_{max} = 0.8868$ and minimum void ratio $e_{min} = 0.4706$. The size of the prismatic soil specimen is $140 \times 40 \times 80$ mm^3, the confining stress is 404 kPa, the void ratio of the specimen before shear was $e_0 = 0.647$.

With the increase of shear strain, the shear stress increases initially and then decreases after the attainment of its peak. The determinants of acoustic tensors det(A_{jk}) becomes negative in both models. As shown in Fig. 4, det(A_{jk}) becomes to zero when the shear strain is 0.684 %, while in non-coaxial model, it becomes

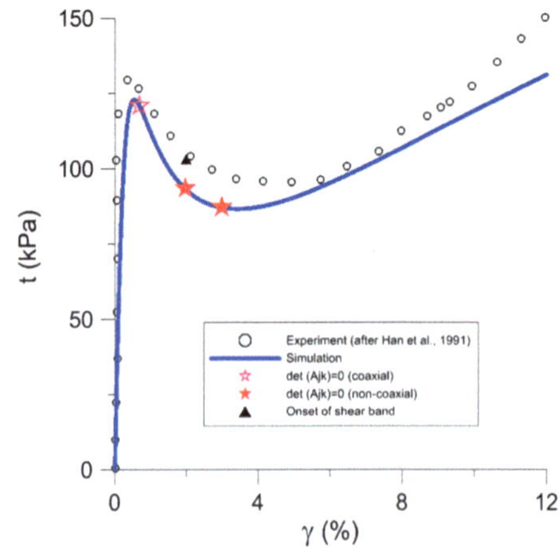

Fig. 4 The stress-strain relationship

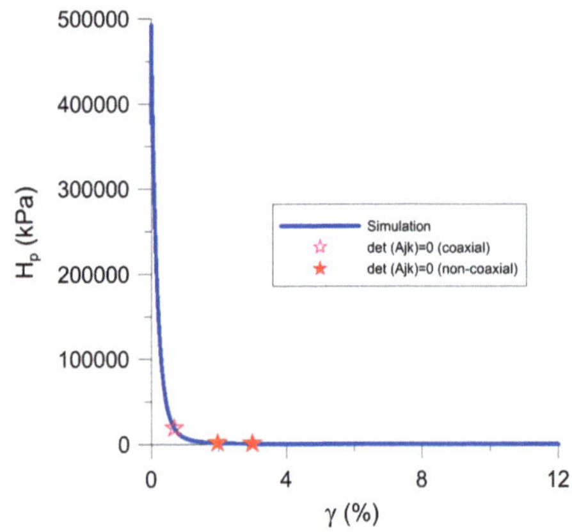

Fig. 5 The evolution of hardening modulus

zero when shear strain are 1.96 and 3.0 %. The shear band would be triggered at the first time when acoustic tensor approaches zero, and the prediction compares well with the experiment. As shown in Fig. 5, the bifurcation point corresponds to the positive value of the hardening modulus.

4.3 True Triaxial Tests

In order to further validate the proposed method, the true triaxial tests of Wang and Lade (2001) on dense sand was used. The soil properties are $D_{50} = 0.25$ mm, $\gamma_d = 15.99$ kN/m^3, $e_0 = 0.63$ and $D_r = 90$ % respectively. The height-to-diameter ratio of the rectangular prismatic specimens was $H/D = 2.47$ and the confining pressure on the test was 49 kPa. By using the material parameters in Table 1, the stress strain under different principal stress ratio condition are shown in Fig. 6.

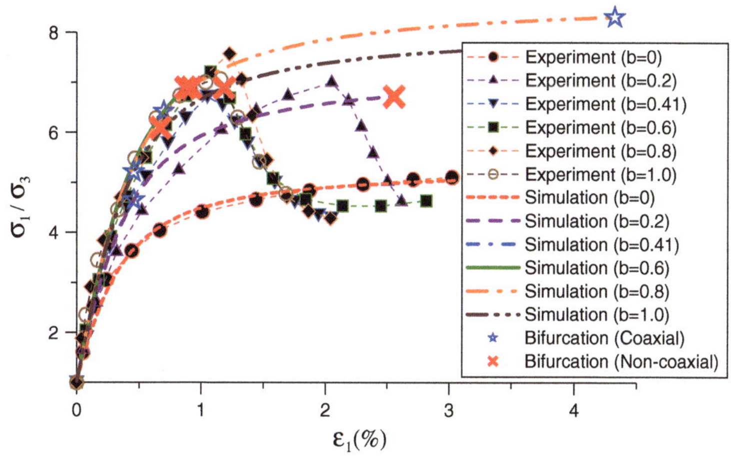

Fig. 6 The prediction of bifurcation points in true triaxial tests (after Lade and Wang 2001)

The adoption of non-claxial flow rule does not change the stress-strain relationships, while the bifurcation points could be changed. In triaxial condition ($b = 0.0$), both models produce no bifurcation point, non-coaxial plasticity gives out bifurcation points except for b is close or equals to zero, while coaxial model gives out bifurcation points only when b in the range of 0.18–0.83. The bifurcation points given out by non-coaxial model are much closer to the experiments than coaxial model.

5 Conclusion

The bifurcation analysis based on 3D non-coaxial plasticity was used to predict the onset of strain localization under plane strain and true triaxial conditions. The comparison between coaxial and non-coaxial model showed the necessity of the introduction of non-coaxial flow rule in accurate prediction of strain localization. In the case of drained conditions, strain localization occurs with shear stress increases, while in the case of undrained conditions, it could occur when shear stress decreases. The decrease of shear stress is induced by the increase of pore water pressure for the material always keeps hardening. The prediction under true triaxial condition by non-coaxial model showed the strain localization occurs when the intermediate principal stress ratio is away from zero value.

Acknowledgments The financial supports by National Program on Key Basic Research Project (grant No. 2014CB049100) and Shanghai natural science foundation (13ZR1443800) are gratefully acknowledged.

References

Cai Y, Yu H-S, Wanatowski D, Li X (2013) Noncoaxial behavior of sand under various stress paths. J Geotech Geoenviron Eng 139(8):1381–1395

Gutierrez M, Ishilhara K, Towhata I (1991) Flow theory for sand during rotation of principle stress direction. Soils Found 31(4):121–132

Han CH, Drescher A (1993) Shear bands in biaxial tests on dry coarse sand. Soils Found 33(1):118–132

Han CH, Vardoulakis I (1991) Plane strain compression experiments on water-saturated fine-grained sand. Géotechnique 41(1):49–78

Huang MS, Lu XL, Qian JG (2010) Non-coaxial elasto-plasticity model and bifurcation prediction of shear banding in sands. Int J Numer Anal Meth Geomech 34(9):906–919

Lade PV, Wang Q (2001)Analysis of shear banding in true triaxial tests on sand. J Eng Mech 127(8):762–768

Lu XL, Huang MS, Qian JG (2011) The onset of strain localization in cross-anisotropic soils under true triaxial condition. Soils Found 51(4):693–700

Miura K, Miura S, Toki S (1986) Deformation behaviour of anisotropic dense sand under principal stress axes rotation. Soils Found 26(1):36–52

Papamichos E, Vardoulakis I (1995) Shear band formation in sand according to non-coaxial plasticity model. Géotechnique 45(4):649–661

Qian JG, Yang J, Huang MS (2008) Three-dimensional noncoaxial plasticity modeling of shear band formation in geomaterials. J Eng Mech 134(4):322–329

Rice JR (1976) The localization of plastic deformation. In: Proceedings of 14th international congress on theoretical and applied mechanics, vol 11. Delft North-Holland Publishing, Delft, pp 371–394

Rudnicki JW, Rice JR (1975) Conditions for the localization of the deformation in pressure sensitive dilatant materials. J Mech Phys Solids 23(6):371–394

Wang Q, Lade PV (2001) Shear banding in true triaxial tests and its effect on failure in sand. J Eng Mech 127(8):754–761

Yang Y, Yu H-S, Kong L (2011) Implicit and explicit procedures for the yield vertex non-coaxial theory. Comput Geotech 38(5):751–755

Micromechanics-Based Constitutive Modeling and DEM Simulation of Localized Failure in Soil

Xiaoqiang Gu, Maosong Huang and Jiangu Qian

Abstract To account for the fabric effect on the localized failure, the fabric tensor describing the soil fabric is related to the stress tensor and a micromechanics-based isotropic-kinematic model is developed in this study. With this model, the effect of fabric anisotropy on the onset of localization and the angle of shear band, is investigated and compared with the experimental results. Meanwhile, discrete element method (DEM) simulation is carried out to study the localized failure of granular soil and the evolution of fabric anisotropy during shearing. The numerical results are compared with the theoretical predictions by the micromechanics-based model. This study successfully illustrates the importance of considering the fabric anisotropy in constitutive modeling of the localized failure in granular soil.

1 Introduction

It is well recognized that many geotechnical failures are attributed to the localized failure of soils which is characterized by the appearance of shear band. In the past decades, tremendous studies have been carried out to predict the occurrence and explore the fundamental mechanism of localized failure (Desrues et al. 1985; Han and Drescher 1993; Papamichos and Vardoulakis 1995; Pietruszczak 1995; Iwashita and Oda 1998; Qian et al. 2013), especially in granular soil. Theoretically, the formation of localized failure can be regarded as a type of instability resulted from the bifurcation of uniformly deformed soil (Rudnicki and Rice 1975). Therefore, the theoretical prediction of localized failure highly depends on the pre-localization constitutive modeling. Extensive studies have showed that

X. Gu · M. Huang (✉) · J. Qian
Department of Geotechnical Engineering and Key Laboratory of Geotechnical and Underground Engineering of the Ministry of Education, Tongji University, Shanghai, China
e-mail: mshuang@tongji.edu.cn

K.-T. Chau and J. Zhao (eds.), *Bifurcation and Degradation of Geomaterials in the New Millennium*, Springer Series in Geomechanics and Geoengineering,
DOI 10.1007/978-3-319-13506-9_33

the fabric anisotropy (inherent or induced) has an essential effect on the macroscopic behavior of granular soil (Arthur et al. 1977; Collins and Muhunthan 2003). Attempts have been made to include the influence of fabric anisotropy in constitutive modeling. However, it is extremely difficult to rationally introduce an anisotropic response into a well-established constitutive model, which is probably due to the difficulty in investigating the anisotropic yield mechanism, particularly under complex loading conditions (Papamichos and Vardoulakis 1995; Qian et al. 2013; Miura et al. 1986; Gutierrez et al. 1991).

To account for the anisotropy effect, micromechanics-based approach has been used as an alternative way to predict the macroscopic behavior of granular soil. In such micromechanics-based approach, the macroscopic parameters (e.g. stress) are linked to the soil microstructure (e.g. fabric anisotropy). Several micromechanics-based theories have been established and used to successfully reveal the anisotropic deformation and shear strength of granular soils (Nemat-Nasser 2000; Yin et al. 2009). However, the localized failure has not been explored by such approaches.

In this study, the fabric tensor describing the soil fabric is linked to the stress tensor and incorporated in the conventional isotropic-kinematic constitutive model to develop a micromechanics-based model. With this model, the effect of fabric anisotropy on the onset of localization and the angle of shear band is investigated and the results are compared with the experiment. Meanwhile, discrete element method (DEM) is used to study the localized failure of granular soil and the evolution of fabric anisotropy during shearing. The numerical results are compared with the predictions by the micromechanics-based model.

2 Micromechanics-Based Modeling

In the conventional isotropic-kinematic hardening theory, the stress can be decomposed into three components as follows:

$$\sigma_{ij} = \tau_{ij} + \alpha_{ij} = p\delta_{ij} + t_{ij} + \alpha_{ij} \tag{1}$$

where τ_{ij} and α_{ij} are the reduced stress and the back stress, and they are known to result in isotropic hardening and kinematic hardening, respectively, p is the hydrostatic pressure, δ_{ij} is the Kronecker delta, and t_{ij} is the deviatoric part of τ_{ij}.

On the other hand, Rothenburg and Bathurst (1989) reported that the stress tensor in a two-dimensional discrete media subjected to boundary forces can be related to the interparticle contact forces and the stress can be expressed as follows:

$$\sigma_{ij} = p\delta_{ij} + s_{ij} = p\delta_{ij} + s^f_{ij} + s^a_{ij} \tag{2}$$

where s_{ij}^{f} and s_{ij}^{a} are the deviatoric stress components related to contact force anisotropy and fabric anisotropy, respectively. The s_{ij}^{a} can be further described by the deviatoric part of the conventional fabric tensor F_{ij} in a matrix form as

$$\left[s_{ij}^{a}\right] = p\left[F_{ij}\right] = p\frac{\omega}{2}\begin{bmatrix} \cos 2\theta_0 & \sin 2\theta_0 \\ \sin 2\theta_0 & -\cos 2\theta_0 \end{bmatrix} \tag{3}$$

where ω and θ_0 denote the degree of anisotropy and the principle direction of the distribution of unit contact normal.

As indicated by Nemat-Nasser (2000), contact force anisotropy changes during shearing, resulting in a change of void ratio and therefore the size of the yield surface, indicating isotropic hardening. On the other hand, the fabric also changes during shearing, resulting in a change of the size and possibly the position of the yield surface (i.e. kinematic hardening) depending on the loading path.

To relate the anisotropic hardening response with the fabric, the loading increment $\dot{s}_{ij}^{a}$ is decomposed into a normal component ($\dot{s}_{ij}^{an}$) and a tangential component ($\dot{s}_{ij}^{at}$) with respect to the unit vector normal to the yield surface N_{ij}. Now, the increment of macroscopic stress tensor can be expressed by

$$\begin{aligned} \dot{t}_{ij} &= \dot{s}_{ij}^{f} + \dot{s}_{ij}^{an} = \dot{s}_{ij}^{f} + N_{ij}\dot{s}_{kl}^{a}N_{kl} \\ \dot{\alpha}_{ij} &= \dot{s}_{ij}^{at} = \dot{s}_{ij}^{a} - N_{ij}\dot{s}_{kl}^{a}N_{kl} \end{aligned} \tag{4}$$

According to Eq. (3), $\dot{s}_{ij}^{a} = p\dot{F}_{ij} + \dot{p}F_{ij}$, where $\dot{F}_{ij}$ is the rate of F_{ij}. Oda's experimental observations (Oda 1993) indicated that F_{ij} tends to evolve with the change of the deviatoric stress ratio s_{ij}/p under proportional loading. Without loss of generality under general loading conditions, $\dot{F}_{ij}$ may be considered as $\dot{F}_{ij} = -\xi\dot{r}_{ij}$, where $r_{ij} = t_{ij}/p$ and ξ quantifies the rate of fabric change during plastic loading. The value of ξ tends to be constant according to Oda's experiments (1993). Finally, the rate of back stress can be given by

$$\dot{\alpha}_{ij} = p\left(\dot{F}_{ij} - N_{ij}\dot{F}_{kl}N_{kl}\right) \tag{5}$$

It indicates that the back stress is linked with the fabric anisotropy.

By replacing the stress $\sigma_{\rm ij}$ with the reduced stress $\tau_{\rm ij}$, the classical isotropic hardening law can be modified into a mixed isotropic-kinematic hardening law. In this study, the yield and plastic potential functions are modified from an isotropic hardening model proposed by Pietruszczak (1995) and given by

$$\begin{aligned} R &= \bar{q} - \eta p = 0 \\ Q &= \bar{q} + \eta_c p \ln(p/p_0) = \text{const} \end{aligned} \tag{6}$$

where $\bar{q} = \sqrt{t_{ij}t_{ij}/2}$, p_0 is the initial confining pressure, $\eta = \frac{\bar{q}}{p} = \eta_f \frac{\varepsilon_s}{A+\varepsilon_s}$ with $\eta_{\rm f} =$ the peak stress ratio at an ideal state, η_c is the critical stress ratio, and A is a material constant that represents the initial shear stiffness in physics. The detail of this micromechanics-based model can be found in Qian et al. (2013).

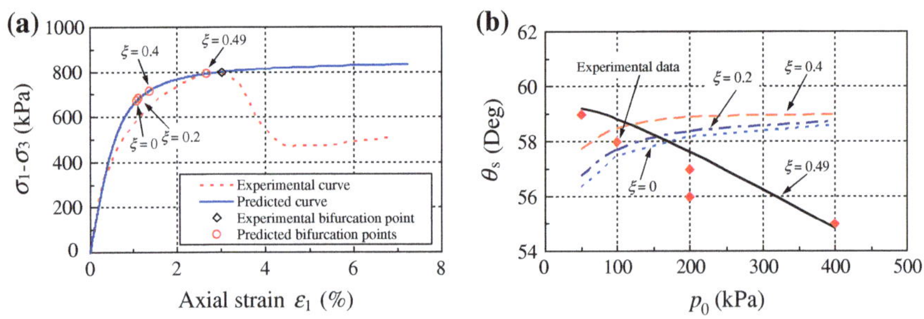

Fig. 1 Effect of fabric anisotropy on the **a** bifurcation point, and **b** inclination angle of shear band in biaxial test ($p_0 = 200$ kPa)

3 Analysis of Experimental Localized Failure

Han and Drescher (1993) performed biaxial tests to investigate the localized failure of dry Ottawa sand. The observed localized failure is analyzed by the proposed micromechanics-based model. Isotropic elasticity is used and the shear modulus and Poisson's ratio are 75 MPa and 0.2, respectively. By fitting the pre-localized behavior, the parameters are determined as $\eta_f = 0.68$, $\eta_c = 0.36$ and $A = 0.0015$. Meanwhile, different ξ values are used to predict the bifurcation state.

Figure 1a shows the experimental stress-strain curve at $p_0 = 200$ kPa, together with the predicted bifurcation points. As seen in Fig. 1a, the localized failure in the experiment is marked by the peak stress state. The predicted bifurcation points significantly depend on the ξ value. The predicted axial strain and deviatoric stress at bifurcation increase as the ξ value increase. It convincingly illustrates that the fabric anisotropy dominants the onset of localized failure.

Another important aspect of the localized failure is the inclination angle of shear band (θ_s). Figure 1b compares the θ_s values from experiments and theoretical predictions. As seen in Fig. 1b, the observed inclination angle of shear band decreases as the initial confining pressure increases. However, the classical isotropic hardening model ($\xi = 0$) yields an opposite trend. On the contrary, as ξ increases to 0.49, the predicted trend approaches the observed one. It again indicates the importance of fabric anisotropy in predicting the characteristics of localized failure.

4 DEM Simulation of Localized Failure

The well recognized DEM program PFC^{2D} is used to simulate the laboratory biaxial test by Han and Drescher (1993). The specimen is of 140 mm height and 40 mm width. The soil particle is represented by a non-circular clump of two

Table 1 Parameters in simulation

Particle density	2,650 kg/m^3	No. of clumps	12,431
Interparticle friction	0.5	Clump aspect ratio	1.5
Wall stiffness	1×10^7 kN/m	Wall-particle friction	0
Particle normal stiffness k_n	Variable[a]	Particle shear stiffness k_t	$k_t = k_n$

[a]The particle normal stiffness $k_n = k_0 \times r$, where $k_0 = 1 \times 10^6$ kN/m^2 and r is the particle's radius

spheres with an aspect ratio of 1.5. For circular particle shape, it is essential to account for the rolling resistance in DEM simulation of localized failure (Zhang et al. 2013; Mohamed and Gutierrez 2010). The top cap and pedestal are simulated by frictionless walls and the membrane is simulated by a flexible boundary with bonded particles. Table 1 lists the basic parameters used.

Figure 2a shows the stress-strain curve of the numerical specimen at a void ratio of 0.126 and confined by $p_0 = 200$ kPa, together with the predicted bifurcation points predicted by the proposed model. By fitting the stress-strain curve before localized failure, the parameters are determined as $\eta_f = 0.66$, $\eta_c = 0.43$ and $A = 0.0035$. It clearly indicates that the predicted bifurcation point significantly depends on the fabric anisotropy. Figure 2b shows the evolution of fabric anisotropy coefficient ω with the stress ratio q/p. It can be seen that the ω value increases with increasing stress ratio and the rate of increase accelerates as the stress ratio approaches the bifurcation point instead of a constant (Zhang et al. 2013). The ω value at bifurcation is 0.42 in DEM simulation, somewhat larger than the predicted value of 0.31. The effect of different fabric anisotropy evolution modes on the localized failure is under investigation. Figure 2c shows the observed shear band in the numerical specimen, indicating the DEM simulation successfully captures the initiation and development of the shear band with the flexible lateral boundaries.

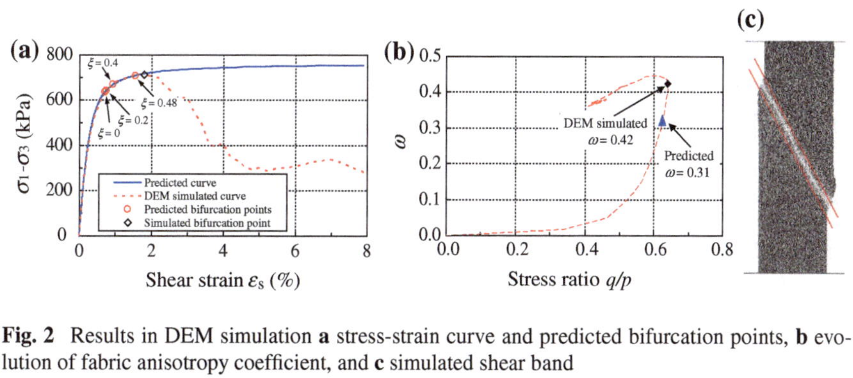

Fig. 2 Results in DEM simulation **a** stress-strain curve and predicted bifurcation points, **b** evolution of fabric anisotropy coefficient, and **c** simulated shear band

5 Conclusions

In this study, the fabric tensor describing the soil fabric is linked to the stress tensor and incorporated in the conventional isotropic-kinematic constitutive model to develop a micromechanics-based model. With this model, the effect of fabric anisotropy on the onset of localization and the angle of shear band is investigated and compared with the experimental results. Meanwhile, discrete element method (DEM) simulation with flexible lateral boundaries successfully captures the localized failure of granular soil. The results show that the fabric anisotropy increases nonlinearly with increasing stress ratio. The agreement between the theoretical predictions, the experimental observations and the numerical simulations convincingly validates the proposed micromechanics-based model and illustrates the importance of considering fabric anisotropy in predicting the localized failure.

Acknowledgments The work presented in this paper is supported by the National Basic Research Program of China (973 program, Grant no. 2012CB719803), National Natural Science Foundation of China (Grant nos. 51308408, 11372228 and 51238009) and Postdoctoral Science Foundation of China (Grant no. 2013M541543).

References

Arthur JRF, Chua KS, Dunstan T (1977) Induced anisotropy in a sand. Geotechnique 27(1):13–30

Collins IF, Muhunthan B (2003) On the relationship between stress-dilatancy, anisotropy, and plastic dissipation for granular materials. Geotechnique 53(7):611–618

Desrues J, Lanier J, Stutz P (1985) Localization of the deformation in tests on sand samples. Eng Fract Mech 21:909–921

Gutierrez M, Ishihara K, Twohata I (1991) Flow theory for sand during rotation of principal stress direction. Soils Found 31(4):121–132

Han C, Drescher A (1993) Shear bands in biaxial tests on dry coarse sand. Soils Found 33(1):118–132

Iwashita K, Oda M (1998) Rolling resistance at contacts in simulation of shear band development by DEM. J Eng Mech ASCE 124:285–292

Miura K, Miura S, Toki S (1986) Deformation behaviour of anisotropic dense sand under principal stress axis rotation. Soils Found 26(1):36–52

Mohamed A, Gutierrez M (2010) Comprehensive study of the effects of rolling resistance on the stress-strain and strain localization behavior of granular materials. Granular Matter 12(5):527–541

Nemat-Nasser S (2000) A micromechanically-based constitutive model for frictional deformation of granular materials. J Mech Phys Solids 48:1541–1563

Oda M (1993) Inherent and induced anisotropy in plasticity theory of granular soils. Mech Mater 16:35–45

Papamichos E, Vardoulakis I (1995) Shear band formation in sand according to noncoaxial plasticity model. Geotechnique 45(4):649–661

Pietruszczak S (1995) On the undrained response of granular soil involving localized deformation. J Eng Mech ASCE 114:1292–1298

Qian JG, You ZP, Huang MS, Gu XQ (2013) A micromechanics-based model for estimating localized failure with effects of fabric anisotropy. Comput Geotech 50:285–292

Rothenburg L, Bathurst RJ (1989) Analytical study of induced anisotropy in idealized granular materials. Geotechnique 39(4):601–614

Rudnicki JW, Rice JR (1975) Conditions for the localization of deformation in pressure sensitive dilatant materials. J Mech Phys Solids 23:371–94

Yin ZY, Chang CS, Hicher PY, Kastunen M (2009) Micromechanical analysis of kinematic hardening in natural clay. Int J Plast 25(8):1413–1435

Zhang WC, Wang JF, Jiang MJ (2013) DEM-aided discovery of the relationship between energy dissipation and shear band formation considering the effects of particle rolling resistance. J Geotech Geoenvironmental Eng ASCE 139(9):1512–1527

Thermally Induced Deformation and Failure in Shale Under Drained and Undrained Heating

Biao Li, Ron C.K. Wong, Bin Xu and Yanguang Yuan

Abstract The clay shale is subjected to high pressures up to 10 MPa and high temperatures up to 300 °C during the recovery process. Very few experimental researches have been conducted to investigate the thermal response of clay shale under such environment. In this paper, the thermal responses of clay shale under different heating rate conditions are investigated by applying X-ray CT scanning. Under drained condition with a slow heating rate, the deformation behavior is the combined effect from the thermal expansion of the solid particle and the thermal dehydration of Stern layer in clay particles. The thermally induced contraction is significant when the thermal dehydration in clay particles dominates the deformation. The thermally induced contraction behavior is more significant in the axial direction (normal to the bedding plane) than that in the radial direction (parallel to the bedding plane) of the shale sample due to the intrinsic oriented fabric. Under fast heating condition, the thermally induced pore pressure generates the failure behavior since the thermal expansion coefficient of water is much higher than that of solid grain in shale.

1 Introduction

Thermal processes have been widely used to recover viscous heavy oil and bitumen from oil sands reservoirs in Canada. During thermal enhanced oil recovery processes, the shale undergoes heating and cooling cycles. Thermally induced deformation and failure in shale can pose an impact on casing and caprock

B. Li · R.C. Wong (✉)
Department of Civil Engineering, University of Calgary, Calgary, Canada
e-mail: rckwong@ucalgary.ca

B. Xu · Y. Yuan
Bitcan Geosciences and Engineering Inc., Calgary, Canada

K.-T. Chau and J. Zhao (eds.), *Bifurcation and Degradation of Geomaterials in the New Millennium*, Springer Series in Geomechanics and Geoengineering,
DOI 10.1007/978-3-319-13506-9_34

Table 1 Physical properties of Colorado shale samples for heating tests

Sample ID	T114S6	T114S4
Testing condition	Slow heating	Fast heating
Height (mm)	166	77
Diameter (mm)	87.13	87.26
Moisture content (%)	15.2	15.5
Bulk density (g/cm^3)	2.218	2.203

integrity (Miyazawa et al. 2008; Yuan et al. 2013). Limited experimental work has been conducted to investigate the thermal response of clay shale under thermal operation conditions. In this paper, the thermal response of clay shale under different heating rates is investigated. Laboratory confined and unconfined tests were conducted on Colorado Shale cores which were retrieved from the overburden shale formation above the oil sands reservoirs in Cold Lake area, Alberta, Canada. Basic properties of the samples for tests are shown in Table 1. The thermally induced deformation and fracturing responses in shale have been captured by X-ray computed tomography scanning (CT scan).

2 Slow Heating Test

The slow heating test was conducted in a specially designed triaxial cell, shown in Fig. 1. The cell is made of 6061-T6 aluminum, which can easily be penetrated by X-ray. During the test, the brine saturated filter paper was wrapped around the sample T114S6 to reduce the drainage time and the Viton membrane was put

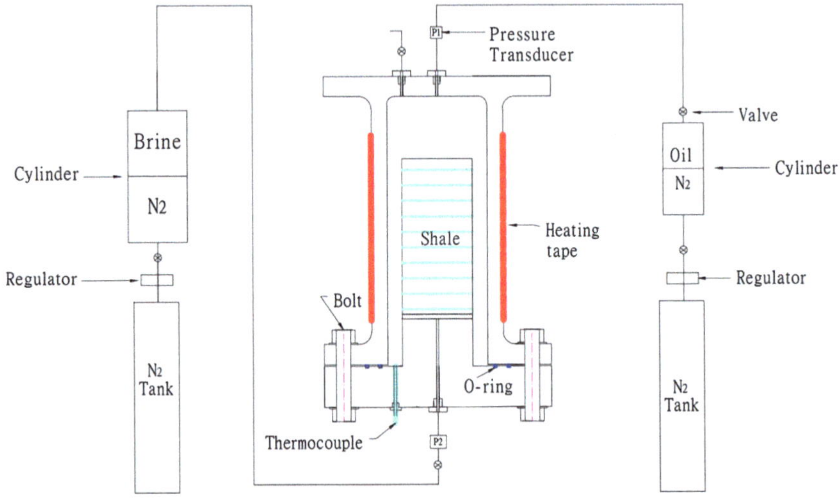

Fig. 1 The schematic of setup for heating test on shale with confining pressure

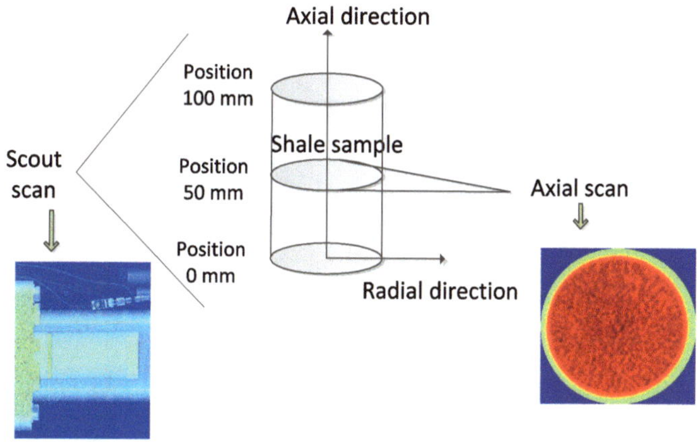

Fig. 2 Sketch showing the sample direction, scout can and axial scan

outside to prevent the invasion of mineral oil. The sample was heated from 20 to 130 °C in 5 days then cooled to 30 °C under confining and pore pressures of 6 and 3 MPa, respectively.

During the test, both scout scans and axial scans were conducted at different temperature conditions. The descriptions of sample directions scout can and axial scan are shown in Fig. 2. The scout scan captures the profile of the sample and only has a resolution of 1 mm. However, we can make use of the difference between the CT numbers in the shale area and the gap area, shown in Fig. 3. The material with the higher density will display the higher CT number. Thereby, a pixel can be divided into a fraction that indicates how much of it is contributed by shale. The dimension change in the axial direction of the sample can be calculated accordingly. The average CT number along the axial line of the sample (the white line in Fig. 3) was derived based on the digital CT image matrix. The deformation and the change in CT number along the axial direction of the sample during the slow heating test are shown in Fig. 4. The shale displays contraction behavior along the axial direction before it is heated beyond 100 °C. To some extent, such nonlinear deformation behavior can be verified by the development of CT number shown in Fig. 4b. After cooling, the sample has significant irreversible contraction.

The axial scan captures the detail information of the sample in slices along the radial direction. Axial scans hold a resolution of 0.195 mm with the beam thickness of 1 mm. The CT images were unified by averaging along the axial direction of the sample. As is indicated in Fig. 5, the diameter of the sample was determined based on the profile of CT number along the while line, and the average CT value was derived by averaging the CT numbers within the black box. The deformation and the change in CT number along the radial direction of the sample under slow heating test are shown in Fig. 6. Under heating, the sample does not display noticeable expansion or contraction along the radial direction until heated beyond 80 °C, where the thermal expansion behavior is significant. After cooling,

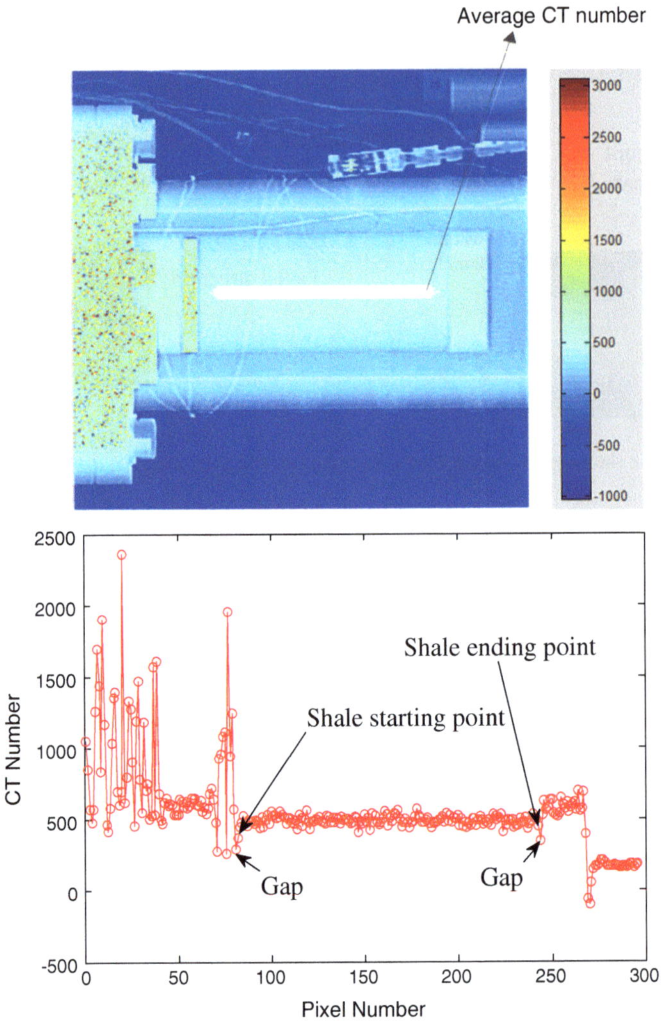

Fig. 3 The approach to estimate the changes in the dimension and the CT number along the axial direction of the sample

the irreversible contraction behavior can be noticed. The difference of deformation behaviors under heating and cooling in Fig. 6a is verified by the different slopes of the CT number-temperature curves in Fig. 6b.

Generally, under slow heating, the shale sample maintains its integrity during the test. The thermally induced contraction behavior is more significant in the axial direction than in the radial direction. The dehydration or degradation of the Stern layer in clay particles can be the affecting factor for such contractile

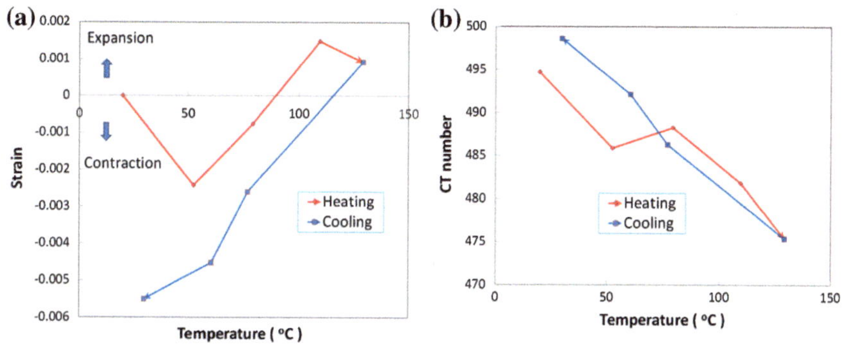

Fig. 4 The deformation (**a**) and the change in CT number (**b**) along the axial direction of Colorado shale sample during the slow heating test

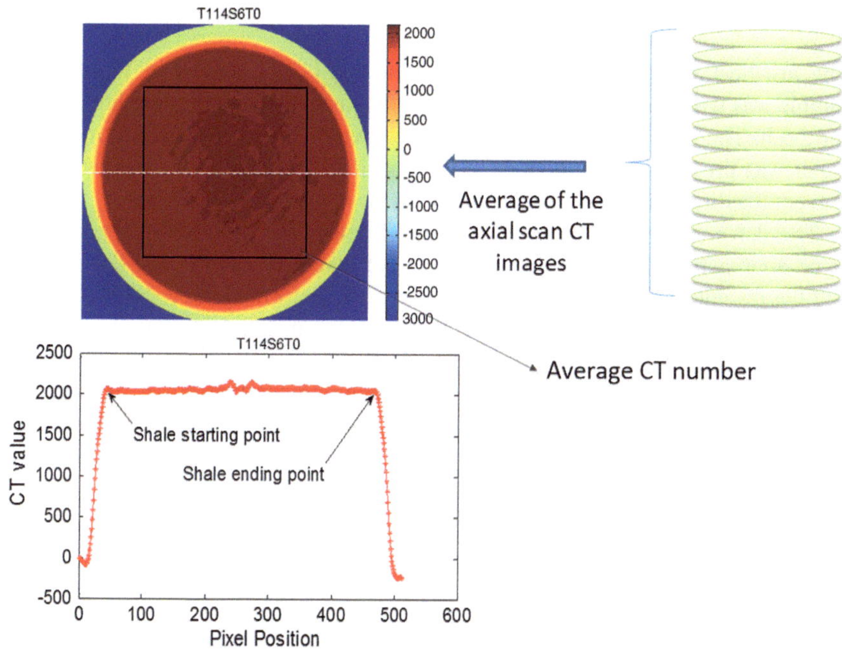

Fig. 5 The approach to estimate the changes in the dimension and the CT number along the radial direction of the sample

behavior (Colten-Bradley 1987; Gonçalvès 2010). Furthermore, the clay shale features in is its intrinsic oriented fabric (Wong and Wang 1997; Hornby et al. 1994; Cholach and Schmitt 2006). The anisotropic deformation behavior can be due to the intrinsic oriented fabric.

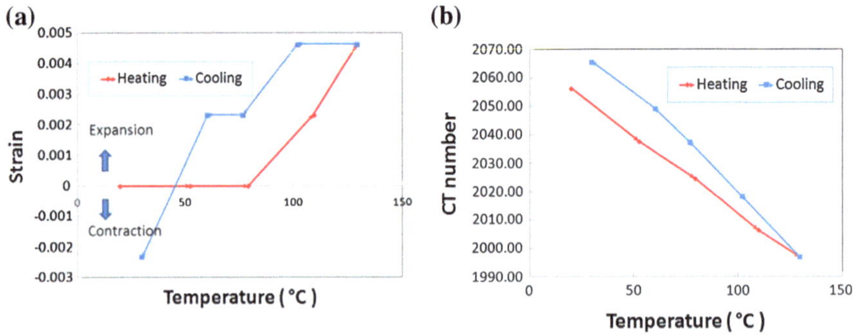

Fig. 6 The deformation (**a**) and the change in CT number (**b**) along the radial direction of Colorado shale sample during the slow heating test

3 Fast Heating Test

The fast heating test was conducted on sample T114S4, shown in Table 1. The sample was wrapped by two layers of vacuum plastic bag and put into boiling bath for 2.5 h. The general layout of the test is shown in Fig. 7. A piece of thick filter paper was provided at the bottom of the container to reduce the disturbance and prevent the direct contact of the sample with the heated steel container.

The deformation and the change in CT number are derived by applying the same approach indicated in Sect. 2. Since the height of sample S114S4 is small and the resolution of the scout scan is limited, the deformation behavior along the axial direction of the sample cannot be monitored in this test. The deformation and the change in CT number along the radial direction of the sample are shown in Fig. 8. Both the strain curve and CT number-temperature curve confirms the significant thermal expansion behavior. After cooling, the dimension of the sample recovers to the initial condition. Under such fast heating condition, the test can be

Fig. 7 General layout of the fast heating test

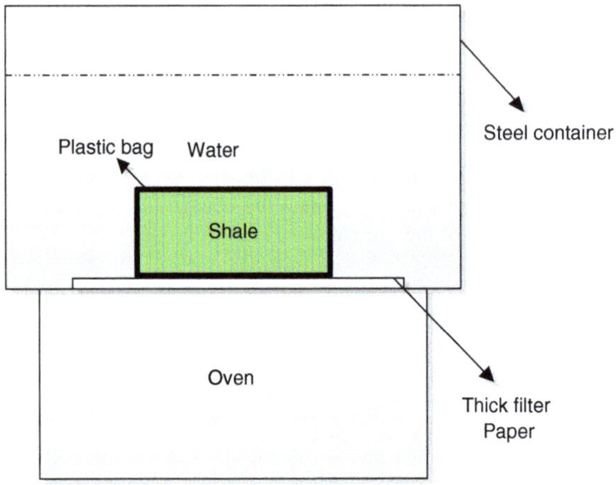

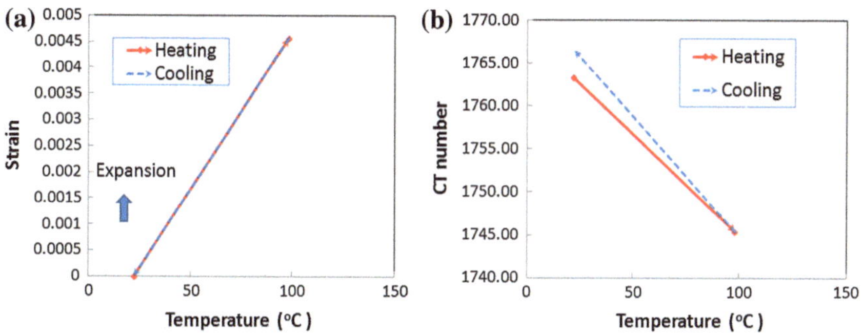

Fig. 8 The deformation (**a**) and the change in CT number (**b**) along the radial direction of Colorado shale sample during the fast heating test

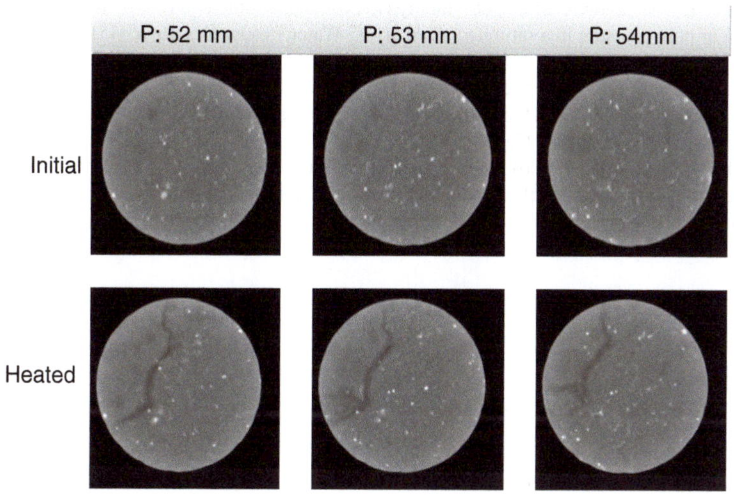

Fig. 9 Generated fractures in sample T114S4 due to fast heating (*P* indicates for position)

treated as undrained heating and the pore fluid dominates the deformation behavior. The hydraulic fracturing happens when the thermally induced pore pressure is higher than the tensile strength of the shale. The thermally induced fracturing was monitored by the X-ray CT scanning. Typical fractures are shown in Fig. 9.

4 Summary

Under drained condition with a slow heating rate, the shale sample maintains its integrity. The thermally induced deformation behavior in clay shale is the combined effect from the thermal expansion of the solid particle and the thermal dehydration of Stern layer in clay particles. The thermally induced contraction is significant when the thermal dehydration in clay particles dominates the deformation. Under undrained condition, the deformation behavior in shale is governed by the thermal expansion of pore fluid. The shale loses its structural integrity when the thermally induced pore pressure is higher than its tensile strength.

References

Miyazawa M, Venkataraman A, Snieder R, Payne MA (2008) Analysis of microearthquake data at Cold Lake and its applications to reservoir monitoring. Geophysics 73:O15–O21

Yuan YG, Xu B, Palmgren C (2013) Design of caprock integrity in thermal stimulation of shallow oil-sands reservoirs. J Can Pet Technol 52:266–278

Colten-Bradley VA (1987) Role of pressure in smectite dehydration; effects on geopressure and smectite-to-illite transformation. AAPG Bull 71:1414–1427

Gonçalvès J, Rousseau-Gueutin P, de Marsily G, Cosenza P, Violette S (2010) What is the significance of pore pressure in a saturated shale layer? Water Resour Res 46:W04514

Wong RCK, Wang EZ (1997) Three-dimensional anisotropic swelling model for clay shale—A fabric approach. Int J Rock Mech Min Sci 34:187–198

Hornby BE, Schwartz LM, Hudson JA (1994) Anisotropic effective-medium modeling of the elastic properties of shales. Geophysics 59:1570–1582

Cholach PY, Schmitt DR (2006) Intrinsic elasticity of a textured transversely isotropic muscovite aggregate: comparisons to the seismic anisotropy of schists and shales. J Geophys Res: Solid Earth 111, B09401

Instability of Dissociation Process of Methane Hydrate Bearing Soil

S. Kimoto, H. Iwai, T. Akaki and F. Oka

Abstract Methane hydrates which exist in both the permafrost sediments and the deep seabed ground are viewed as a new energy resource. It is, however, well known that dissociation process of gas hydrate such as methane hydrate may lead to instability such as large deformation, uncontrollable gas production etc. In the present study, a linear instability analysis was performed in order to investigate which variable has a significant effect on the onset of the instability of methane hydrate bearing geo-materials subjected to methane hydrate dissociation. Stability analysis shows that the onset of the growing instability of the material system mainly depends on the hydrate dissociation rate and the strain hardening-softening parameters. In addition, the stability depends on the wave number of the fluctuation. For larger value of the wave number of the fluctuation, the magnitude of viscoplastic parameter affects the instability, as well as the hardening-softening parameter.

1 Introduction

Recently, methane hydrates have been viewed as a potential energy resource since a large amount of methane gas is trapped within ocean sediments and regions of permafrost. A unit volume of methane hydrate dissociates into approximately 160–170 times of volume (at 0 °C and 1 atmosphere) of methane gas, although the exact amount varies depending on the measuring environment. We do not have, however, enough knowledge about the behaviors of sediment induced by dissociation of hydrates in the ground. Some researchers have pointed out that gas hydrates may be a trigger of a submarine geohazard which impacts the global

S. Kimoto (✉) · H. Iwai · T. Akaki · F. Oka
Department of Civil and Earth Resources Engineering, Kyoto University,
Kyoto 615-8540, Japan
e-mail: kimoto.sayuri.6u@kyoto-u.ac.jp

K.-T. Chau and J. Zhao (eds.), *Bifurcation and Degradation of Geomaterials in the New Millennium*, Springer Series in Geomechanics and Geoengineering,
DOI 10.1007/978-3-319-13506-9_35

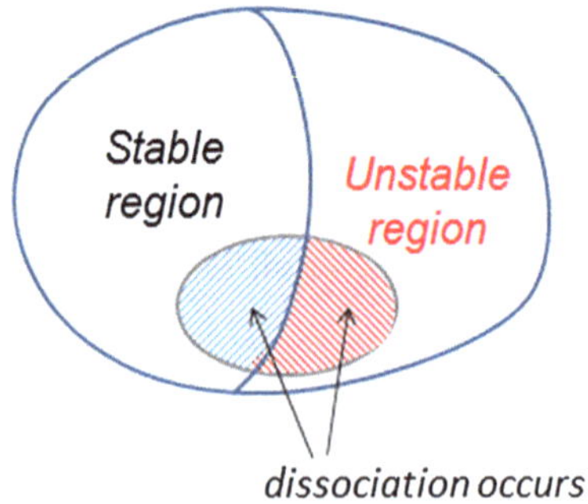

Fig. 1 Stable and Unstable regions of methane hydrate bearing sediments

climate change. For example, it has been indicated that dissociation of methane hydrates is a possible reason of the giant Storegga Slide on the Norwegian margin (Sultan et al. 2004). The laboratory dissociation tests by Wu and Grozic (2008) show the possibility that the dissociation of hydrates causes significant increase in the pore pressure and thus significant decrease in the effective stress under undrained conditions. In addition, many experimental studies (e.g., Waite et al. 2009) indicate that the loss of hydrate bonding in sediments will lead to the decrease in the stiffness and the strength. Thus, the dissociation process may lead to unstable behavior such as large deformation, uncontrollable gas production etc.

In the present study, we have conducted a linear stability analysis to investigate the onset of instability during dissociation process. Figure 1 shows an illustration of the stable and unstable regions of methane hydrate bearing sediments with and without hydrates dissociation. We discuss which parameters or variables have a significant effect on the instability of methane hydrates bearing materials when they are subjected to a dissociation process. In the linear stability analysis, we have extended the method by Oka et al. (1995) and Garcia et al. (2010) to the chemo-thermo-mechanically coupled material considering hydrates dissociation.

2 One-dimensional Instability Analysis

In this section, the linear stability analysis of methane hydrate bearing soil considering dissociation is shown. We followed the method by Garcia et al. (2010), and extended the method by considering energy balance and hydrates reaction process in order to deal with the dissociation phenomenon. The governing equations for the chemo-thermo-mechanically coupled behavior are based on Kimoto et al. (2010), and a viscoplastic constitutive model for the soil skeleton is used. The details of the governing equations for the stability analysis are shown below.

Stress variables are defined in the following one-dimensional form. The total stress σ is obtained from the sum of the partial stresses, namely:

$$\sigma = \sigma^S + \sigma^H + \sigma^W + \sigma^G \tag{1}$$

where superscripts, *S, H, W, G* indicate soil, hydrate, water and gas phase, respectively. The partial stresses for the fluid phases can be written as

$$\sigma^W = -n^W P^W \tag{2}$$

$$\sigma^G = -n^G P^G \tag{3}$$

where P^G, P^G, n^W and n^G are the pore water pressure, the pore gas pressure, the volume fraction of the water phase and volume fraction of the gas phase respectively. Tension is positive for the stresses. For simplicity, we assume that soil and hydrate are in the same phase, namely, solid phase. Thus the partial stress of solid phase is defined as:

$$\sigma^{SH} = \sigma^S + \sigma^H \tag{4}$$

$$\sigma^{SH} = \sigma' - n^{SH} P^F \tag{5}$$

$$n^{SH} = n^S + n^H \tag{6}$$

where σ' is called skeleton stress in the present study; it acts on the solid phase and is used as the stress variable in the constitutive equation. P^F is the average fluid pressure given by

$$P^F = sP^W + (1-s)P^G \tag{7}$$

where s is the water saturation. Substituting Eqs. (2)–(7) into Eq. (1), the skeleton stress is obtained as:

$$\sigma' = \sigma + P^F \tag{8}$$

The conservation of mass for the soil, the water, the gas, and the hydrate phases are given by the following equations:

$$\frac{D^\alpha}{Dt}\left(n^\alpha \rho^\alpha\right) + n^\alpha \rho^\alpha \frac{\partial v^\alpha}{\partial x} - \dot{m}^\alpha = 0 \quad (\alpha = S, W, G, H) \tag{9}$$

in which ρ^α is the material density for α phase, v^α is the velocity vector of each phase, and $\dot{m}^\alpha$ is the mass increasing rate per unit volume due to hydrate dissociation. In order to describe the changes in gas density, the equation for ideal gas is used. Darcy type's law is adopted for the flow of the water and the gas. The one-dimensional equilibrium equation can be written as:

$$\frac{\partial \sigma}{\partial x} + \rho\bar{F} = \frac{\partial \sigma'}{\partial x} - \frac{\partial P^F}{\partial x} + \rho\bar{F} = 0 \tag{10}$$

In the present study, we consider heat conductivity and heat sink rate associated with the hydrate dissociation. The one-dimensional equation of energy conservation is written as:

$$(\rho c)\dot{\theta} = -\frac{\partial h}{\partial x} + \sigma'\dot{\varepsilon} + \dot{Q}^H \tag{11}$$

$$\rho c = \sum_{\alpha} n^{\alpha} \rho^{\alpha} c^{\alpha} \quad (\alpha = \text{S, W, G, H}) \tag{12}$$

where c^{α} is the specific heat of phase α, θ is the temperature for all phases; $\dot{Q}^H$ is time rate of dissociation heat per unit volume due to hydrate dissociation:

$$\dot{Q}^H = (a - b\theta)\frac{\dot{N}_H}{V} \quad a = 56599,\, b = 16.744 \tag{13}$$

The heat flux h follows Fourier's law as $h = -k^{\theta}\partial\theta / \partial x$ in which k^{θ} is the thermal conductivity for all phases. Methane hydrates dissociate into water and methane gas with the chemical reaction. For the methane hydrate dissociation rate $\dot{N}_H$, we use Kim-Bishnoi's equation (Kim et al. 1987):

$$\dot{N}_H = -D_H \exp\left(-\frac{9{,}400}{\theta}\right)\left(P^e - P^F\right) N_{H0}^{\frac{1}{3}} N_H^{\frac{2}{3}}, \quad D_H = 5.85 \times 10^{12} \tag{14}$$

where N_H is the moles of hydrates in the volume V, N_{H0} is the moles of hydrates in the initial state, P^F is the average pore pressure and θ is the temperature. When the dissociation occurs, the dissociation rate is negative, i.e. $\dot{N}_H < 0$. In the analysis, a simplified viscoplastic constitutive model is used. The stress-strain relation can be expressed as:

$$\sigma' = H\varepsilon + \mu\dot{\varepsilon} \tag{15}$$

where ε is the strain, $\dot{\varepsilon}$ is the strain rate, H is the strain hardening-softening parameter and μ is the viscoplastic parameter. We neglect the dependency of the hardening-softening parameter H on the skeleton stress σ', namely, we assume that the strain hardening-softening parameter H is a function of suction P^C and hydrate saturation S_r^H for simplicity. The viscoplastic parameter μ is a function of the temperature θ.

$$H = H\left(P^C, S_r^H\right), \quad \mu = \mu(\theta) \tag{16}$$

The relation between the suction and the effective water saturation is given by the equation (Van Genuchten 1980). In order to estimate the instability of the material system, we consider the equilibrium equation, the continuity equation, the energy balance equation, the constitutive equations, and the equation of hydrate dissociation rate in perturbed configuration. In the governing equations above, unknowns are the pore water pressure P^W, the pore gas pressure P^G, the strain ε, the temperature θ, and the moles of hydrate N_H. For each unknown, we suppose that

$$\begin{aligned} &P^W = P^W_{(0)} + \tilde{P}^W, \quad P^G = P^G_{(0)} + \tilde{P}^G, \quad \varepsilon = \varepsilon_{(0)} + \tilde{\varepsilon}, \\ &\theta = \theta_{(0)} + \tilde{\theta}, \quad N^H = N^H_{(0)} + \tilde{N}^H \end{aligned} \tag{17}$$

$$\left\{\tilde{P}^W, \tilde{P}^G, \tilde{\varepsilon}, \tilde{\theta}, \tilde{N}_H\right\}^T = \left\{P^{W*}\, P^{G*}\, \varepsilon^*\, \theta^*\, N_H^*\right\}^T \exp\left(\omega t + iqx\right) \tag{18}$$

where the first terms in right side in Eq. (17) indicate the values which satisfy the governing equations and second terms are the perturbations of each variables. For the perturbations, we assume the periodic form given in Eq. (18) where q is the wave number ($= 2\pi/l, \quad l:\ wave\ length$), ω is the rate of the fluctuation growth, and superscript "(*)" indicates the amplitude of each variable. By substituting the variables into the governing equations and retaining only the terms of the first order of perturbation quantities, simultaneous linear equations for the perturbed quantities are obtained.

$$[A]\{y\} = \{0\} \tag{19}$$

$$\{y\}^T = \left\{P^{W*},\ P^{G*},\ \varepsilon^*,\ \theta^*,\ N_H^*\right\} \tag{20}$$

$$[A] = \begin{bmatrix} 0 & 0 & 0 & \omega + \frac{k^\theta}{\rho c} q^2 + \frac{\dot{N}_H b}{\rho c V} & -\frac{a - b\theta}{\rho c V}\omega \\ -B_N\left(B_C P^C + s\right) & B_N\left(B_C P^C + s - 1\right) & 0 & A_N & \omega + C_N \\ -\left(\varepsilon H_{PC} + B_C P^C + s\right) & \varepsilon H_{PC} + B_C P^C + s - 1 & \mu\omega + \left(H - \varepsilon H_{SH}\frac{n^H}{n^2}\right) & -\alpha_\mu \dot{\varepsilon} & \varepsilon H_{SH}\frac{m_H}{n} \\ B_C n^F \omega & -\left(B_C n^F - \frac{n^F(1-s)}{P^G}\right)\omega + \frac{k^G}{\gamma^G} q^2 & (1-s)\omega & -\frac{n^F(1-s)}{\theta}\omega & (m_G - (1-s)m_H)\omega \\ -B_C n^F \omega + \frac{k^W}{\gamma^W} q^2 & B_C n^F \omega & s\omega & 0 & (m_W - s m_H)\omega \end{bmatrix} \tag{21}$$

For nonzero solutions of P^{W*}, P^{G*}, ε^*, θ^*, and N_H^* in Eq. (19), the determinant of the matrix $[A]$ has to be equal to zero. From the condition that $\det[A] = 0$, we have a polynomial function of ω as:

$$a_5\omega^5 + a_4\omega^4 + a_3\omega^3 + a_2\omega^2 + a_1\omega + a_0 = 0 \tag{22}$$

If the growth rate of the perturbations ω, which is the root of Eq. (22), has a positive real part, the material system is unstable. The necessary and sufficient conditions that the all roots have negative real parts are given by the Routh-Hurwitz criteria. At first, the necessary and sufficient conditions do not meet when the sign of a_0 becomes negative. If Eq. (23) is satisfied, a_0 becomes negative.

$$\left(H - \varepsilon H_{SH}\frac{n^H}{n^2}\right)\left(k^\theta q^2 + \dot{N}_H b\right) < 0 \tag{23}$$

The term of $k^\theta q^2 + \dot{N}_H b$ is negative when the magnitude of hydrate dissociation rate $\left|\dot{N}_H\right|$ is larger than $k^\theta q^2 / b$ The condition $H - \varepsilon H_{SH} n^H / n^2 > 0$ is satisfied in the case of viscoplastic hardening $H > 0$ with the compressive strain $\varepsilon < 0$, or with the expansive strain which satisfies the following inequality, $0 < \varepsilon H_{SH} n^H / n^2 < H$. In the case of viscoplastic softening $H < 0$, the compressive strain which satisfies $\varepsilon H_{SH} n^H / n^2 < H < 0$, gives the instability condition. In contrast, the term of $k^\theta q^2 + \dot{N}_H b$ is positive when the hydrate dissociation rate $\left|\dot{N}_H\right|$ is smaller than $k^\theta q^2 / b$. The condition $H - \varepsilon H_{SH} n^H / n^2 < 0$ is satisfied in the case of the viscoplastic softening $H < 0$ with the expansive strain $\varepsilon > 0$ or with the compressive strain $\varepsilon < 0$ which satisfies the inequality,

$H < \varepsilon H_{SH} n^H / n^2 < 0$. In the case of the viscoplastic hardening $H > 0$, the positive strain which satisfies, $H < \varepsilon H_{SH} n^H / n^2 < 0$, may lead to the material instability.

Since it is rather difficult to discuss the sign of coefficients $a_1 \sim a_4$ due to the complexity, we will consider the extreme condition of the wave number $q \to \infty$ When the wave number of fluctuation q is very large, the sign of coefficients are determined by the highest order of q. It is possible that the terms of the highest order of q in a_0 and a_1 can be negative in the following cases.

$$\mu C_N + H - \varepsilon H_{SH} \frac{n^H}{n^2} < 0 \quad \text{or} \quad H - \varepsilon H_{SH} \frac{n^H}{n^2} < 0 \tag{24}$$

When the term $H - \varepsilon H_{SH} n^H / n^2$ is negative, the highest order of q in a_0 is always negative, and the highest order of q in a_1 may also become negative depending on the magnitude of the term μC_N. If the term μC_N is smaller than negative value of $H - \varepsilon H_{SH} n^H / n^2$, the highest order of q in a_1 can be negative, since both C_N and the viscoplastic parameter μ are always positive.

3 Conclusions

A linear stability analysis was performed in order to investigate the effects of the parameters on the onset of instability of hydrate bearing sediments induced by hydrate dissociation. In the analysis we used a linearized viscoplastic constitutive model. The main conclusions obtained in this study are as follows:

1. The parameters related to the hydrate dissociation and the material degradation affect the instability of the material system.
2. When the hydrate dissociation rate is very large, the material system is possible to be unstable even in the viscoplastic hardening region regardless the strain is compressive or expansive.
3. Even when the dissociation rate is very small, namely, the dissociation effect is small; the viscoplastic softening may lead to the large deformation.
4. The stability depends on the wave number of the fluctuation. For the larger value of the wave number of the fluctuation, the magnitude of viscoplastic parameter affects the instability, as well as the hardening-softening parameter.

References

Garcia E, Oka F, Kimoto S (2010) Instability analysis and simulation of water infiltration into an unsaturated elasto-viscoplastic material. Int J Solids Struct 47:3519–3536

Kim HC, Bishnoi PR, Rizvi SSH, Engineering P (1987) Kinetics of methane hydrate decomposition. Chem Eng Sci 42:1645–1653

Kimoto S, Oka F, Fushita T (2010) A chemo-thermo-mechanically coupled analysis of ground deformation induced by gas hydrate dissociation. Int J Mech Sci 52:365–376

Oka F, Adachi T, Yashima A (1995) A strain localization analysis using a viscoplastic softening model for clay. Int J Plast 11:523–545
Sultan N, Cochonat P, Foucher JP, Mienert J (2004) Effect of gas hydrates melting on seafloor slope instability. Mar Geol 213:379–401
Van Genuchten MT (1980) A closed-form equation for predicting the hydraulic conductivity of unsaturated soils. Soil Sci Soc Am J 44:892–898
Wu L, Grozic JLH, Eng P (2008) Laboratory analysis of carbon dioxide hydrate-bearing sands. J Geotech Geoenvironmental Eng 134:547–550
Waite WF, Santamarina JC, Cortes DD, Dugan B, Espinoza DN, Germaine J, Jang J, Jung JW, Kneafsey TJ, Shin H, Soga K, Winters WJ (2009) Physical properties of hydrate-bearing sediments. Rev Geophys 47:1–38

Development of Image Analysis Tools to Evaluate In-Situ Evolution of the Grain Size Distribution in Sand Subjected to Breakage

Marios Gkiousas-Kapnisis, Edward Andò, Alessandro Tengattini, Itai Einav and Gioacchino Viggiani

Abstract Grain crushing is a phenomenon of pivotal importance in the inelastic deformation of granular materials. The progressive evolution of the grain size distribution is known to play a major role in a number of geotechnical engineering problems. There is, however, a lack of experimental work tackling the quantification of the three dimensional evolution of the grain size distribution of materials undergoing grain crushing. The technological advancements in X-ray computed tomography now allow in situ, 4 dimensional (3D + time) images of geomaterials to be obtained as they evolve. While recent investigations of the kinematics of persistent grains have allowed a deeper experimental understanding of some inelastic micro-mechanisms to be obtained, a further effort is required when interpreting tomographic images in which grains are not persistent (i.e., they can break). In this contribution, a novel image-analysis technique under development is proposed to quantify the evolution of the grain size distribution as grain crushing proceeds in an experiment. This technique is applied to the analysis of 3D tomographic images of sand sheared at high confinement.

1 Introduction

New micro-mechanics based models have started integrating our understanding of how micro (e.g., grain scale in the case of sand) processes affect the macro (REV scale) behaviour. Some micro-mechanics based models explore the possibility

M. Gkiousas-Kapnisis · E. Andò (✉) · A. Tengattini · G. Viggiani
Univ. Grenoble Alpes, 3SR, Grenoble F-38000, France
e-mail: edward.ando@3sr-grenoble.fr

M. Gkiousas-Kapnisis · E. Andò · A. Tengattini · G. Viggiani
CNRS, 3SR, Grenoble F-38000, France

A. Tengattini · I. Einav
School of Civil Engineering, University of Sydney, Sydney, Australia

K.-T. Chau and J. Zhao (eds.), *Bifurcation and Degradation of Geomaterials in the New Millennium*, Springer Series in Geomechanics and Geoengineering,
DOI 10.1007/978-3-319-13506-9_36

of predicting the evolution of the micro-structure (e.g., de-bonding of cemented grains, or grain breakage, (see Einav 2007a, b), with clear practical implications for geotechnical engineering (e.g., permeability evolution in oil production). The need for information regarding micro-mechanical processes pushes experimental mechanics to strive for a more quantitative understanding of the observed processes to assess the predictions of these models.

X-ray tomography is proving to be a very powerful tool for the experimental investigation into the mechanics of granular media—allowing a range of scales to be imaged non-destructively. Recent work in Grenoble (Ando et al. 2012a, b) has used X-ray tomography to image the shearing of small cylindrical specimens (22 mm height by 11 mm diameter) of different types of sand in triaxial compression. Analysis of the acquired images allows most of the 50,000 grains of a specimen to be identified and followed during a test, allowing in turn the identification of some grain-scale processes responsible for the different specimen-scale responses of these materials.

In an even more recent experimental campaign using the same specimens and imaging techniques, triaxial compression tests were carried out at relatively high confining pressures for sand (in a range from 100 to 7,000 kPa, see Alikarami et al. 2015). In the high confinement tests, x-ray imaging reveals that grain crushing occurs during the shearing of the specimen (as suggested in Ando et al. 2013, whose final figure is reproduced here as Fig. 1), since pores can clearly be seen to fill with fragments

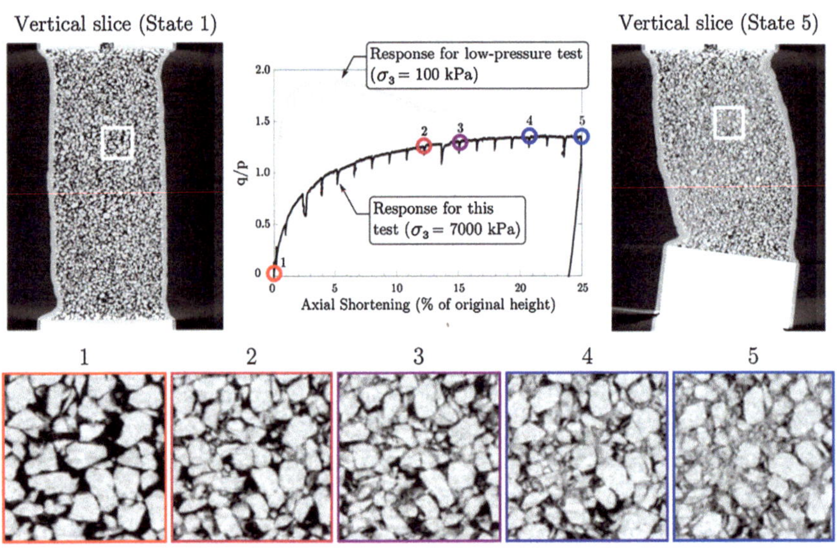

Fig. 1 Figure showing the production of fines during a triaxial compression test on Hostun sand at a relatively elevated confining pressure of 7,000 kPa. *Top left* and *right*: *vertical slices* through the 3D images of the specimen after isotropic compression and at the end of deviatoric loading, respectively. *Top, middle* q/p *versus* axial shortening for a test on Hostun sand at 7,000 kPa (*black*) and at 100 kPa cell pressure (*grey*). *Bottom*: Following the neighbourhood around a grain which breaks, for 5 different imaged states (noted on the graph). Entire Figure adapted from Ando et al. (2013)

of grains. The quantification of the grain crushing occurring in these time-resolved images is a key step in the dialogue between experiments and theory of breakage.

The objective of this paper, therefore, is to obtain a grain size distribution from images of a breaking granular medium, by image analysis.

2 Image Analysis

Figures 1 (top and bottom, left) show vertical cross sections of 3D x-ray tomography images of a sand specimen before shearing. In such an image it is clear that this is a two-phase system (pores and grains). If these two phases can be clearly identified and separated (the simplest way being with a threshold value of x-ray attenuation), grains can then be identified and counted in what amounts to a virtual sieving process. If, under shear, this system evolves without grains crushing, the techniques presented in Ando et al. (2012a, b) can be used to measure grain kinematics since grains are permanent objects.

However, the series of five vertical sections shown at the bottom of Fig. 1 (each centered on the same grain, and taken from X-ray images acquired at different points during shearing) tell a clear story: as the specimen is sheared, grains break and these fragments of grains fill the pores. The system remains a two-phase system even after crushing (pores and potentially broken grains). However, from an image analysis point of view, some grain fragments become smaller than, or around the size of one voxel (a 3D pixel) in the x-ray tomography (a voxel in the case of these images measures $15.6 \times 15.6 \times 15.6\ \mu m^3$); their partial occupancy of the voxel means that they appear to have an x-ray attenuation value between the value for pores (i.e., air in this case) and the sand grains (quartz). Although these "partial volume" voxels are also necessarily present along the grain-pore interfaces in the images before shear, their number grows as grains reduce in size, giving what appears to be a three phase system by the end of the test.

Recalling that the objective of the analysis is to obtain a grain size distribution, it is clear that this information will necessarily be partial when grains start to become smaller than the resolution at which the images are acquired. As a first concrete step in the analysis of this type of image, the technique presented herein attempts to identify the coarsest grains (those bigger than a few voxels) in order to reconstruct the coarser part of the grain size distribution using digital sieving of individual grains. The technique, which for the moment requires a manual selection of global threshold values, starts from a very conservative initial threshold in order to capture the voxels with the highest x-ray attenuation values. The voxels thus selected should in principle be inside grains sufficiently large to have at least one non-partial-volume voxel. Since some noise is inherent in these images, some voxels within the grains are not captured by this first threshold; a dilate procedure is put in place to locally expand and therefore join the voxels selected by this first threshold. The dilation operation (which is far from shape-preserving) is combined with information of the highest local greyscale variance in the image in order to have better-resolved edges. The result of the procedure at this stage is a series of markers—zones that the technique considers to be certainly within grains. Observation of these markers shows that they select the insides of the grains with a fair reliability; however they fall short of selecting the edges of these grains. This information is recovered by going back to the original data with the markers subtracted and making a second thresholding operation on the remaining data; this selects some additional material, which is added to the markers, and which describes the coarse grains. Further technical details of this technique can be found in Gkiousas-Kapnisis (2013). Currently the different thresholds are operator-chosen, which can cause a certain amount of subjectivity in the results.

Figure 2 shows that when grain crushing has occurred in a test, the technique described above allows the coarse grains to be extracted with a good degree of success. These coarse grains are then separated from each other with a commercial watershed code, which in turn (after due labelling of each grain) allows access to the volume of each grain. This is converted into the diameter of a sphere of equal volume. These discrete data can then be shown with a normalised, cumulative histogram (normalising with respect to the mass or volume of grains, and not the *number* of grains), also known as a grain size distribution.

Figure 3 shows the evolution of the sample-wide grain size distributions obtained by the digital sieving of a few of the images of the test presented in

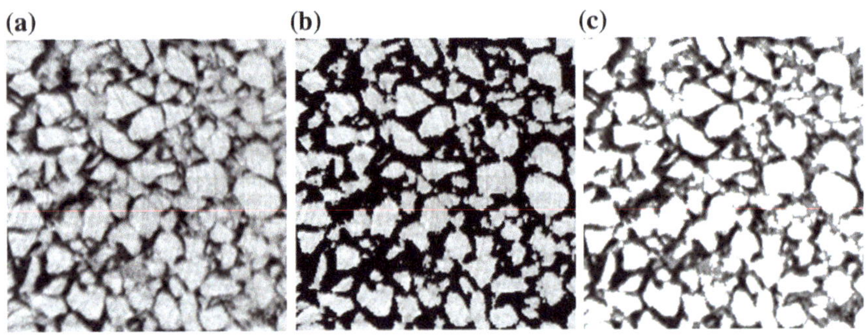

Fig. 2 *Horizontal* sections of an x-ray tomography image showing a zone with grain crushing. **a** raw data, **b** voxels identified as the coarse grain phase, **c** a–b, the remaining two phases, with the coarse grain phase replaced by *white*

Fig. 3 Specimen-wide grain size distributions obtained by the digital sieving of all of the identified coarse grains in several images acquired during progressive shearing of specimen of Hostun sand at 7,000 kPa cell pressure

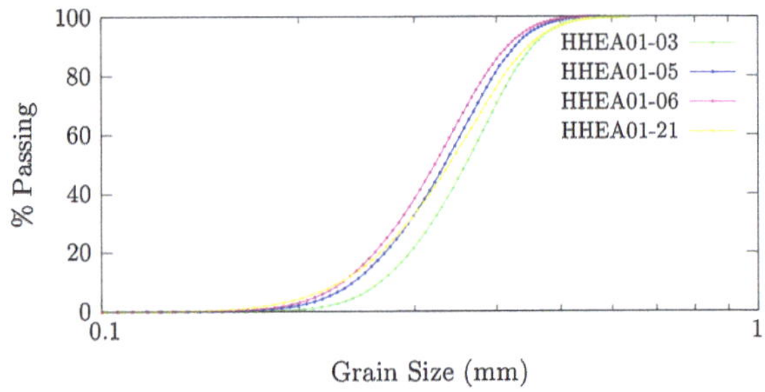

Fig. 1. We recall that a voxel is 0.016 mm wide in these images. The technique is clearly able to pick up the evolution of the grain size distribution to some extent, although the final state presented (HHEA01-21) appears to have a grain-size distribution describing larger grains than previous steps, which is considered to be physically impossible and is almost certainly due to the subjective choice of threshold values.

3 Conclusion

This paper has presented the work in progress on an image analysis technique able to extract grain-scale information from x-ray tomography images of a granular system which is experiencing crushing. Although at the moment the technique requires manual intervention (particularly in the choice of thresholds, which are manually chosen for each image) the first results using only the coarse grains seem promising.

There is a limit—the image's spatial resolution—to the amount of quantitative information that can be obtained about the smallest grains from such an image. However, it is likely that some information about the distribution of grain sizes can be obtained from the "crushed grains" phase. Over and above the considerations of the preservation of mass, further information about the expected distribution could be built into the technique as an *a-priori*.

This *a-priori* is the sort of information that can come from a rich dialogue with micro-mechanical models of breakage, see Tengattini et al. (2014).

Once a more complete image-analysis measurement technique is put in place, and duly verified, one clear path for investigation of these images is the assessment of the distribution (diffuse or localised) of breakage within a specimen. Once obtained, this field of grain breakage can be understood in comparison with other maps, incremental shear/volumetric strain or porosity for example.

References

Andò E, Hall SA, Viggiani G, Desrues J, Bésuelle P (2012a) Grain-scale experimental investigation of localised deformation in sand: a discrete particle tracking approach. Acta Geotech 7(1):1–13

Andò E, Hall SA, Viggiani G, Desrues J, Bésuelle P (2012b) Experimental micromechanics: grain-scale observation of sand deformation. Géotech Lett 2(3):107–112

Andò E, Hall SA, Viggiani G, Desrues J (2013) Experimental micro-mechanics of granular media studied by X-ray tomography: recent results and challenges. Géotech Lett 3:142–146

Alikarami R, Andò E, Gkiousas-Kapnisis M, Torabi A, Viggiani G (2015) Strain localisation and grain breakage in sand under shearing at high mean stress: insights from in-situ x-ray tomography. (accepted)

Einav I (2007a) Breakage mechanics-part I: theory. J Mech Phys Solids 55(6):1274–1297

Einav I (2007b) Breakage mechanics-part II: modelling granular materials. J Mech Phys Solids 55(6):1298–1320

Gkiousas-Kapnisis M (2013) Strain localisation in sand at high mean stress studied by x-ray tomography and DIC. Masters thesis, Université Joseph Fourier

Tengattini A, Andò E, Nguyen GD, Viggiani C, Einav I (2014) Double-scale assessment of micro-mechanics based constitutive models for granular materials undergoing mechanical degradation. (this volume)

FEM × DEM Multi-scale Analysis of Boundary Value Problems Involving Strain Localization

J. Desrues, T.K. Nguyen, G. Combe and D. Caillerie

Abstract The paper presents a FEM × DEM multiscale modeling analysis of boundary value problems involving strain localization in cohesive granular materials. At the microscopic level, a discrete element method (DEM) is used to model the granular structure. At the macroscopic level, the numerical solution of the boundary value problem (BVP) is obtained via a finite element method (FEM) formulation. In order to bridge the gap between micro- and macro-scale, the concept of representative volume element (REV) is applied: the average REV stress and the consistent tangent operators are obtained in each macroscopic integration point as the results of DEM simulation. The numerical constitutive law is determined through the DEM modeling of the microstructure to take into account the discrete nature of granular materials. The computational homogenization method is described and illustrated in the case of a hollow cylinder made of cohesive-frictional granular material, submitted to different internal and external pressures. Strain localization is observed to occur at the macro scale in this simulation.

1 Introduction

When modeling boundary value problems encountered in geotechnical engineering, the designer has often to consider the risk of localized failure. Besides long-standing works on experimental characterization of shear banding in laboratory tests and physical models, trying to catch as realistically as possible the localized failure in computational geomechanics has been the subject of theoretical and numerical works for a long while in the IWBDG community.

Recently, multi-scale analysis using a numerical approach of the homogenization of the microstructural behavior of materials to derive the constitutive response at

J. Desrues (✉) · T.K. Nguyen · G. Combe · D. Caillerie
Laboratoire 3SR, UJF-Grenoble 1, Grenoble-INP, CNRS UMR 5521, BP 53,
38041 Grenoble Cedex 09, France
e-mail: jacques.desrues@3sr-grenoble.fr

K.-T. Chau and J. Zhao (eds.), *Bifurcation and Degradation of Geomaterials in the New Millennium*, Springer Series in Geomechanics and Geoengineering,
DOI 10.1007/978-3-319-13506-9_37

the macro scale has become a new trend in numerical modeling. In recent years, different authors have proposed multi-scale approaches (Kouznetsova et al. 2001; Miehe and Dettmar 2004; Meier et al. 2008) to investigate the behavior of materials by using informations from the micro level. As for granular media, a two-scale fully coupled approach can be defined by using FEM at the macroscale, together with DEM at the microscale (Nitka et al. 2009, 2011; Guo and Zhao 2013). Despite an evident computational cost penalty with respect to mono-scale approaches like FEM and DEM, two-scale FEM × DEM approach allows one to perform real-size grain micro-structure modeling on real-size macroscopic problems, without facing the intractable problem of dealing with trillions of grains in a fully DEM mapped full field problem. Using this approach, microscale related features such as the inherent and induced anisotropy of the material, or material softening/hardening with strain, flow naturally from the microscale DEM model to the macroscale FEM model. An implementation of the FEM × DEM method in the FEM code Lagamine (ULg) (University of Liège, Belgium) is presented, and representative results are discussed, including aspects related to strain localization.

2 Multi-scale Coupling Method

A two-scale numerical homogenization approach by FEM × DEM is considered, Fig. 1. At the microscopic scale level (for each FEM Gauss point), the constitutive equation $\sigma = \Theta(\varepsilon)$ is numerically obtained by a DEM simulation on a

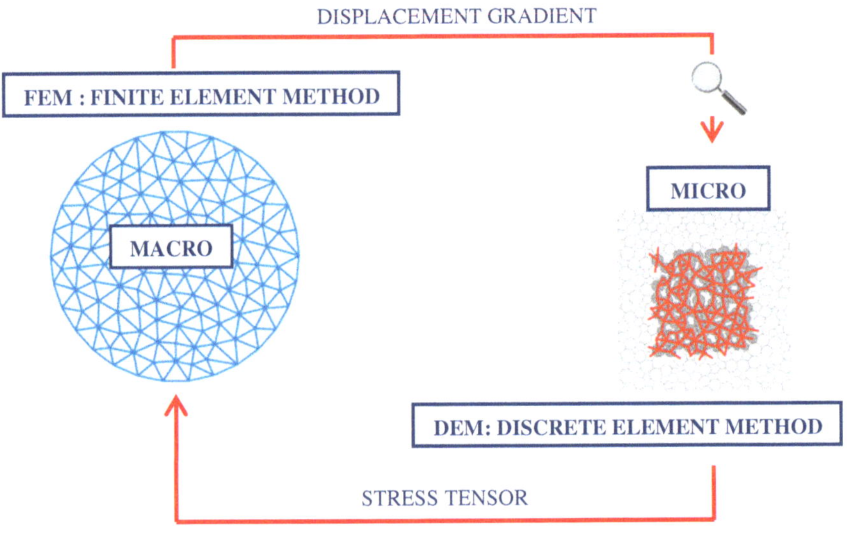

Fig. 1 Computational homogenization scheme

representative elementary volume (REV). The stress response of the REV is computed using the classical homogenization formula defined in Weber (1966):

$$\sigma_{ij} = \frac{1}{S} \sum_{c=1}^{N_c} f_i^c \cdot l_j^c \tag{1}$$

where S is the volume of the REV (area in 2D). f_i^c and l_j^c are respectively the component i of the contact forces acting in contact c and the component j of the branch vector l joining the mass centers of two grains in contact.

At the macroscopic level, a numerical solution for the BVP is obtained using FEM. In every Gauss point of all the elements of the mesh, a specific REV is attached and followed all along the computation, and the stress in this Gauss point at the time t results from the whole deformation history of the REV. A consistent tangent stiffness matrix C_{ijkl} is computed by numerical perturbation, giving the stress increment as a function of the displacement gradient:

$$d\sigma_{ij} = C_{ijkl} \cdot \frac{\partial du_k}{\partial x_l} \tag{2}$$

3 Micro-scale (DEM) Model

The numerical model of the granular material behavior is herein obtained by a DEM approach (soft-contact dynamics type) using periodic boundary conditions (PBC), following (Radjaï and Dubois 2011). The specific REV associated to each Gauss point is made of a dense packing of 400 polydisperse disks, as shown in Fig. 2. Comparing the response of different REVs involving an increasing number of disks, this choice leads to a response reasonably close to the asymptotic one. All grains interact via linear elastic laws and Coulomb friction when they are in contact (Cundall and Strack 1979). Accordingly, the normal repulsive contact force f_{el} is related to the normal apparent interpenetration δ of the contact as $f_{el} = k_n \cdot \delta$, where k_n is a normal stiffness coefficient ($\delta > 0$ if a contact is present, $\delta = 0$ if there is no contact). The tangential component f_t of the contact force is proportional to the tangential elastic relative displacement, with a tangential stiffness coefficient k_t. In order to model cohesive-frictional granular materials, a local cohesion is introduced at the level of each pair of particles by adding an attractive force f_c to f_{el}; f_c is constant for each contact. The overall normal force for two grains in contact is $f_n = f_{el} + f_c$. The Coulomb condition $\|f_t\| \leq \mu \cdot f_{el}$ requires an incremental evaluation of f_t in each time step, which leads to some amount of slip each time one of the equalities $f_t = \pm\mu \cdot f_{el}$ is imposed. In that study, k_n is such that $\kappa = k_n / \sigma_2 = 1{,}000$ (Combe and Roux 2003), where σ_2 is the 2D isotropic pressure. The stiffness ratio is $k_n / k_t = 1$. The adhesion force f_c is defined by reference to the mean level of pressure as suggested by Gilabert et al. (2007): $p^* = f_c / (a \cdot \sigma_2)$ where a is the typical diameter of grains. So p^* is a ratio scaling the attractive part of the mean stress in the sample with the repulsive part due to particle overlap. Hereafter, $p^* = 1$. The intergranular friction angle is $\mu = 0.5$.

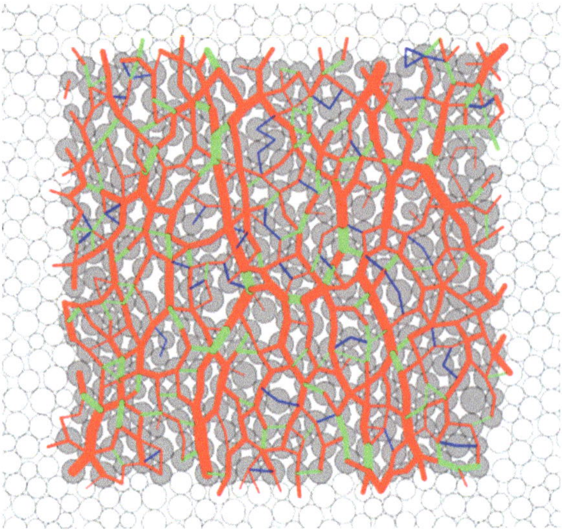

Fig. 2 A REV of 400 particles with PBC. The contact forces are displayed with the following conventions: the width of the lines joining the centers of two particles in contact is proportional to the amplitude of the normal force. *Red*, *green*, *blue lines* distinguish respectively compressive forces with cohesion ($f_n > 0, f_c < 0$), cohesionless contacts ($f_n > 0, f_c = 0$) and attractive contacts ($f_n < 0, f_c < 0$)

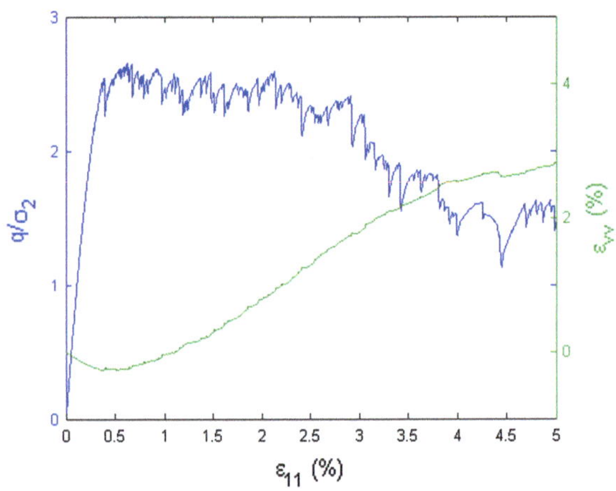

Fig. 3 Mechanical response of the REV of 400 cohesive-frictional discs submitted to a biaxial loading. The *blue curve* corresponds to the evolution of the normalized vertical deviatoric stress $(q/\sigma_2), q = \sigma_1 - \sigma_2$ versus the vertical strain ε_{11}. The *green curve* displays the evolution of the volumetric strain $\varepsilon_{vv} = tr(\varepsilon)$ along the biaxial compression. Strain softening is obtained as the result of the degradation of the contact strength and distribution

A degradation of the cohesion is taken into account by considering a vanishing of f_c at a contact when sliding or separation occurs. This corresponds to a simple model of granular materials with brittle cemented contacts. The mechanical response of the REV exhibits strain softening (Fig. 3).

4 FEM × DEM Simulation of BVPs

The FEM × DEM approach was implemented in the FEM code Lagamine that is able to perform finite strain analysis. The implementation consisted in inserting the DEM modeling code as a new constitutive law, and solving some specific difficulties linked to the determination of the consistent tangent operator. Different BVPs were studied, showing strain localization due to the inherently strain-softening behavior of the micro-scale model. The case of a biaxial test has been presented in Nguyen et al. (2013). Due to lack of space, the present paper concentrates on the results of the simulation of a pressurized hollow cylinder, using 400 eight-nodes quadrilateral order-2 elements with 4 integration points, with the geometry shown in Fig. 4 and the loading conditions in Fig. 5: starting from an homogeneous state of isotropic compression, first the internal pressure is decreased to zero then the external pressure is increased up to 4 times the initial isotropic stress.

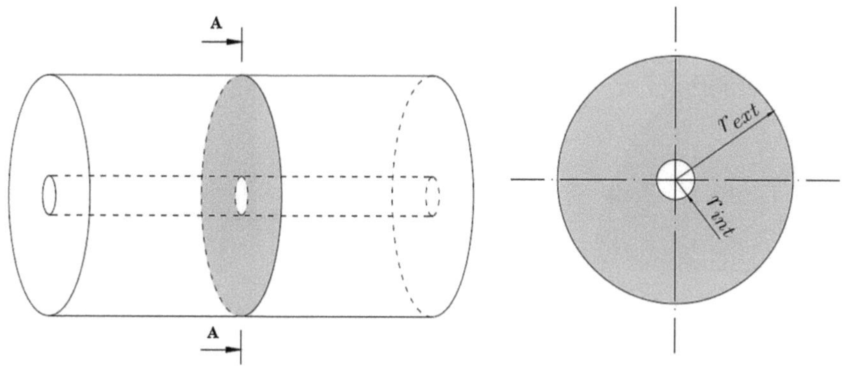

Fig. 4 2D model of a hollow cylinder

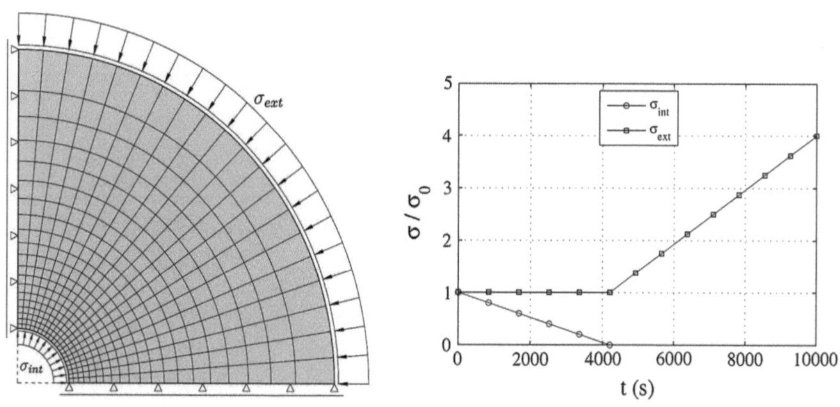

Fig. 5 Modeling of 2D hollow cylinder: discretization in finite elements and loading conditions

Figure 6 shows the deformation mode in the model at different stages of the loading: (a) after internal pressure drop, (b) and (c) after large external pressure increase. In (a), the deformation is more or less axisymmetric with a strong gradient in the radial direction in the immediate vicinity of the internal wall; but in (b) and (c) strain localization has taken place, organized in spiral shear bands originated at the internal wall and progressing significantly inside the cylinder, as illustrated in more details in the zoom at t = 10,000 s in Fig. 7. This is the result of the inherent strain softening exhibited by the material as shown in Fig. 3. This result shows the ability of the FEM × DEM scheme to produce complex and realistic computations in BVPs. On the other hand, it is well known that implementing strain softening constitutive laws in FEM produces mesh dependency: the deformation concentrates in zones as narrow as the mesh permits, independently of any material parameter. Such pathologic response is observed here, as in the biaxial test simulation in Nguyen et al. (2013). In order to restore a mesh independent behavior in such computations, higher order constitutive models can be introduced, as in Chambon et al. (2001); Matsushima et al. (2002) in which a second gradient model is used with success.

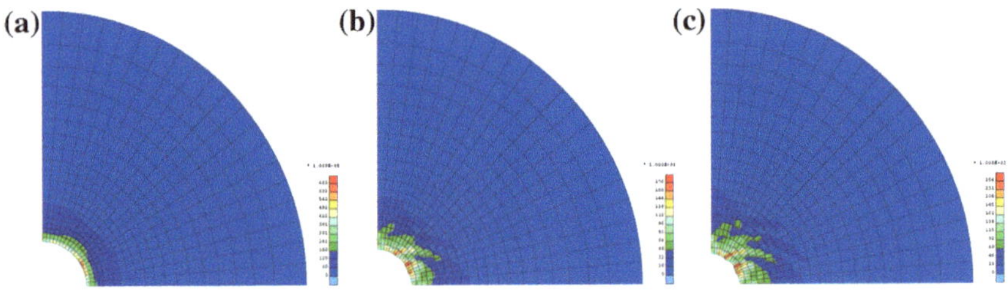

Fig. 6 Strain localization: 2nd invariant of strain tensor. **a** t = 4,400 s. **b** t = 9,000 s. **c** t = 10,000 s

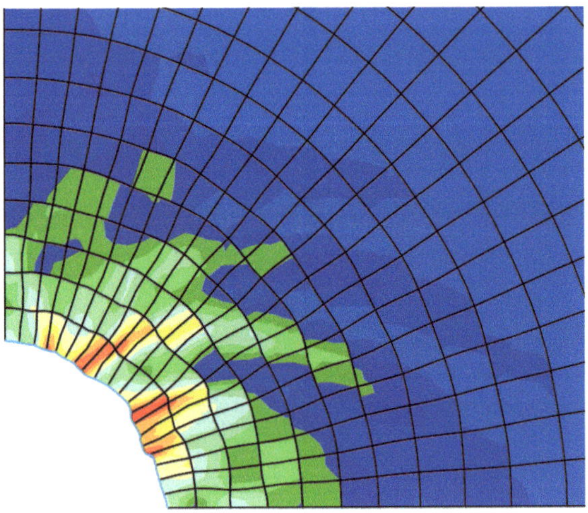

Fig. 7 Strain localization at t = 10,000 s

5 Conclusion

A two-scale approach to investigate the behavior of cohesive granular materials has been presented, combining DEM at the micro-scale with FEM modeling at macroscopic level. The FEM × DEM mechanical response of a cohesive-frictional granular material submitted to a hollow cylinder pressurization test was analyzed and strain localization was detected. The results obtained allow to validate the approach and open new perspectives. Further developments will concern the use of a second gradient extension of the model in the FEM formulation, to overcome mesh dependency and restore objectivity of the post-localization simulations.

Acknowledgments This work was carried out as part of *GeoBridge* research project at 3SR lab, Grenoble, France, which is funded by the French Agence Nationale de la Recherche (ANR).

References

Chambon R, Caillerie D, Matsushima T (2001) Plastic continuum with microstructure, local second gradient theories for geomaterials: localization studies. Int J Solids Struct 38(46):8503–8527

Combe G, Roux JN (2003) Discrete numerical simulation, quasistatic deformation and the origin of strain in granular materials. In: 3rd international symposium on deformation characteristics of geomaterials, Lyon, France, pp 1071–1078

Cundall PA, Strack ODL (1979) A discrete numerical model for granular assemblies. Geotechnique 29(1):47–65

Gilabert F, Roux JN, Castellanos A (2007) Computer simulation of model cohesive powders: influence of assembling procedure and contact laws on low consolidation states. Phys Rev E 75:011303

Guo N, Zhao J (2013) A hierarchical model for cross-scale simulation of granular media. AIP Conf Proc 1542:1222–1225

Kouznetsova V, Brekelmans WAM, Baaijens FPT (2001) An approach to micro-macro modeling of heterogeneous materials. Comput Mech 27(1):37–48

Matsushima T, Chambon R, Caillerie D (2002) Large strain finite element analysis of a local second gradient model: application to localization. Int J Numer Methods Eng 54(4):499–521

Meier HA, Steinmann P, Kuhl E (2008) Technische Mechanik Band 28. Heft 1:32–42

Miehe C, Dettmar J (2004) A framework for micro–macro transitions in periodic particle aggregates of granular materials. Comput Methods Appl Mech Eng 193(3):225–256

Nguyen TK, Combe G, Caillerie D, Desrues J (2013) Modeling of a cohesive granular materials by a multi-scale approach. AIP Conf Proc 1542:1194–1197

Nitka M, Bilbie G, Combe G, Dascalu C, Desrues J (2009) A DEM—FEM two scale approach of the behaviour of granular materials. Powders and grains 2009. Golden CO Colorado School of Mines, USA, pp 443–446

Nitka M, Combe G, Dascalu C, Desrues J (2011) Two-scale modeling of granular materials: a DEM-FEM approach. Granular Matter 13(3):277–281

Radjaï F, Dubois F (2011) Discrete-element modeling of granular materials. Wiley, New York, pp 139–157

Weber J (1966) Recherches concernant les contraintes intergranulaires dans les milieux pulvérulents. Bul. liaison P. et Ch. N°20, juillet–août

Evolution of Packing Structure in Cyclic Mobility and Post-liquefaction of Granular Soils

Jiangtao Wei and Gang Wang

Abstract Micromechanical change in packing structures can provide significant insights to better understand the cyclic mobility and post-liquefaction behaviors of granular soils. In this study, Discrete Element Method (DEM) is used to investigate the evolution of the packing structure under undrained cyclic loading. The coordination number is used to indicate the formation and destruction of a load-carrying structure in the post-liquefaction stage. A new index, termed as centroid distance, is proposed to quantify the effect of void and particle redistribution during cyclic loading. The new index is found to have a strong correlation with the mobilized maximum cyclic strain in post-liquefaction deformation.

1 Introduction

Understanding the behaviors of cyclic mobility and liquefaction in granular soils has been a subject of intensive study for many decades (e.g., Seed and Lee 1966). Laboratory experiments have observed significant changes in soil behaviors before and after the initial liquefaction. For very loose sands, flow-type failure often occurs after the triggering of liquefaction. On the other hand, the mobilized strain in medium to dense sands continues to accumulate progressively in subsequent loading cycles, which is termed as "cyclic mobility" (Castro 1975). To date, there is no direct experimental observation regarding how the microscopic structure of the granular packing changes before and after the liquefaction. Researches on the post-liquefaction behaviors of granular soils are quite limited.

In this study, Discrete Element Method (DEM) is used to simulate the cyclic mobility and post-liquefaction behaviors of granular soils under undrained cyclic loading. The numerical investigation provides detailed microscopic information of

J. Wei · G. Wang (✉)
Department of Civil and Environmental Engineering,
The Hong Kong University of Science and Technology, Clear Water Bay,
Kowloon, Hong Kong
e-mail: gwang@ust.hk

K.-T. Chau and J. Zhao (eds.), *Bifurcation and Degradation of Geomaterials in the New Millennium*, Springer Series in Geomechanics and Geoengineering,
DOI 10.1007/978-3-319-13506-9_38

the granular packing that cannot be obtained from conventional laboratory tests (e.g. Ng and Dobry 1994; Thornton 2000; Sitharam et al. 2009). Indexes will be proposed to quantitatively characterize the evolution of the granular packing structure. These indicators are found to have strong correlations with the cyclic mobility and post-liquefaction behaviors of the granular packing.

2 Discrete Element Simulation

In this study, an open source DEM code, Yade, is used to conduct the numerical simulations. 4,000 disk-shaped particles are randomly generated in a square representative volume element (RVE). Periodic boundary is prescribed on this RVE to eliminate the non-uniformity caused by RVE boundary. The radius of particles ranges from 0.15 to 0.45 mm and the mean radius $R_{50} = 0.3$ mm. A nonlinear Hertz-Mindlin model was used to describe the particle contact behavior. The following material properties are assigned to all the particles: Young's modulus of 70 GPa, Poisson's ratio of 0.3, friction coefficient of 0.5. After particles generation, the packing was isotropically consolidated under an initial confining pressure $p = 100$ kPa to reach a void ratio of 0.228.

The simulation was stress controlled quasi-static simple shear loading with constant volume prescribed and a cyclic stress ratio (CSR) of 0.2. The pore water pressure is determined by the difference of stress between the vertical total stress and the vertical effective stress. As shown in Fig. 1, the soil sample reached the initial liquefaction after 17 cycles. The simulated result is quantitatively similar to the laboratory test results of a dense granular sample, such as the gradual decrease of effective vertical stress in each load cycle till liquefaction, increase of shear strain with the number of load cycles, and phase transform from contraction to dilation in each load cycle. The double-amplitude maximum strain of each cycle before initial liquefaction increases slowly to 2.1 % at cycle No. 17 when the initial liquefaction occurs. From cycle No. 17 to 22, the maximum strains increase dramatically. After cycle No. 22, the mobilized maximum strain is saturated to around an ultimate value, which will be further discussed later.

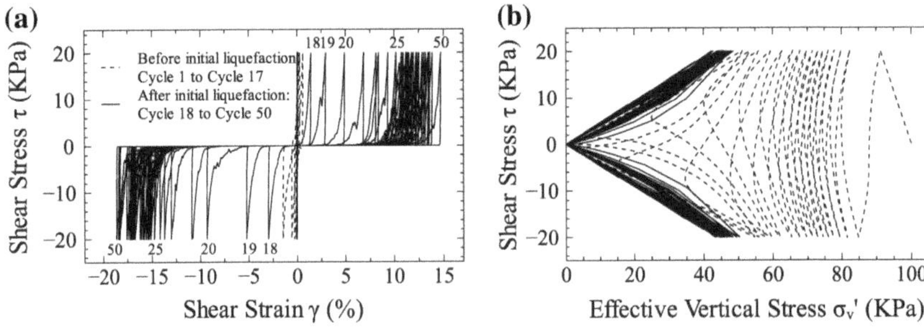

Fig. 1 Macroscopic behaviors of granular packing from DEM. **a** Shear stress and shear strain relation. **b** Shear stress and effective vertical stress relation

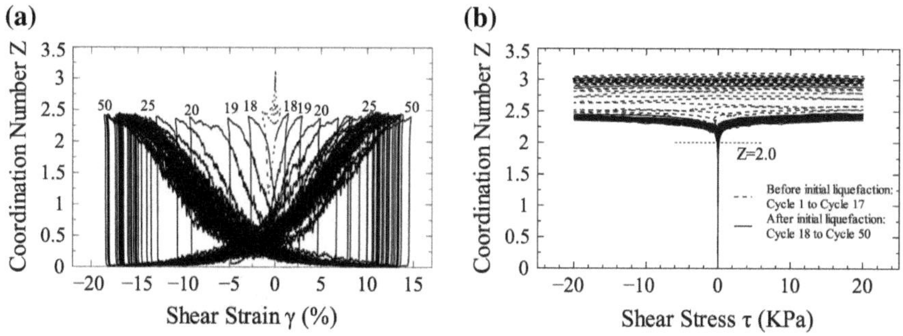

Fig. 2 **a** Coordination number and shear strain relation. **b** Relation between coordination number and shear stress

3 Micromechanical Structure During Post-liquefaction Stage

3.1 Evolution of the Coordination Number

The coordination number is a good indicator of the micromechanical load-carrying structure since it represents the average number of contacts for each particle, which is defined as $Z = 2N_c/N_p$ (where N_c is the total number of contacts and N_p is the total number of particles). Figure 2 shows the evolution of Z during the cyclic loading process. Initially, the coordination number is 3.1, and the material behaves almost elastically. The contact number decreases in each subsequent loading cycle. During post-liquefaction, the circulate loop like butterfly wings in Fig. 2a.

Liquefied sand experiences a large flow-type deformation and eventually regains strength under continued shear deformation, which is regarded as shear-induced dilatancy, i.e., the tendency of the granular matter to dilate under shear deformation. Figure 3 shows the stress-strain relationship and evolution of the coordination number versus the shear stress during cycle No. 19 (post-liquefaction). The shear stress remains almost zero at point 0 and starts to increase when the shear strain $\gamma = 1.13\ \%$ at point 1, where the shear stress reaches 0.1 kPa. This stage is termed as the "flow stage". The flow stage is followed by a "hardening stage" when the shear stress starts to grow substantially to 20 kPa. The coordination number reaches 2.39 from point 1 to 2. Comparing with the peak stress point 2, there are 84 % of the contacts established at the end of the flow stage (i.e. point 1). These contacts form a load-carrying structure that permits stress to increase during further shear deformation. Figure 2b clearly demonstrates a threshold contact number ($Z = 2$) needed for establishment of such a loading-carry structure. Another interesting phenomenon observed from the Fig. 3 is that immediately upon unloading (from point 2 to 3), the coordination number decreases dramatically to reach a minimum value ($Z = 0.1$), implying that the load-carrying structure is completely destroyed upon stress reversal.

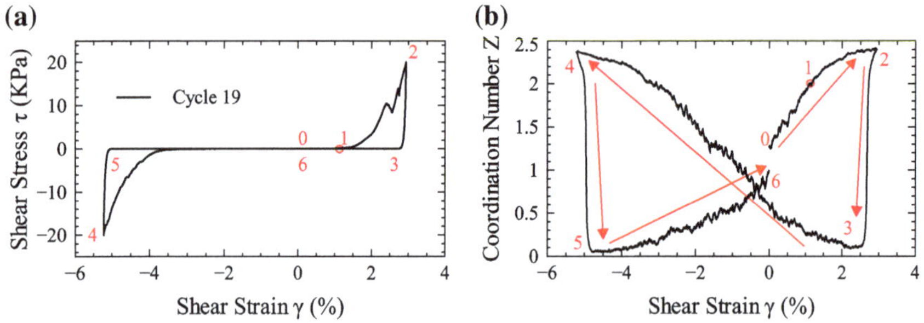

Fig. 3 Stress-strain relation and evolution of coordination number during cycle number 19 (**a**, **b**)

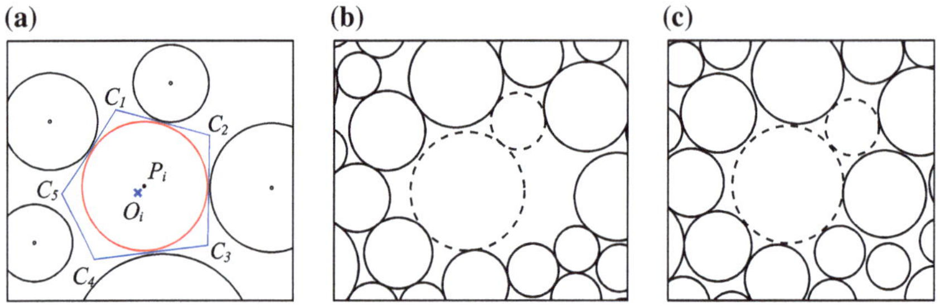

Fig. 4 **a** Definition of D_c^i. **b** Packing configuration before cyclic loading. **c** Packing configuration after 50th loading cycle

3.2 Definition of Centroid Distance (D_c)

For 2D DEM simulation, Voronoi cell can be conveniently used to divide the void space around each particle. As shown in Fig. 4a, the Voronoi cell for particle *i* is a convex polygon enclosed by C_1–C_2–C_3–C_4–C_5. The mass center of the Voronoi cell and the mass center of the particle are denoted as O^i and P^i, respectively. The centroid distance for particle *i* can be defined as $D_c^i = \left|P^i - O^i\right| / R_{50}$, where R_{50} is the average radius of particles in the packing. Snapshots of particle configuration before cyclic loading (under effective confining stress of 100 kPa) and after 50 loading cycle (effective stress is almost zero) are illustrated in Fig. 4b, c. A relatively large pore is surrounded by particles in dotted line in the initial configuration. After cyclic loading, the large pore diminishes, and D_c^i of the particles in dotted line decreases considerably. The above observation corroborates the suggestion by Youd (1977), who attributed the pore-pressure buildup in saturated soils to collapse of unstable particles within the packing. The centroid distance appears to be an effective indicator to quantify the pore space redistribution in undrained cyclic loading.

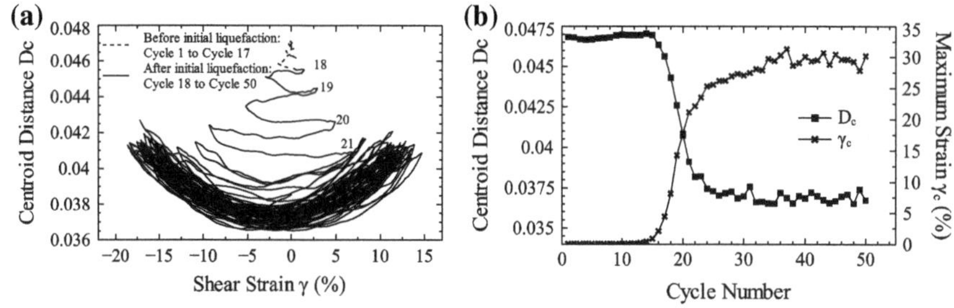

Fig. 5 Simulated results for stress-controlled test with CSR = 0.2: **a** D_c versus shear strain, **b** evolution of γ_c and D_c with the number of cycles

3.3 Evolution of D_c and Cyclic Mobility in Post-liquefaction

The centroid distance (D_c) of the whole packing can be defined as the average of D_c^i over all particles: $D_c = (\sum_i^N D_c^i)/N$, where N is the number of particles. Through the DEM simulation, it is observed that the cyclic mobility of the granular packing is strongly correlated to the evolution of D_c.

Figure 5a shows the evolution of D_c with the shear strain γ during 50 loading cycles. In the first 25 cycles, D_c decreases almost monotonically. In the following load cycles, D_c increases in loading and then decreases upon unloading, with a net effect of decrease in value after a full loading cycle. After approximately 30 loading cycles, D_c reaches a lower-bound limit.

Here, we use γ_c to measure the maximum mobilized (double-amplitude) strain in each load cycle. For example, γ_c is calculated as the shear strain between point 2 and 4 in Fig. 3a for load cycle No. 19. The relation between γ_c and D_c is illustrated in Fig. 5b, where D_c in the figure refers to its minimum value within a loading cycle. Interestingly, significant change in D_c and γ_c occurs simultaneously from cycle No. 17 (the initial liquefaction) to cycle No. 22. When D_c reaches its lower limit, the mobilized maximum shear strain γ_c also stabilizes around a constant value. The maximum shear strain ceases to increase under further loading cycles, and the stress–strain behavior of the soil is eventually saturated.

The evolution of D_c indicates change in granular packing in cyclic loading. Due to the friction between granular particles and complex particle shape, a local arching structure can be formed during initial consolidation, which preserves relative large pores. More large pores can be observed in the loose packing compared with the dense or medium dense packing under the same stress condition. However, the arching structure is not stable and can be progressively destroyed during the cyclic loading. As large pores are redistributed during cyclic loading, the packing is more homogeneous and D_c decreases accordingly.

4 Conclusions

In this study, DEM is used to simulate the cyclic mobility and post-liquefaction behaviors of granular soil under undrained cyclic loading. The coordination number is used to quantify the change in the load-carrying structure of the granular pacing. Upon initial liquefaction, the coordination number significantly reduces to almost zero when particles lose contact, and it gradually increases when the packing continues to deform. A load-carrying structure can be gradually formed in the post-liquefaction stage to sustain considerable load when the coordination number exceeds a threshold value. The granular packing exhibits dilative behavior and strain hardening. However, the load-carrying structure can be completely destroyed upon stress reversal, as indicated by the immediate reduction of the coordination number to near zero. To quantify the change in microscopic configuration during cyclic loading, centroid distance is defined as an effective indicator to quantify the physical process of pore and particle redistribution. It is found that the centroid distance of the packing is strongly correlated to the cyclic mobility in the post-liquefaction stage. The numerical simulation reveals the existence of an ultimate stage for the centroid distance of the packing, and correspondingly, saturation of the maximum shear strain that can be mobilized in post-liquefaction deformation.

Acknowledgments The study was financially supported by Research Project Competition (UGC/HKUST) grant No. RPC11EG27 and DAG11EG03G.

References

Castro G (1975) Liquefaction and cyclic mobility of saturated sands. J Geotech Eng Div ASCE 101(GT6):551–569

Ng TT, Dobry N (1994) Numerical simulations of monotonic and cyclic loading of granular soil. Int J Geotech Eng 120(2):388–403

Seed HB, Lee KL (1966) Liquefaction of saturated sands during cyclic loading. J Soil Mech Found Eng ASCE 92:105–134

Sitharam TG, Vinod JS, Ravishankar BV (2009) Post-liquefaction undrained monotonic behavior of sands: experiments and DEM simulations. Géotechnique 59(9):739–749

Thornton C (2000) Numerical simulations of deviatoric shear deformation of granular media. Géotechnique 50(1):43–53

Youd TL (1977) Packing changes and liquefaction susceptibility. J Geotech Eng Div ASCE 103(8):918–922

Experimental Versus Finite Element Approach to Study Scale Dependent Features in Granular Materials' Stress and Deformation Fields

Danuta Lesniewska and Magdalena Pietrzak

Abstract This paper presents selected results of physical model tests on large granular assembly and finite element analysis of the same configuration. Standard Mohr-Coulomb model was employed as a soil model. The comparison between experimental strain fields and FEM simulations was performed. It was found that the general alignment of strain localization pattern (dilation bands) belongs to macro scale features rather and is dominated by a test boundary conditions and the separated compaction bands may result entirely from granular nature of the sample.

1 Introduction

The two competing approaches evolved historically in granular materials science: the first—traditional, which could be called 'continuous', typical for engineering applications and originally embedded in classical continuum mechanics, and the second, which could be called 'discreet'—more basic, representing discrete systems' mechanics and starting directly from fundamental laws of physics, applied to a system of individual grains.

Each of the two approaches has its limitations. 'Continuum' approach is based on phenomenological or semi-empirical models coming from soil mechanics or on not always physically justified extensions of elasticity and plasticity. The approach proved to be helpful in finding many solutions of practical problems, but its predictive power is regarded as limited.

In spite of this apparent lack of physical justification, there are some macro scale features of granular materials behavior, which were successfully captured by continuum approach, like for example the failure modes coming from plastic solutions of soil mechanics boundary value problems, confirmed by numerous physical model tests.

D. Lesniewska (✉) · M. Pietrzak
Koszalin University of Technology, Koszalin, Poland
e-mail: danuta.lesniewska@tu.koszalin.pl

K.-T. Chau and J. Zhao (eds.), *Bifurcation and Degradation of Geomaterials in the New Millennium*, Springer Series in Geomechanics and Geoengineering,
DOI 10.1007/978-3-319-13506-9_39

As to the 'discreet' approach, the main barrier of its significant progress seems to be the lack of sufficient experimental knowledge. It is extremely difficult however to model granular material in laboratory conditions in a way allowing to control the state of each individual grain and also the whole granular assembly.

As a result, only small collections of artificial 2D grains are usually investigated, being in a number of 10^3 range at most, compared to a range of 10^6–10^9 for typical soil mechanics tests and billions of grains in reality.

In spite of the mutual skepticism, both approaches have some strong sides: discreet approach is able to explain some micro scale phenomena, apparently ruling an overall behavior of large natural grain assemblies, when the 'continuous' approach can offer macro scale view, including numerous extremely interesting experimental observations, resulting from some tens of years of research.

This paper tries to identify, using an example of physical model tests on large granular assembly (~10^6 grains) and finite element analysis, what features of 'strain' field may come from the granular nature of the material and which of them can be regarded as a product of 'continuum' behavior and how in this light some meso-scale phenomena could be defined.

2 Experimental Basis

The experimental set-up applied in this work was described in detail by Lesniewska and Muir Wood (2009, 2010), Muir Wood and Lesniewska (2011) and also Lesniewska et al. (2012) and Niedostatkiewicz et al. (2011).

Small-scale tests on granular samples retained by a movable rigid wall were performed in a glass-sided box (Fig. 1).

The glass sides were 20 mm thick and the model was loaded vertically by constant pressure. The particular configuration included a smooth and rigid vertical wall, 180 mm high, supported by rods that were able to slide horizontally through

Fig. 1 View of the test box filled in with substitute granular material (glass granules)

the box. An active earth pressure mode was investigated, wherein a retaining wall moved away from the backfill.

Such an experimental configuration was selected, as it should produce, according to a previous research, simple mode of strain localization.

The mode of the test was quasi static with a constant wall displacement increment equal to 0.0625 mm (1/20 of the supporting screw lead).

Starlit beads 1000 glass granules were used as a surrogate granular material. Glass grains were mainly of a spherical shape and have d_{50} equal to ~1.1 mm.

The choice of glass as granular material was dictated by the demands of photoelastic method, which was used in combination with PIV deformation analysis, but is not discussed here.

3 Experiment Versus FEM Simulation

The main purpose of the study presented in this paper was to check up whether it is possible, using relatively simple tools, to draw a boundary between macro and micro scale characteristics of a granular material's deformation.

To achieve this goal, primary comparison between experimental strain fields and FEM simulations of the same problem was performed. Software by White and Take (geoPIV, White et al. 2003) was employed for PIV analysis and Plaxis 2010 for finite element calculations (standard Coulomb-Mohr model was assumed to simulate granular material behavior).

The results of this preliminary comparison are presented in Figs. 2, 3 and 4.

Figure 2 (column A), presents the PIV analysis of the selected experimental step (single wall movement increment), Fig. 2 (columns B and C), FEM calculations of the same step, for simplified modelling of the external loading system (uniformly distributed load, Fig. 2-1B) and for realistic model of the loading system, which consisted of five metal segments (Fig. 2-1C).

Figure 2-2A and 2C show that accurate modelling of the test setup may produce striking resemblance between experimental and calculated strain fields, what means that the general alignment of strain localization pattern (dilation bands) belongs to macro scale features rather and is dominated by a test boundary conditions.

The different conclusion can be lead looking at Fig. 3, comparing fields of volumetric strains produced by an external load increment equal to 0.2 MPa (no wall displacement).

Figure 3A shows the experimental field, clearly indicating the existence of separated compaction bands, Fig. 3B corresponding calculation results, showing constant volumetric strains within the whole sample, what suggests that the compaction bands result entirely from granular nature of the sample.

Figure 4 presents some composite experimental strain fields, extracted by PIV, which show an interesting feature: clear splitting of localized deformation into

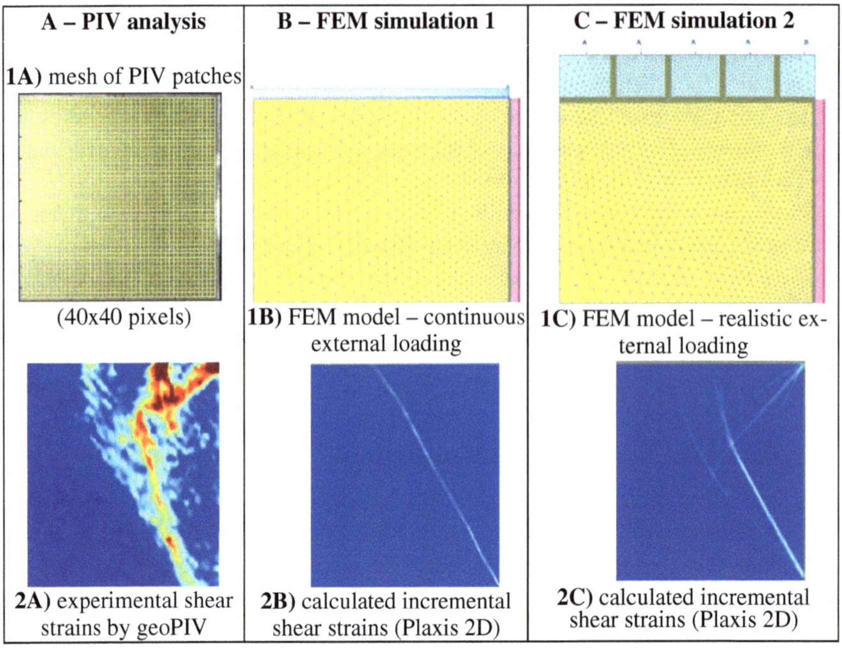

Fig. 2 Comparison of experimental (column **A**) and calculated (columns **B** and **C**) shear strain fields evolved due to a vertical wall (*right boundary*) incremental horizontal displacement

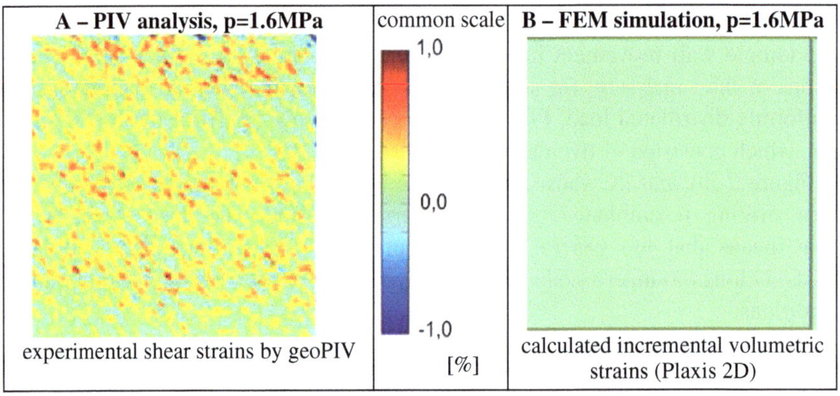

Fig. 3 Pure external load effect—measured (**A**) and calculated (**B**) volumetric strains

two families: dilation bands (Fig. 4-1B and 2B, blue color) and compaction bands (Fig. 4-1B and 2B, red and yellow).

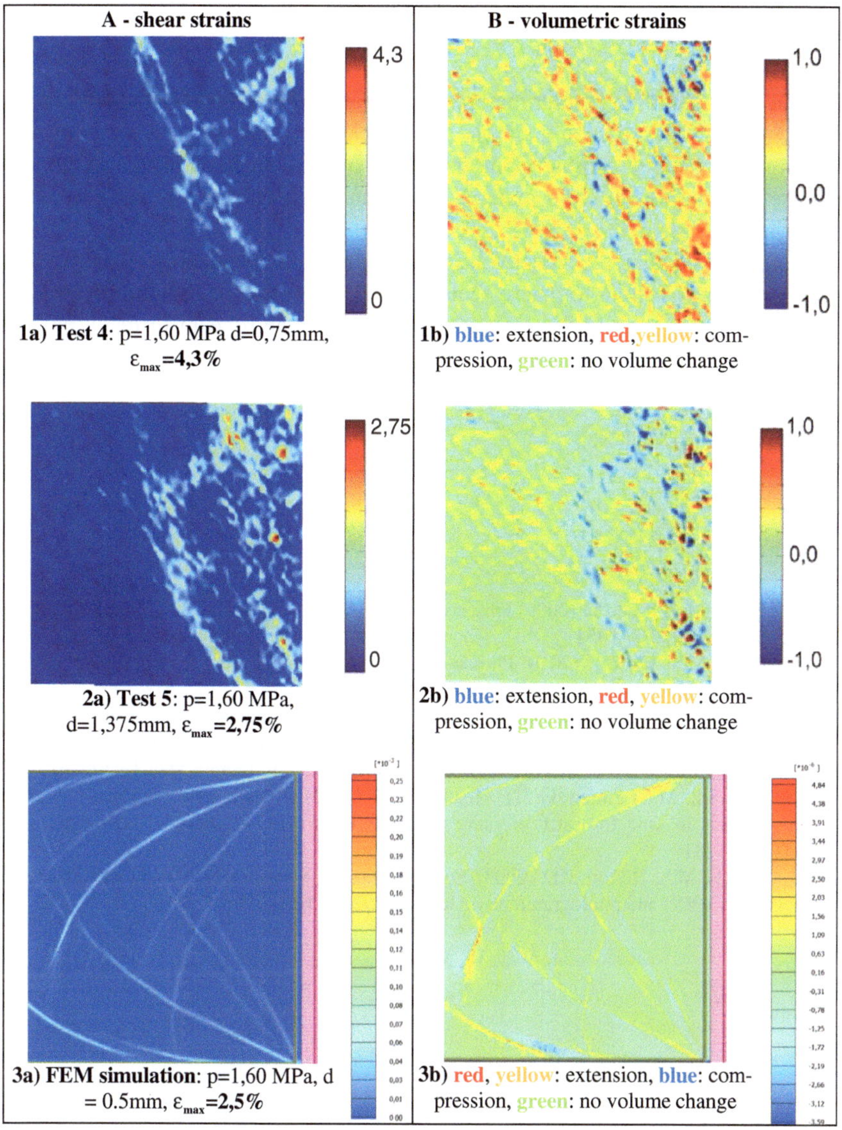

Fig. 4 Qualitative comparison of experimental (rows **1** and **2**) and calculated (row **3**) shear and volumetric strain fields evolved due to a vertical wall (*right boundary*) incremental horizontal displacement

FEM simulation (Fig. 4, row 3) shows some geometrical similarity to the pattern of dilation bands but no clear trace of separate compaction bands system, confirming the previous finding.

4 Conclusions

The mechanisms of a granular material deformation are still not sufficiently recognized, even in case of such a simple test arrangement, like vertical retaining wall moving horizontally. Simple Coulomb-Mohr model can be helpful in recognizing the meaning of different material scales and their role in a granular material behavior.

It was found that the general alignment of strain localization pattern (dilation bands) may belong to macro scale features rather and be dominated by a test boundary conditions and the separated compaction bands seem to result entirely from the granular nature of the sample.

Acknowledgments This research was funded by Polish National Science Centre, decision nr DEC-2011/03/B/ST8/05865.

References

Lesniewska D, Muir Wood D (2009) Observations of stresses and strains in a granular material. J Eng Mech 135:1038–1054
Lesniewska D, Muir Wood D (2010) Photoelastic and photographic study of a granular material. Geotechnique 60:903–911
Lesniewska D, Niedostatkiewicz M, Tejchman J (2012) Experimental study on shear localization in granular materials within combined strain and stress field. Strain 48:430–444
Muir Wood D, Lesniewska D (2011) Stresses in granular materials. Granul Matt 13:395–415
Niedostatkiewicz M, Lesniewska D, Tejchman J (2011) Experimental analysis of shear zone patterns in sand for earth pressure problems using particle image velocimetry. Strain 47(s2):218–231
White DJ, Take WA, Bolton MD (2003) Soil deformation measurements using particle image velocimetry (PIV) and photogrammetry. Geotechnique 53:619–631

Onset of Strain Localization in Fiber Reinforced Composites Subjected to Plane Stress Loading

Marta Miletić and Dunja Perić

Abstract The main objective of this study was to find analytical solutions for the onset of strain localization in fiber reinforced composites subjected to plane stress loading. In particular, elastic multidirectional fibers are embedded into an elastic-plastic matrix. A macroscopic tangent stiffness tensor of the fiber-reinforced composite is obtained by consistently homogenizing the contribution of fibers in a cylindrical representative volume element. Upon deriving analytical solutions their properties are further illustrated on the example of Drucker–Prager model. Results show that fibers decrease the critical hardening modulus, thus inhibiting the onset of strain localization. The main fiber parameters that control the stress level at the inception of strain localization are their volumetric content and their stiffness modulus.

1 Introduction

The main objective of this study was to quantify the role of fibers in inhibiting the inception of strain localization in fiber reinforced composites subjected to plane stress loading. A macroscopic stiffness tensor of the composite is obtained by consistently homogenizing the contribution of fibers in a representative volume element (RVE).

2 Fiber Contribution

The RVE is selected to be a cylinder shown in Fig. 1. Due to plane stress state the height t of RVE is assumed to be half of its radius, which corresponds to a length of fiber L_f. Fibers are also assumed to have a cylindrical shape with a diameter d_f,

M. Miletić · D. Perić (✉)
Department of Civil Engineering, Kansas State University, 2111 Fiedler Hall,
Manhattan KS 66506-5000, USA
e-mail: peric@ksu.edu

K.-T. Chau and J. Zhao (eds.), *Bifurcation and Degradation of Geomaterials in the New Millennium*, Springer Series in Geomechanics and Geoengineering,
DOI 10.1007/978-3-319-13506-9_40

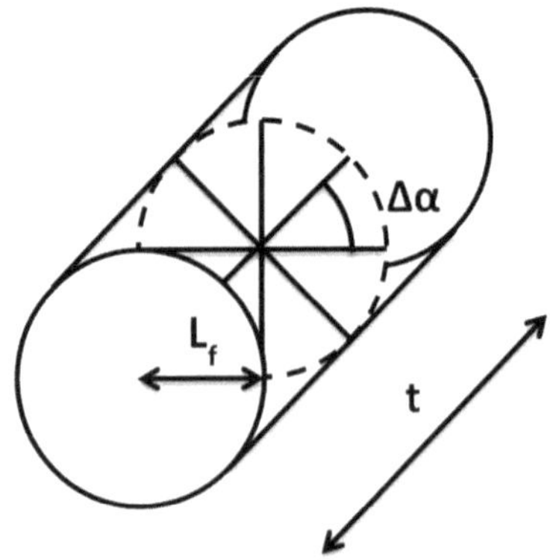

Fig. 1 Fiber distribution in RVE for $n_f = 8$

length L_f and an aspect (length to diameter) ratio η_f. Volumetric fiber content χ_f then can be expressed as:

$$\chi_f = \frac{V_f}{V} = \frac{n_f}{2\eta_f^2}, \tag{1}$$

where V_f is the volume of fibers contained in RVE, V is the volume of the RVE and n_f is the total number of fibers in the RVE.

All fibers are located in the same plane, which is parallel with the base of the cylinder. The angle between any two neighboring fibers is denoted by $\Delta\alpha$, and it is equal to $2\pi/n_f$.

The contribution of fibers to the tangent stiffness tensor $D^e_{(\mathrm{f})ijkl}$ is given by:

$$D^e_{(f)ijkl} = \frac{K_f}{n_f}\sum_{m=1}^{n_f} N_{i(m)}N_{j(m)}N_{k(m)}N_{l(m)} \tag{2}$$

where K_f is an elastic modulus of a fiber, N_i is a unit vector in the direction of the fiber and m is the summation index.

3 Tangent Stiffness Moduli Tensor

The stress-strain relationship for an elastic-plastic fiber reinforced material involving a non-associated flow rule under plane stress is described by:

$$\dot{\sigma}_{ij} = D_{ijkl}\dot{\varepsilon}_{kl} \tag{3}$$

where indices i, j, k, l refer to in-plane components and take values 1 and 2.

The elastic part of D_{ijkl} is given by

$$D^e_{ijkl} = (1 - \chi_f)D^e_{(m)ijkl} + \chi_f D^e_{(f)ijkl} \tag{4}$$

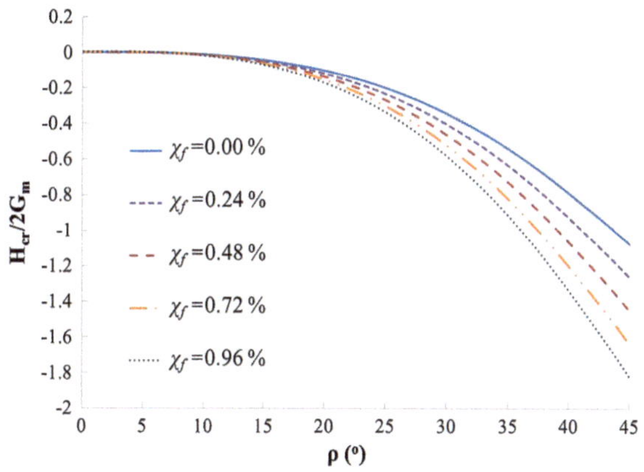

Fig. 3 $H_{cr}/2G_m$ versus ρ for $K_f = 5{,}000$ GPa

In Figs. 2 and 3 the aspect ratio of fibers is constant while n_f is changing to obtain different values of χ_f. By comparing Figs. 2 and 3 it can be seen that the increase of stiffness moduli tensor of fibers produces a significant decrease in the critical hardening modulus, thus delaying the onset of strain localization.

6 Conclusions

The explicit analytical expression for the critical hardening modulus of the fiber reinforced composites subjected to plane stress loading is derived. The results obtained for Drucker–Prager model indicate that addition of fibers decreases the critical hardening modulus, thus delaying the onset of strain localization in the region where both principal stresses are tensile.

Reference

Runesson K, Ottosen NS, Perić D (1991) Discontinuous bifurcations of elastic-plastic solutions at plane stress and plane strain. Int J Plasticity 7:99–121

Using Shear Strain Localisation to Model the Fracturing Around Gallery in Unsaturated Callovo-Oxfordian Claystone

B. Pardoen, S. Levasseur and F. Collin

Abstract Galleries drilling leads to damage propagation, fracturing and properties modifications in the surrounding medium. The prediction of the damaged zone behaviour is an important matter and needs to be properly assessed. To do so the fractures can be modelled using shear strain localisation. The coupled local second gradient model is used under unsaturated conditions to correctly model the strain localisation behaviour. The permeability evolution and the rock desaturation due to air ventilation in galleries are considered. Finally, a hydro-mechanical modelling of a gallery excavation in Callovo-Oxfordian claystone is performed leading to a fairly good representation of the damaged zone.

1 Introduction

Owing to its low permeability, the Callovo-Oxfordian claystone is nowadays considered as a favorable medium for nuclear waste repository. In this rock, the gallery excavation induces damage propagation, fractures and drastic properties changes leading to the development of the excavation damaged zone (Fig. 1). The safety function of the host rock may be altered thus predicting the fracturing structure and behaviour of this zone is crucial.

Numerous experimental works have emphasize that strain localisation in shear band mode appears prior to fractures, therefore we propose to model the damaged zone by considering this mode of strain localisation. Among the existing enhanced models with regularisation methods that allow a proper modelling of the strain localisation behaviour, we use the coupled local second gradient model under unsaturated conditions. Moreover, it is known that hydraulic permeability is not homogeneous in the damaged zone (Tsang et al. 2005), then its modification is considered through a dependency with shear band properties.

B. Pardoen (✉) · S. Levasseur · F. Collin
ArGEnCo Department, University of Liège, 4000 Liège, Belgium
e-mail: b.pardoen@ulg.ac.be

K.-T. Chau and J. Zhao (eds.), *Bifurcation and Degradation of Geomaterials in the New Millennium*, Springer Series in Geomechanics and Geoengineering,
DOI 10.1007/978-3-319-13506-9_41

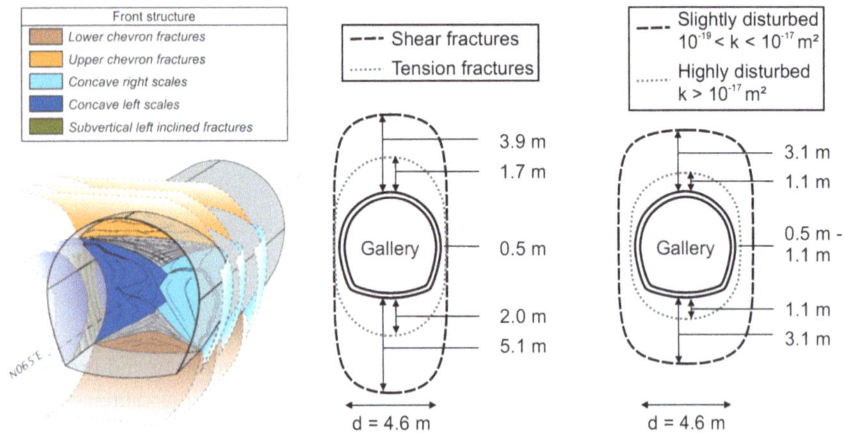

Fig. 1 In situ fractures and permeability in Callovo-Oxfordian claystone (Armand et al. 2014)

2 Coupled Local Second Gradient Model

Within the framework of classical finite elements, the strain localisation depends on the mesh size and orientation (Collin et al. 2009). An enhanced model, introducing an internal length scale, is thus needed to correctly model the post peak and localisation behaviour. Among the different regularization methods, the second gradient local model (Chambon et al. 1998) is used. In the latter, the continuum is enriched with microstructure effects: the kinematics includes the classical ones (macro) and the microkinematics (Mindlin 1964). For coupled second gradient model, the two balance equations to be solved (Collin et al. 2006), for every kinematically admissible virtual displacement field u_i^* and virtual pore water pressure field p_w^*, read in a weak form:

$$\int_{\Omega} \left(\sigma_{ij} \frac{\partial u_i^*}{\partial x_j} + \Sigma_{ijk} \frac{\partial^2 u_i^*}{\partial x_j \partial x_k} \right) d\Omega = \int_{\Gamma_\sigma} \left(\bar{t}_i u_i^* + \bar{T}_i D u_i^* \right) d\Gamma \tag{1}$$

$$\int_{\Omega} \left(\dot{M} p_w^* - m_i \frac{\partial p_w^*}{\partial x_i} \right) d\Omega = \int_{\Omega} Q p_w^* d\Omega + \int_{\Gamma_q} \bar{q} p_w^* d\Gamma \tag{2}$$

where σ_{ij} is the total stress field, Σ_{ijk} is the double stress dual of the (micro) second gradient, which needs an additional constitutive law, and Ω denotes the current solid configuration (volume). t_i is the external traction (classical) forces per unit area, T_i is an additional external (double) force per unit area, both applied on a part Γ_σ of the boundary of Ω and Du_i is u_i normal derivative. Further, M is the time derivative of the water mass inside Ω, m_i is the mass flow, Q is a sink term and Γ_q is the part of the boundary where the input water mass per unit area $\bar{q}$ is prescribed.

The total stress field is defined according to Bishop's postulate, corresponding to Biot's definition and taking into account partial saturation: $\sigma_{ij} = \sigma'_{ij} - b\, S_{r,w}\, p_w\, \delta_{ij}$ where σ'_{ij} is the effective stress, b is Biot's coefficient, $S_{r,w}$ is the water degree of saturation and δ_{ij} is the Kronecker symbol.

3 Constitutive Models

The constitutive mechanical law used for the clayey rock is an elastoplastic model with a Drucker-Prager yield surface. It includes friction angle hardening and cohesion softening as a function of the Von Mises equivalent plastic strain. The second gradient law gives Σ_{ijk} as a function of the (micro) second gradient. It is a linear elastic law with isotropic linear relationship deduced from Mindlin (1964) and it depends only on one elastic parameter D. The shear band width is proportional to this elastic parameter (Chambon et al. 1998) and Σ_{ijk} has no link with p_w.

A flow model is used to reproduce water transfer in partially saturated porous media. The advection of the liquid phase is modelled by Darcy's flow. The retention and the water relative permeability curves are given by van Genuchten's model (van Genuchten 1980). Because hydraulic properties are not homogeneous in the damaged zone, a modification of the intrinsic hydraulic permeability tensor k_{ij} is considered through a dependency with a mechanical parameter. Considering an exponential formulation depending on the porosity n (Chavant and Fernandes 2005) or on the total equivalent strain ε_{eq}:

$$k_{ij} = k_{ij,0}\left(1 + \alpha(n - n_0)^{\beta}\right) \qquad or \qquad k_{ij} = k_{ij,0}\left(1 + \alpha(\varepsilon_{eq} - \varepsilon^{t}_{eq})^{\beta}\right) \tag{3}$$

where $k_{ij,0}$ is the initial intrinsic hydraulic permeability tensor, α and β are two coefficients of the permeability evolution, n_0 is the initial porosity, ε_{eq} is the total equivalent strain $\varepsilon_{eq} = 1.5\sqrt{\hat{\varepsilon}_{ij}\hat{\varepsilon}_{ij}}$, $\hat{\varepsilon}_{ij}$ is the total deviatoric strain $\hat{\varepsilon}_{ij} = \varepsilon_{ij} - (\varepsilon_{kk}/3)\,\delta_{ij}$, ε^{t}_{eq} is a threshold value from which the permeability increases, below this value no permeability change occurs. The mechanical and hydraulic parameters for the Callovo-Oxfordian claystone are detailed in the Table 1.

4 Numerical Modelling and Results

The major issue of the simulations is the prediction of the damaged zone behaviour and fracturing structure that develops due to excavation. A hydro-mechanical modelling of a gallery excavation is performed in two-dimensional plane strain state. It takes into account the hydraulic permeability anisotropy and the initial anisotropic stress state corresponding to a gallery of the Andra URL drilled in Callovo-Oxfordian claystone and oriented in the direction of the minor horizontal principal stress. A schematic representation of the model, the mesh, the initial conditions and the boundary conditions is detailed in Fig. 2.

Table 1 Mechanical and hydraulic parameters for the Callovo-Oxfordian claystone

Parameter	Symbol	Name	Value	Unit
Mechanical	E	Young's modulus	4,000	MPa
	ν	Poisson's ratio	0.3	–
	b	Biot's coefficient	0.6	–
	ψ	Dilatancy angle	0.5	°
	φ_0	Initial friction angle	10	°
	φ_f	Final friction angle	20	°
	c_0	Initial cohesion	3	MPa
	c_f	Final cohesion	0.3	MPa
	D	Second gradient elastic modulus	5,000	N
Hydraulic	$k_{xx,0}$	Initial horizontal intrinsic water permeability	4×10^{-20}	m^2
	$k_{yy,0}$	Initial vertical intrinsic water permeability	1.33×10^{-20}	m^2
	n_0	Initial porosity	0.18	–
	M	van Genuchten coefficient	0.33	–
	N	van Genuchten coefficient	1.49	–
	P_r	van Genuchten air entry pressure	15	MPa

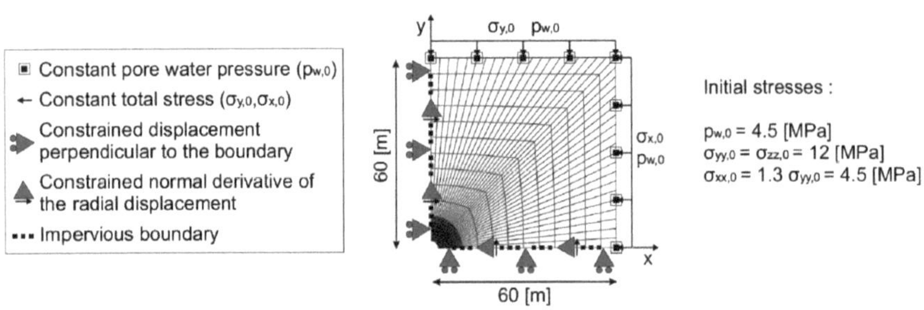

Fig. 2 Representation of the model used for the modelling of the gallery excavation

During the excavation, which lasts 5 days, the radial stresses and the pore water pressure at gallery wall decrease from their initial values to one atmospheric pressure. Then, gallery ventilation is performed. The air inside the gallery is considered to have a relative humidity of 80 %, which corresponds by Kelvin's law to a suction of $p_w = -30.7$ MPa. This constant suction is applied on the gallery wall after the excavation.

The results provide information about the damaged zone extension, structure and behaviour. A chevron fracture pattern develops during the drilling (Fig. 3) with an extension in the rock mass corresponding to in situ experimental measurements of shear fractures (Fig. 1).

Following Eq. 3, a cubic evolution of intrinsic permeability with porosity is considered ($\alpha = 2 \times 10^{12}$, $\beta = 3$). In Fig. 4, one can observe that the porosity increases close to the gallery. The horizontal intrinsic permeability evolution along the x and y-axis indicates that even if this increase is higher in the shear bands, it remains quite diffuse in the gallery vicinity because porosity is linked to volumetric strain.

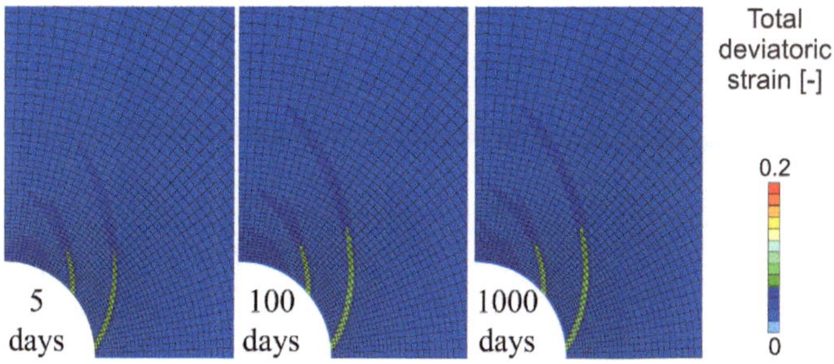

Fig. 3 Evolution of strain localisation bands pattern after the excavation

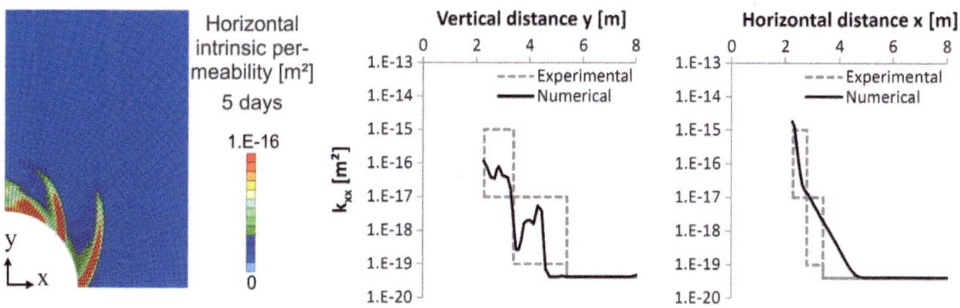

Fig. 4 Evolution of horizontal intrinsic permeability with porosity at the end of excavation (*left*) 2D view, (*middle*) vertical cross-section on y-axis, (*right*) horizontal cross section on x-axis

To better represent the permeability increase in the fractures represented by shear bands, a cubic evolution with the total equivalent strain can be considered (Eq. 3, $\alpha = 2 \times 10^8$, $\beta = 3$). In fact, this evolution only considers a dependence with the deviatoric strain, i.e. shear strain, and the threshold value $\varepsilon_{eq}^{t} = 0.01$ allows to restrict the permeability increase in the shear bands as indicates in Fig. 5. Furthermore, the in situ experimental measures of permeability are well reproduced (Figs. 1, 4 and 5). The damaged zone desaturation, due to gallery ventilation and

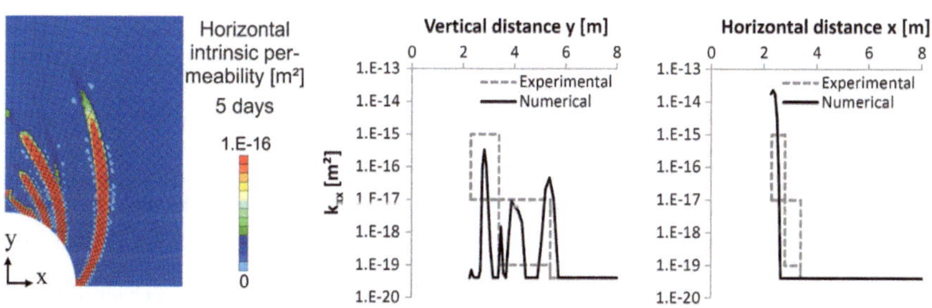

Fig. 5 Evolution of horizontal intrinsic permeability with total equivalent strain at the end of excavation (*left*) 2D view, (*middle*) vertical cross-section on y-axis, (*right*) horizontal cross section on x-axis

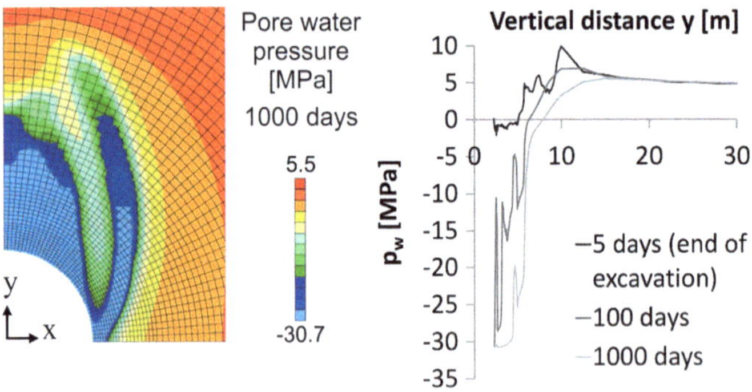

Fig. 6 Pore water pressure evolution when permeability depends on total equivalent strain (*left*) 2D view after 1,000 days of ventilation, (*right*) vertical cross-section on y-axis

rock-atmosphere interaction, is also well reproduced (Fig. 6) thanks to the permeability increase in the fractures. The results along the y-axis in Fig. 6 highlight a strong desaturation, which extends a few meters in the rock, its evolution during time and the influence of the shear strain localisation bands on the pore water pressure.

5 Conclusion

The excavation damaged zone around a gallery and its behaviour is successfully reproduced. First, the fractures are properly represented using shear strain localisation and the coupled local second gradient model. Second, the permeability increase within the fractures corresponds to in situ measurements. Finally, the rock desaturation resulting from gallery ventilation is well reproduced.

References

Armand G, Leveau F, Nussbaum C, de La Vaissiere R, Noiret A, Jaeggi D, Landrein P, Righini C (2014) Geometry and properties of the excavation-induced fractures at the Meuse/Haute-Marne URL drifts. Rock Mech Rock Eng 47(1):21–41

Chambon R, Caillerie D, Hassan NE (1998) One-dimensional localisation studied with a second grade model. Eur J Mech A-Solid 17(4):637–656

Chavant C, Fernandes R (2005) Evaluating the reliability of hydro-mechanical simulation: a benchmark of numerical techniques carried out by Research Group of MoMas. In: Second international meeting clays in natural & engineered barriers for radioactive waste confinement, Tours, 14–18 March 2005, pp 249–250

Collin F, Chambon R, Charlier R (2006) A finite element method for poro mechanical modelling of geotechnical problems using local second gradient models. Int J Numer Meth Eng 65(11):1749–1772

Collin F, Levasseur S, Chambon R (2009) Numerical post failure methods in multiphysical problems. Eur J Environ Civ Eng 13(7–8):983–1004

Mindlin RD (1964) Micro-structure in linear elasticity. Arch Ration Mech Anal 16(1):51–78
Tsang CF, Bernier F, Davies C (2005) Geohydromechanical processes in the excavation damaged zone in crystalline rock, rock salt, and indurated and plastic clays—in the context of radioactive waste disposal. Int J Rock Mech Min 42(1):109–125
van Genuchten MT (1980) A closed-form equation for predicting the hydraulic conductivity of unsaturated soils. Soil Sci Soc Am J 44(5):892–898

Experimental Study on the Behavior of a Plasticine Material

Wei-Qiang Feng, Fei Tong, Xiao-Ming Tao and Jian-Hua Yin

Abstract This paper presents an experimental study, which includes the oedometer test and isotropic creep test, on the compressibility and the creep effect of the plasticine materials. The Compression Index ($C_{c\varepsilon}$), Rebounding Index ($C_{r\varepsilon}$) and creep coefficient ($C_{\alpha\varepsilon}$) are obtained from the test results. However, it can be observed that there is almost no loop in unloading-reloading stages of stress-strain relationship, which is different from most soil test results. By means of the Scanning Electron Microscope, it could be found that the compression of the plasticine is mainly due to the porosity decrease and the structure flocculence, which is nonrecoverable.

1 Introduction

Plasticine is a material for analog modeling, which was studied by many researchers since it is widely used in modeling geological structures (Zulauf and Zulauf 2004). The compressional and rheological behaviors were examined by uniaxial compression test and constant strain rate test (McClay 1976; Sofuoflu and Rasty 2000). And the thermomechanical behavior and deformation characteristic were also investigated (Ji et al. 2009). However, two dimensional tests of plasticine are

W.-Q. Feng (✉) · J.-H. Yin
Department of Civil and Environmental Engineering, The Hong Kong Polytechnic University, Hung Hom, Kowloon, Hong Kong
e-mail: fengweiqiang2015@gmail.com

F. Tong
College of Architecture and Civil Engineering, Taiyuan University of Technology, Taiyuan, China

X.-M. Tao
Department of Institutive of Textile and Clothing, The Hong Kong Polytechnic University, Hung Hom, Kowloon, Hong Kong

K.-T. Chau and J. Zhao (eds.), *Bifurcation and Degradation of Geomaterials in the New Millennium*, Springer Series in Geomechanics and Geoengineering,
DOI 10.1007/978-3-319-13506-9_42

limited and the loading-unloading behavior of plasticine is not examined. This paper proposes to present the loading, unloading-reloading behavior of the plasticine by a series of laboratory tests.

2 The Material and Test

The plasticine was plate shaped with a green color, and the density is 1.525 g/cm^3. The pore size of the plasticine is in the range of 10–50 μm, and the structure of the plastine is agglutinate and flocculence at the micro-scale by means of SEM (Scanning Electro Microscope). All the tests were carried out under a constant temperature (20 ± 1 °C) in order to exclude the temperature effect.

The conventional multi-staged oedometer test was conducted with a specimen of 70 mm in diameter and 19 mm in height. The loading schedule includes two cycles of unloading-reloading process. It should be noted that all the loading was applied suddenly on the specimen and the duration was up to 7 days for each loading.

The double cell triaxial system, developed by Yin (2003), was used for isotropic creep test by applying a constant cell pressure on the specimen. The details and reliability of double cell triaxial system could be referred in Yin (2003). The measurable variables in double cell triaxial apparatus are mainly the volume change of the inner cell water, which is also regarded as the volume change of the specimen.

3 Test Results and Discussion

Figure 1 presents the relationships of log(stress) and vertical strain under loading, unloading and reloading conditions for the time t = 0.01 min and t = 64 min. It could be found that there is nearly a linear relationship between log(σ_z) and vertical strain, meanwhile, the unloading line and reloading line are nearly overlapped in the stress-strain curves, which is different from most soil oedometer test results (Yin 1999; Tong and Yin 2011). In order to quantitatively analyze the results,

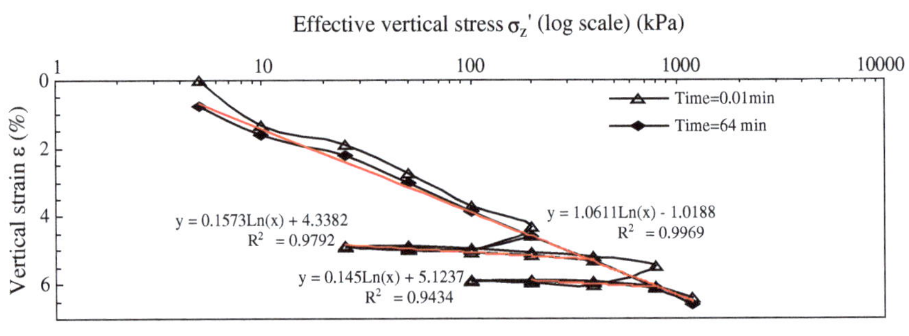

Fig. 1 Relationships of log(stress) and strain under oedometer condition

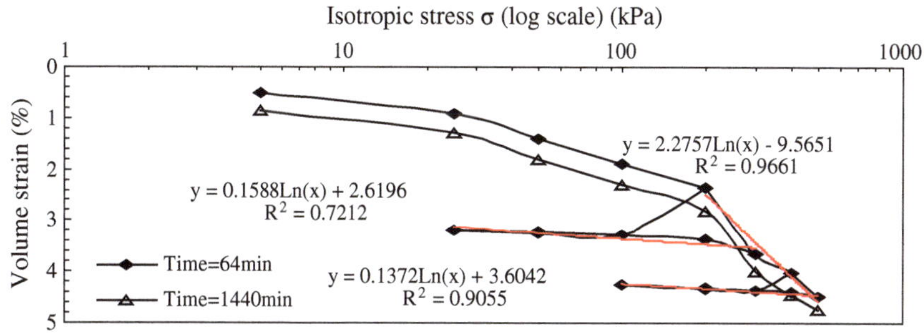

Fig. 2 Relationships of log(stress) and strain under isotropic condition

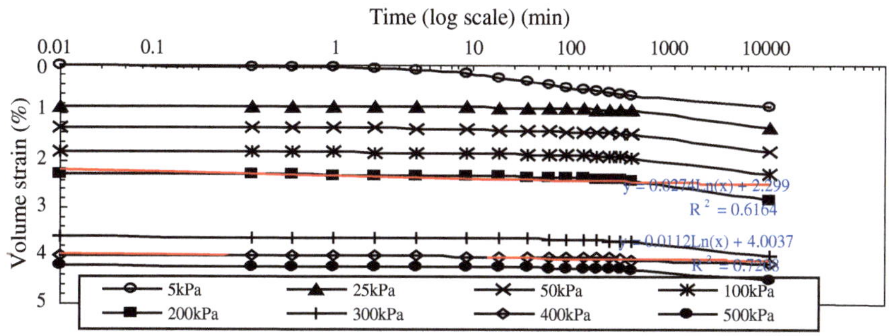

Fig. 3 Relationships of strain-log(time) under isotropic condition

Table 1 The parameters in oedometer test and isotropic creep test

Test	$C_{r\varepsilon}$	$C_{c\varepsilon}$	$C_{r\varepsilon}/C_{c\varepsilon}$	Loading (kPa)	$C_{\alpha\varepsilon}$	$C_{\alpha\varepsilon}/C_{c\varepsilon}$
Oedometer test	0.00362	0.0244	0.148	200	0.00119	0.049
	0.00334		0.137	1,200	0.00110	0.045
Isotropic test	0.00366	0.0593	0.0617	200	0.000631	0.011
	0.00316		0.0533	400	0.000258	0.004

Compression Index ($C_{c\varepsilon}$), Rebounding Index ($C_{r\varepsilon}$) can be obtained by the fittings for time = 0.01 min. And the creep coefficient ($C_{\alpha\varepsilon}$) can also be obtained by fitting the log(time)-vertical strain curves under the loading of 200 and 1,200 kPa.

Similarly, the isotropic creep test results are illustrated as Figs. 2 and 3. And the parameters were also determined by fitting the curves. All the values are listed in Table 1, and it should be noted that there is a constant ratio between fitting values (λ, K) and parameter values ($C_{c\varepsilon}$, $C_{r\varepsilon}$) since the fitting values are corresponding to the ln scale while the parameter values are related to the log scale:

$$C_{c\varepsilon} = \lambda/0.4343 \tag{1}$$

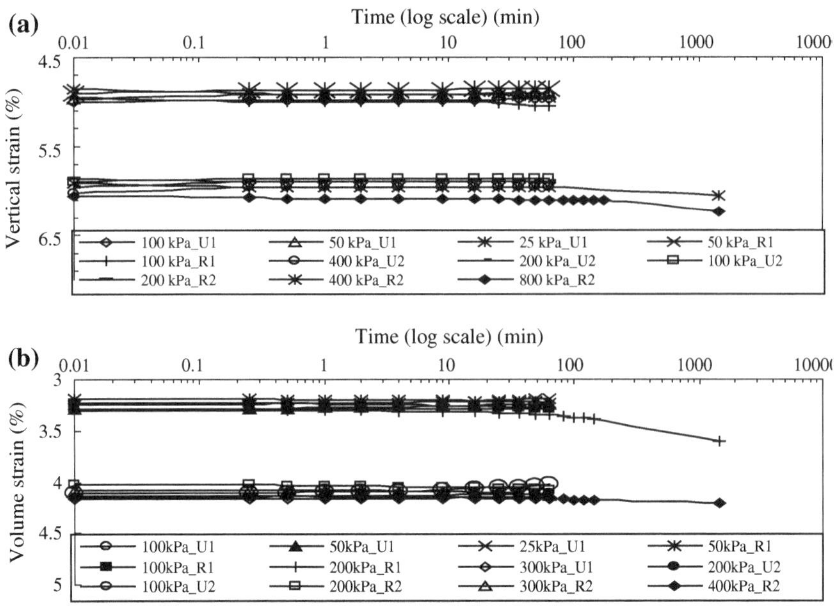

Fig. 4 The relationship of strain and log(time) in unloading-reloading stages: **a** under oedometer condition **b** under isotropic condition

Compared with the soil results of oedometer test and isotropic consolidation test, it could be noticed that the stress-strain curves of the plasticine are similar to those of soils except that the unloading-reloading stages. Meanwhile, the creep coefficient is also suitable to the plasticine under the loading stages. It indicates that these parameters of soils could be used to describe the plasticine behavior. In this part, the mechanism and reasons on the differences of unloading-reloading stages are discussed and interpolated.

Yin (1999) reported that there is a loop in the unloading-reloading stages of stress-strain curves for Hong Kong Marine Deposits, which is also mentioned by Craig (2004). Kawabe et al. (2009) explained the unloading reloading loop as the negative creep with the assumption of isotach viscous property. Yin and Tong (2011) illustrated that the unloading-reloading loop is related to the swelling behavior for the bentonite clayey soils. According to the results of the plasticine, it could be found that there are nearly horizontal lines in strain and log(time) curves, as shown in Fig. 4, which results in a line in unloading-reloading stage of Figs. 1 and 5. As a result, it is reasonable that the loop is related to the time-dependent behavior in the unloading-reloading stages.

In order to interpolate the reason of the behavior more accurately, the SEM photo of the plasticine is compared with that of the HKMD, shown as Fig. 6. It could be observed that the structure of the plasticine is different from the HKMD: The plasticine is agglutinate while the HKMD is open honeycomb with the particles. When the loading is unloaded, the agglutinate structure of the plasticine could keep stable while the open honeycomb structure of HKMD would deform

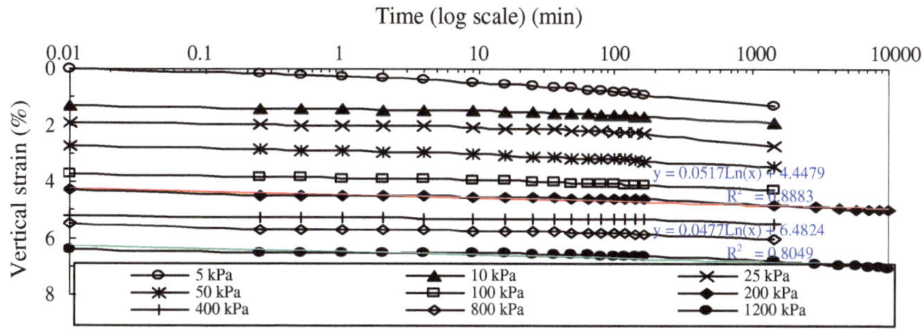

Fig. 5 Relationships of vertical strain-log(time) under oedometer condition

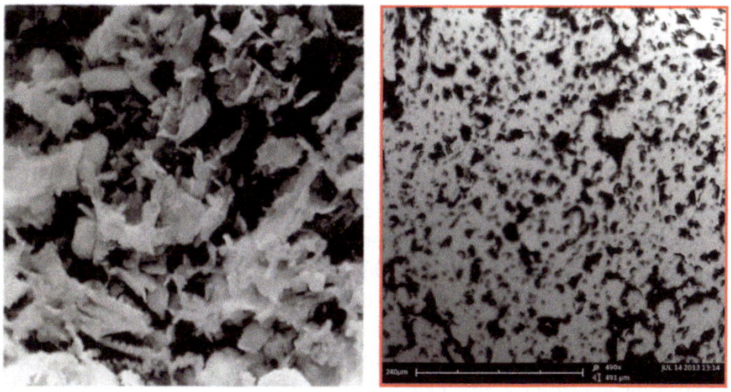

Fig. 6 The micrograph of the HKMD and the plasticine **a** the open honeycomb of HKMD (Bennett et al. 1994). **b** the agglutinate of the plasticine

gradually, therefore, the different structures should be related to the different behaviors in unloading-reloading stages.

4 Conclusions

Based on the plasticine results of oedometer test and isotropic creep test, conclusions could be drawn as following:

(a) The creep behavior of the plasticine material is nearly linear in log(time)-strain space. And the Compression Index, Rebounding Index and creep coefficient are suitable for the plasticine.
(b) There are nearly horizontal lines in the unloading-reloading stages of the stress-strain curves for the plastcine, which results in the unloading line and reloading line are overlapped in the stress-strain curves.
(c) Based on the SEM photos, it could be regarded that the structure of the plasticine is agglutinate while the HKMD is open honeycomb with the particles.

Acknowledgments The work in this paper is supported by a research grant (project No. 51278442) from National Natural Science Foundation of China (NSFC), PolyU Shenzhen Research Institute, and The Hong Kong Polytechnic University, China.

References

Bennett RH, Bryant WR, Hulbert MH (1994) Microstructure of fine-grained sediments from mud to shale. Springer-Verilag, NewYork, p 527

Craig RF (2004) Craig's soil mechanics. Spon Press, London and New York

Ji H, Robin E, Rouxel T (2009) Compressive creep and indentation behavior of plasticine between 103 and 353 K. Mech Mater 41:199–209

Kawabe S, Kongkitkul W, Hirakawa D, Tatsuoka F (2009) Rate-dependent behaviour of clay during cyclic 1D compression. In: Proceedings of 17th international conference on SMGE, Alexandria, Egypt, pp 303–307

McClay KR (1976) The rheology of plasticine. Tectonophysics 33:T7–T15

Sofuoglu H, Rasty J (2000) Flow behavior of plasticine used in physical modeling of metal forming processes. Tribol Int 33:523–529

Tong F, Yin JH (2011) Nonlinear creep and swelling behaviour of bentonite mixed with different sand contents under oedometric condition. Marine Geores Geotech 29:346–363

Yin JH (1999) Non-linear creep of soils in oedometer tests. Géotechnique 49:699–707

Yin JH (2003) A double cell triaxial system for continuous measurement of volume changes of an unsaturated or saturated soil specimen in triaxial testing. Geotech Test J 26:1–6

Yin JH, Tong F (2011) Constitutive modeling of time-dependent stress–strain behavior of saturated soils exhibiting both creep and swelling. Can Geotech J 48:1870–1885

Zulauf J, Zulauf G (2004) Rheology of plasticine used as rock analogue: the impact of temperature, composition and strain. J Struct Geology 26:725–737

Thermodynamic Analysis of Instability in Unsaturated Soils

R. Tamagnini and W. Wu

Abstract The paper deals with the description of mechanical instability in unsaturated soils. The problem is studied by applying the thermodynamics of open continua. The localization and bifurcation in plane strain condition is studied as a boundary value problem. The paper shows that classic Finite Element Method (FEM) can be successfully applied in the analysis of instability induced by the distortion of the original reference volume. The paper also shows that classic elastoplasticity can be successfully applied for the description of the continuum without the need of any length scale. The ill-posedness of the problem is resolved by introducing into rate-independent constitutive equations the effect of the enthalpy of the gas phase.

1 Introduction

The research on unsaturated soils has been focused on the collapse of porous matrix due to the variation of the water content (Jennings and Burland 1962). The conclusions drawn were that the extension of the principle of the effective stress to the case of two fluid phases was not possible for the important feature of collapse that in the researchers' opinion could not be reproduced by a single stress tensor (Jennings and Burland 1962; Alonso et al. 1990). This consideration has reduced unsaturated soils mechanics to a complex discipline. The large amount of formulated constitutive models using many different tensors is an evidence of the confusion in the field (Alonso et al. 1990; Gallipoli et al. 1962). Even though the instability due to the water content changes had a deep influence in the developed researches it has not been studied from a theoretical point of view

R. Tamagnini (✉) · W. Wu
Universität Für Bodenkultur, Institut Für Geotechnik (IGT),
Feistmantelstraße 4A, 1180 Vienna, Austria
e-mail: roberto.tamagnini@boku.ac.at

K.-T. Chau and J. Zhao (eds.), *Bifurcation and Degradation of Geomaterials in the New Millennium*, Springer Series in Geomechanics and Geoengineering,
DOI 10.1007/978-3-319-13506-9_43

but only experimentally. Few exceptions can be found in literature (Schrefler et al. 2006; Borja et al. 2013). On the other hand, the research on instability of fully saturated granular materials has been approached transposing the work of Hill (1958) to multiphase materials without any critical consideration on the thermodynamic nature of the studied system. In fact Hill's criterion was obtained for metals in which there is no fluid filtration. Rudnicki and Rice (1975) concluded that an appropriate constitutive equation for instability should be based on non-associate flow rule. This idea can be found also in recent works on unsaturated soils (Buscarnera and di Prisco 2012; Lignon el al. 2009). The weakness of these scientific productions, which are based on the analysis of the second order work, consists in the identification of the first order work. In fact the expression of the internal energy for an open system is not the same expression of the internal energy of a closed system. Particularly, granular materials are characterised by the presence of fluids into the porous network that can migrate. These fluids can move and can change phase during shearing even in undrained conditions (Mokni and Desrues 1999). The disregard of the enthalpy in the analysis of instability of granular materials with fluid-flow led to an increasing complexity of the mathematical models. The paper shows that mechanical instability of unsaturated soils can be modelled by a classic rate-independent Cam-clay model by introducing the enthalpy contribution. This feature allows modelling "undrained" biaxial test avoiding mesh dependency and without the introduction of any length scale.

2 Thermodynamics

In a closed system the work performed by a pressure p on a system that could be a mixture of gas, water and a solid matrix composed of incompressible grains can be defined at the macroscopic or local level as:

$$\delta W_{shaft} = -pdV \quad \text{or} \quad \delta W_{shaft} = \sigma_{ij} d\varepsilon_{ij} \tag{1}$$

where V is the volume of the closed system, σ_{ij} is the total stress tensor and ε_{ij} is the strain tensor. Note that the Eq. (1) is true only if the system is perfectly sealed and the fluids can not escape from the reference volume. If the system is not perfectly closed, the internal energy of the reference volume is the sum of the work performed by the pressure p (that is called the shaft work), plus the work performed on the system and performed by the system to make room for the new entering mass (of fluids) and to push out the leaving mass. The internal energy U of an open system is defined as:

$$\begin{aligned} \dot{U} - \left(U_g \frac{1}{V_g} \frac{\partial V_g}{\partial t} + U_w \frac{1}{V_w} \frac{\partial V_w}{\partial t} \right) &= \dot{W}_{shaft} - p_g V_g \frac{1}{V_g} \frac{\partial V_g}{\partial t} \\ &\quad - p_w V_w \frac{1}{V_w} \frac{\partial V_w}{\partial t} + \dot{Q} \end{aligned} \tag{2}$$

in which the subscript g and w indicates the gas and the fluid phase respectively; $\partial V_\alpha/\partial t$ is the gross amount of the fluid α that is stored or expelled from the elementary volume and Q is the sum of the total amount of heat produced inside the volume and the heat due to the mass exchange. The Eq. (2) can be rearranged as:

$$dU = \delta W_{shaft} + G_g \nabla m_g + G_w \nabla m_w + \delta Q_{in} \tag{3}$$

in which G_α and m_α are respectively the Gibbs free energy and the mass of the fluid α. The Eq. (3) can be found in Coussy (1995). Note the Eq. (3) differs from the closed system expression for the two terms related to the fluid phases. These energies represent the work done to make room for the fluid α in the elementary volume. The mass and Gibbs free energy of the fluids can be expressed at the local level as:

$$m_\alpha = \rho_\alpha S_\alpha \phi \quad \text{and} \quad G_g = \frac{p_\alpha}{\rho_\alpha} + \psi_\alpha \tag{4}$$

in which ρ_α, S_α and ψ_α are respectively the density, the degree of saturation and the free energy of the fluid α. ϕ is the porosity. Substitution of Eqs. (1) and (4) in Eq. (3) leads to the expression:

$$\begin{aligned} dU = {} & \left\{\sigma_{ij} + \left(S_w p_w + S_g p_g\right)\delta_{ij}\right\} d\varepsilon_{ij} + \left(\rho_g S_g \psi_g\right)\nabla\phi \\ & + \phi\left\{\left(p_g - p_w\right) - \rho_g \psi_g + \rho_w \psi_w\right\}\nabla S_w + \delta Q_{in} \end{aligned} \tag{5}$$

in which the first term is the Bishop's stress as defined in reference (Lewis and Schrefler 1998), the third term is related to the retention properties of the soils, the water retention curve and the second term is the variation of the gas phase free energy due to the variation in the porosity. This term is not accounted for in the second order work analyses of unsaturated soils (Buscarnera and di Prisco 2012; Lignon et al. 2009). The variation of this second term can be expressed using the ideal gas law.

$$p_g = \frac{nRT}{V_g} \tag{6}$$

where n is the number of moles, R is the gas constant and T is the temperature. At constant temperature the variation of the free energy of the gas can be expressed as:

$$\rho_g S_g \psi_g = \int_V \rho_g \frac{\partial \psi_g}{\partial \rho_g} d\rho_g = \int_V nRT d\rho_g \tag{7}$$

Even in undrained condition and in the hypothesis of incompressible grains the second term in Eq. (5) can change due to the compressibility of the gas phase. This result in a second gradient term that can be expressed in the form:

$$\begin{aligned} dU = {} & \left\{\sigma_{ij} + \left(S_w p_w + S_g p_g\right)\delta_{ij}\right\} d\varepsilon_{ij} + \int_V nRT \nabla\phi d\rho_g \\ & + \phi\left\{\left(p_g - p_w\right) - \rho_g \psi_g + \rho_w \psi_w\right\} dS_w + \delta Q_{in} \end{aligned} \tag{8}$$

3 Numerical Analyses

The numerical analyses are performed with the model by Tamagnini (2004) and its implementation in the Finite Element (FE) code Abaqus. The analysed sample is modelled with 3 different meshes and the refinement is obtained doubling the number of the elements (see Fig. 1).

The numerical analyses simulate a stress path that is composed by a drained isotropic compression till the isotropic total stress of 2.5 MPa (the initial preconsolidation pressure of the sample is equal to 4.8 MPa) followed by a shearing at quasi-constant volume. The shearing is obtained with a strain driven step that preserves the original area of the sample with a ramped application of the displacements. Figure 1 reports the localization of the strains into the meshes. The analyses are performed in two ways. The first set of analyses is performed with the original Modified Cam clay. The ill-posedness of the problem is shown in Fig. 2,

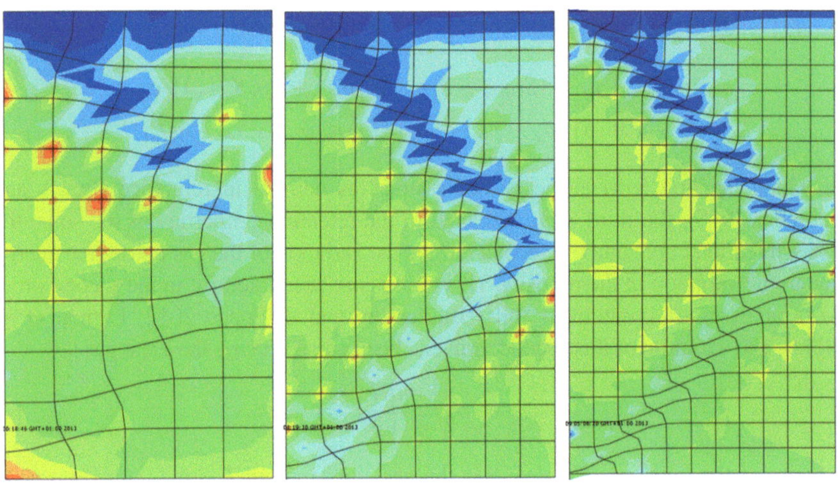

Fig. 1 The shear test with the Tamagnini model for the 3 different mesh refinements

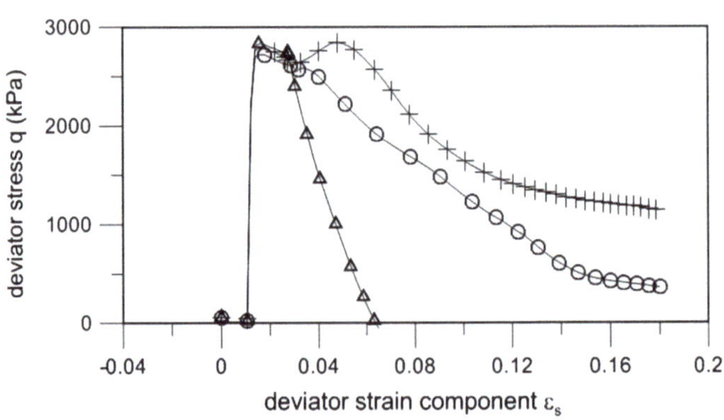

Fig. 2 Shear tests with the original modified cam clay (mesh dependency)

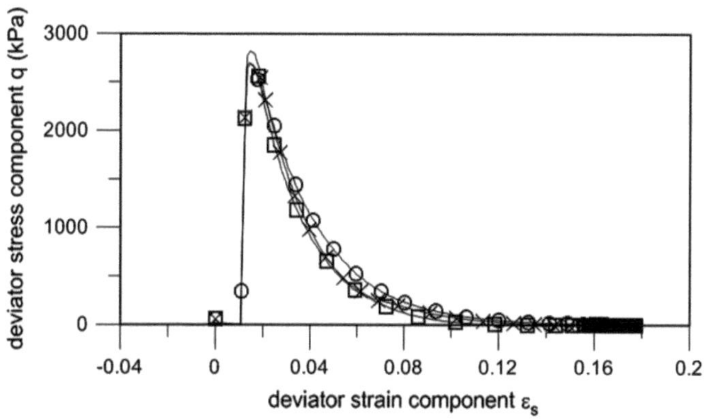

Fig. 3 Shear tests with the Tamagnini model (well posed problem)

in which the three curves correspond to the three different meshes. The effect of the mesh size has not physical meaning. The second set of analyses is performed with the model by Tamagnini (2004) that accounts for the gas effects in the modified hardening law. Figure 3 shows that the mesh dependency is not present in the finite element analyses results. The contours of the deviator stress component obtained with the model by Tamagnini (2004) are reported in Fig. 1 and they shows that the geometrical localization of the shear bands is well captured for each of the three refinements. In the second set of analyses (the ones with the model by Tamagnini (2004)) the work performed by the gas enthalpy is obtained allowing for a small variation of the fluid pressure at the boundary.

4 Conclusions

The paper has presented a preliminary study on the localization and bifurcation in unsaturated granular materials. The analysis has been performed applying the thermodynamics of open continua. The results have shown that the correct identification of the thermodynamic boundary conditions implies a simpler formulation of the local and discretised continuum.

Acknowledgments The paper is part of a larger research on the analyses of landslides and the role of vegetation and evapotranspiration, funded by the European Commission in the 7th Framework Program for Industry-Academia Partnerships and Pathways FP7-PEOPLE-2012-IAPP Grant no: 324466.

References

Alonso E, Gens A, Josa A (1990) Constitutive model for partially saturated soils. Geotechnique 40(3):405–430

Borja RI, Song X, Wu W (2013) Critical state plasticity. Part VII: triggering a shear band in variably saturated porous media. Comput Meth Appl Mech Eng 261–262:66–82

Buscarnera G, di Prisco C (2012) Discussing the definition of second-order work in unsaturated soils. Int J Numer Anal Meth Geomech 36(1):36–49

Coussy O (1995) Mechanics of porous continua. Wiley, England

Gallipoli D, Gens A, Sharma R, Vaunat J (1962) An elasto-plastic model for unsaturated soil incorporating the effects of suction and degree of saturation on mechanical behaviour. Geotechnique 53(1):123–136

Hill R (1958) General theory of uniqueness and stability in elastic-plastic solids. J Mech Phys Solids 6:236–249

Jennings JEB, Burland JB (1962) Limitations of the use of the effective stress in partly saturated soils. Geotechnique 12(2):125–144

Lewis RW, Schrefler BA (1998) The finite element method in the static and dynamic deformation and consolidation of porous media. Wiley, England

Lignon S, Laouafa F, Prunier F, Khoa HDV, Darve F (2009) Hydro-mechanical modelling of landslides with a material instability criterion. Geotechnique 59(6):513–524

Mokni M, Desrues J (1999) Strain localisation measurements in undrained plane-strain biaxial tests on Hostun RF sand. Mech Cohesive-frictional Mater 4:419–441

Rudnicki JW, Rice JR (1975) Conditions for the localization of deformation in pressure-sensitive dilatant materials. J Mech Phys Solids 23:371–394

Schrefler BA, Zhang HW, Sanavia L (2006) Interaction between internal length scales in strain localization analysis of fully and partially saturated porous media—the 1-D case. Int J Numer Anal Meth Geomech 30:45–70

Tamagnini R (2004) An extended Cam-clay model for un-saturated soils with hydraulic hysteresis. Geotechnique 54(3):223–228

Plastic Deformations in Granular Materials with Rotation of Principal Stress Axes

Richard Wan and Nejib Hadda

Abstract The paper examines the nature of plastic flow in granular materials under loading that involves rotation of principal stress axes. Examples in engineering practice involve offshore foundations under wave loading, earthquake loading, pavement and railway track foundations, among others. Micro-scale investigations through discrete element modelling (DEM) with a focus to non-coaxiality of incremental plastic strains to stress tensors are presented. Results can enter constitutive models for numerical applications through an amended plasticity flow rule.

1 Introduction

The effect of principal stress rotation on the behaviour of granular media is known to induce plastic strains, despite the principal stresses (hence stress invariants) are kept constant as demonstrated on sand in simple shear tests (Roscoe 1970). By contrast, classic plasticity flow rule predicts zero plastic deformations under such a particular but pertinent loading process. This is strongly linked to micro-structural changes in fabric as well as sliding characteristics during loading history, whereas continuum plasticity theories are devoid of any micro-scale information. This study describes a systematic micro-mechanical investigation of granular media deformations under a rotation of principal stress directions so that findings can be incorporated into a micro-structurally enriched flow rule which accounts for such effects.

The issue of principal stress rotation on sand behaviour has been long studied in simple shear and torsional shear (hollow cylinder) tests, see Cai (2010) and Roscoe (1970), among others. The principal strain increment direction and

R. Wan (✉) · N. Hadda
University of Calgary, Calgary, Canada
e-mail: wan@ucalgary.ca

N. Hadda
e-mail: nhadda@ucalgary.ca

K.-T. Chau and J. Zhao (eds.), *Bifurcation and Degradation of Geomaterials in the New Millennium*, Springer Series in Geomechanics and Geoengineering,
DOI 10.1007/978-3-319-13506-9_44

principal stress direction are clearly different and this phenomenon is coined as non-coaxial behaviour (Yu and Yuan 2006). The paper examines the deformational response of a granular assembly subjected to various magnitudes of rotation of principal stress directions through discrete element modelling (DEM). A series of numerical experiments of pure stress rotations have been conducted on a DEM sample, herein considered as a representative element volume (REV). The plastic strain response is thereafter studied with respect to non-coaxial behaviour.

2 Discrete Element Results and Discussions

In order to perform principal stress rotation, a 3-D discrete element (Cundall and Strack 1979) specimen is considered using the open-source software YADE (Smilauer et al. 2010). The model consists of a granular assembly made of 10,000 spherical particles following a uniform size distribution enclosed within six frictionless rigid walls. Frictionless walls are considered so that non-coaxiality is assumed to be totally produced from the rotation of principal stresses.

Particle diameters range from 4 to 16 mm. The contact law used to govern the inter-particle interaction within the granular assembly is noncohesive and involves three mechanical parameters; the normal contact stiffness k_n, the tangential contact stiffness k_t and the friction coefficient μ which is incorporated at the contact level by the Coulomb friction law. These parameters are introduced such that nearly rigid particles are considered with $k_n/D = 356$ MPa, $k_t/k_n = 0.42$ and $\mu = 0.7$, where D is the mean diameter of the two particles in contact.

In the first stage, following the isotropic compression of the specimen at a confining pressure $\sigma_3 = 100$ kPa, a classic axi-symmetric drained triaxial test was performed by applying a strain rate $\dot{\varepsilon}_1 = 0.01\ \mathrm{s}^{-1}$ in the axial direction and using a servo-control in the lateral directions to maintain a constant confining pressure. The difference between the axial stress σ_1 and the lateral (confining) stress σ_3 gives rise to the deviatoric stress q.

The model shows a typical behaviour of a dense dry sand as presented in Fig. 1 where both deviatoric stress q and volumetric strain ε_v are plotted in terms of the axial strain ε_1. The deviatoric stress q passes through a peak with concomitant dilation along the triaxial compression as observed from the variation of ε_v.

In the second stage, for a given initial principal stress state, various principal stress rotations were carried out by sweeping the 3-D space following Euler angles θ and φ as illustrated in Fig. 2. In order to evaluate the influence of the initial stress level on the material response under such rotations of principal stresses, two different stress-strain states were chosen at stress ratios $\eta = q/p = 0.63$ and 0.79 with p being the mean stress within the specimen; see Fig. 1. As shown in Fig. 2, starting from a chosen stress state $\mathbf{S}^{\mathrm{R}}$ with principal axes rotated by Euler angles θ and φ, the stress tensor $\mathbf{S}$ in the fixed reference frame (x_1, x_2, x_3) is readily found knowing $\mathbf{R}$, the rotation matrix obtained as a product of two rotation matrices $\mathbf{R}_\theta$ and $\mathbf{R}_\varphi$ about two different principal axes. The angles θ and φ were set to vary

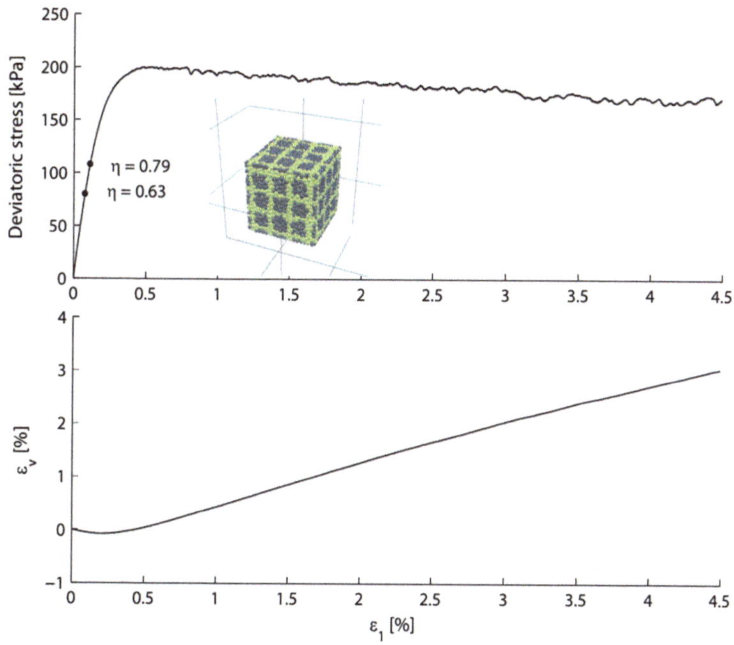

Fig. 1 Deviatoric stress q and volumetric strain ε_v in terms of the axial strain ε_1 at a confining pressure of 100 kPa

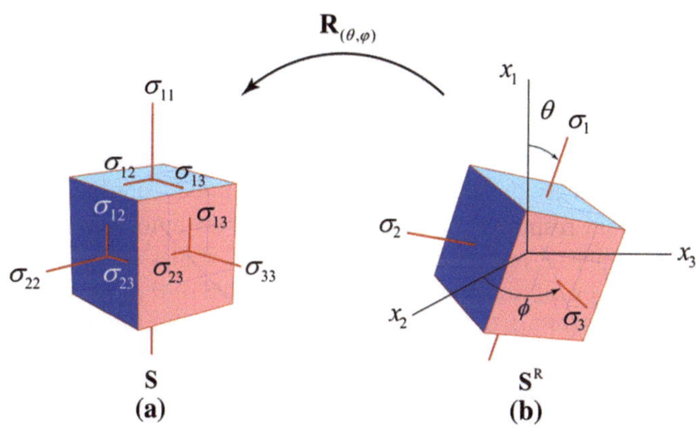

Fig. 2 Pure stress rotation on DEM specimen (REV) parametrized by Euler angles φ and θ

every 10° over the ranges [0, 90°] and [0, 360°] respectively in order to sweep all possible rotations in 3-D space. Except for the isotropic stress case as initial state, any rotation of principal stresses gives rise to the appearance of shear stress components in the stress tensor **S** expressed in the fixed reference frame (x_1, x_2, x_3). As such a principal stress rotation induces a stress variation $\Delta\sigma$ and $\Delta\tau$ along the normal and shear directions respectively from initial stress values.

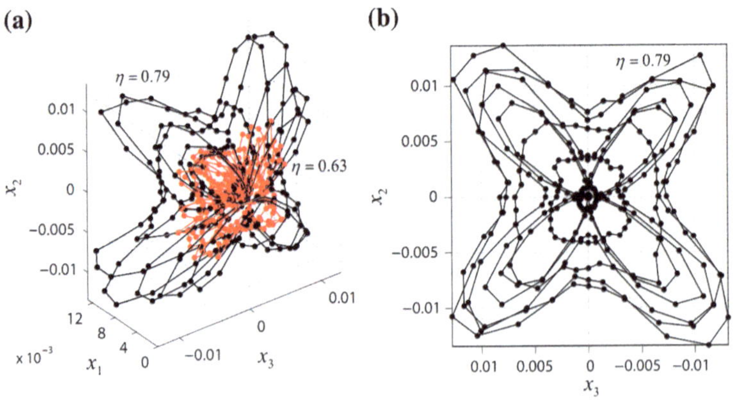

Fig. 3 Plastic shear strain $||\Delta\gamma^p||$ resulting from principal stresses rotation performed on the stress states defined by $\eta = 0.79$ (*black*) and $\eta = 0.63$ (*red*): **a** 3D view, **b** projected view

The normal stress loading can be performed numerically using either a stress or a strain control. However, since both normal and shear stresses are to be incremented simultaneously, a stress control (servo-control) is considered in this study to ensure the proper transformation of the initial principal stress components to the rotated ones. For the shear loading, a strain control is considered by applying a shear strain in each direction. Each pair of parallel walls are rotated along a different principal direction with a different rotation velocity in such a way that the shear stresses reach the targeted shear components required by the rotated stress tensor at almost the same time together with the normal stresses. Once the rotated stress state is reached, a force equilibrium stabilization is performed.

Figure 3 shows the directional plastic response under all possible principal stresses rotations for the two stress ratio values. Basically, each stress rotation is characterized by angles φ, and θ with the norm of the plastic strain increment $||\Delta\gamma^p||$ being plotted as the radius r in spherical polar coordinates. As anticipated, plastic strain response increases with the stress ratio level η. In fact, the contours refer to plastic responses for fixed rotation angles θ revealing two main φ-directions of major plastic shear strains irrespective of θ as clearly shown in a 2-D projected view in Fig. 3b. This indicates the existence of two specific orientations for which the deviation between the principal fabric (initial and induced) and principal stress directions is maximum so as to cause the most sliding between particles. This issue seems also to be connected to the rate of change of fabric with respect to rate at which shear stresses are being mobilized and distributed within the specimen. This interpretation, albeit crude and speculative, can be also be made from Fig. 4 where the norm of the plastic shear strain $||\Delta\gamma^p||$ is plotted against both the angles θ and φ. Besides the clear evidence of preferred values of rotations giving major plastic deformations, it is seen that the plastic shear strain response becomes more prominent for θ-angles around 50°.

It appears that the various observations made in the above suggest that the plastic response under principal stress rotation rest on microstructural aspects

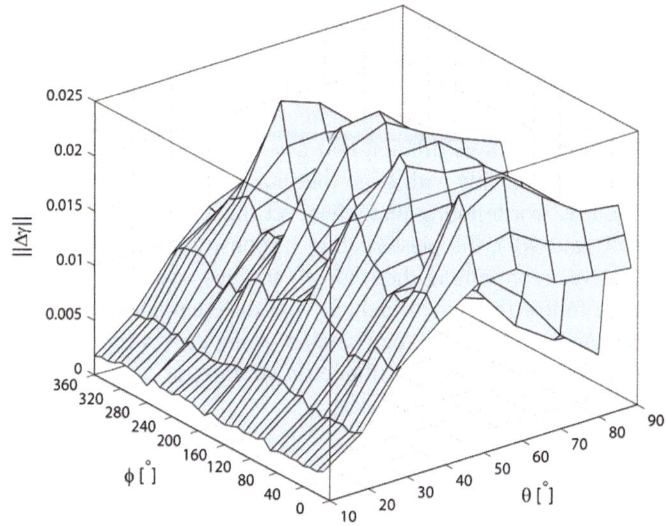

Fig. 4 Plastic shear strain $||\Delta\gamma^p||$ in terms of θ and φ resulting from principal stresses rotation performed on the stress states defined by $\eta = 0.79$

relative to principal stress directions. Thus, we next analyze the DEM results in the special 2-D case by setting $\varphi = 0°$ and $50°$ and restricting principal stress rotations to occur only about the vertical axis through various angles θ.

Figure 5 shows respectively the evolutions of plastic shear strain $||\Delta\gamma^p||$, volumetric strain ε_v, and cosine of the angle of non-coaxiality (α) with principal stresses rotations θ performed at the stress state defined by $\eta = 0.79$. The angle α

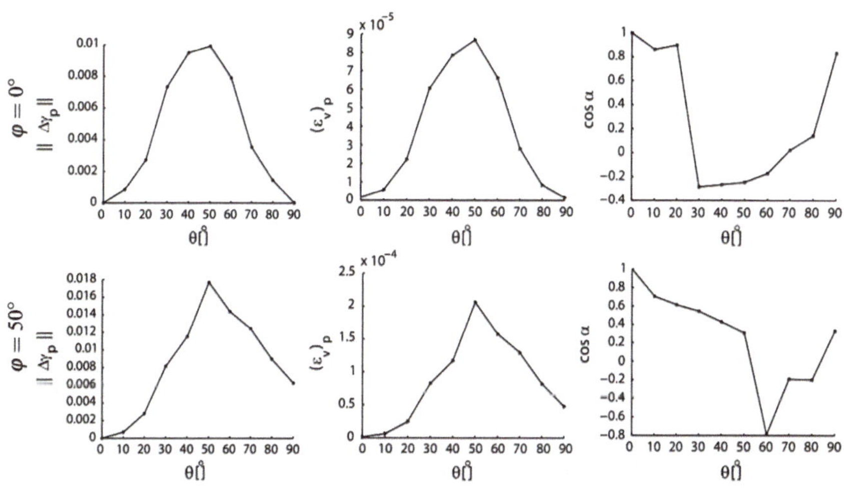

Fig. 5 Plastic shear strain $||\Delta\gamma^p||$, volumetric strain ε_v and cosine of the angle of non-coaxiality (α) in terms of θ for $\varphi = 0$ and $50°$ respectively

denotes the non-coaxiality between the principal stress directions and those of the principal strains.

It turns out that for both $\varphi = 0$ and $50°$, the maximum plastic shear strain response corresponds to the angle $\theta = 50°$. This maximum plastic response is characterized by the maximal dilation that can be observed along all θ-directions which also seems to coincide with large (but not necessarily the largest) angles of non-coaxiality. It is worth noting that whenever the angle of non-coaxiality is the smallest ($\theta = 0$ and $90°$), the plastic strain response is shown to be the smallest as well. The curves representing these three variables resulting from the principal stresses rotation follow the same trend but with different magnitudes according to the values of θ and φ. Also, the "bell" shape of the curves representing both the plastic shear strain $\|\Delta\gamma^p\|$ and the volumetric strain ε_v at $\varphi = 0°$ tends to be lost from the right side as soon as φ increases.

3 Conclusions

This short numerical study provides some insights in the plastic deformation of granular media under a rotation of principal stress direction. The anisotropy of the granular fabric can be revealed by the signature of the plastic strain response of a granular assembly under such special loading. There is a close relationship between the angle of non-coaxiality and the plastic response which can be exploited to formulate a plastic flow rule which accounts for this type of plasticity mechanism. Further DEM studies are being conducted to characterize plastic strain response with respect to fabric anisotropy, internal structure rotation and fabric of sliding contacts.

Acknowledgments This work is jointly funded by the Computer Modelling Group (CMG) and the Natural Science and Engineering Council of Canada through a CRD grant.

References

Cai Y (2010) An experimental study of non-coaxial soil behaviour using hollow cylinder testing. PhD Dissertation, Department of Civil Engineering, University of Nottingham, Nottingham

Cundall PA, Strack ODL (1979) A discrete numerical model for granular assemblies. Géotechnique 29(1):47–65

Roscoe KH (1970) The influence of strains in soil mechanics. Géotechnique 20(2):129–170

Smilauer V, Catalano E, Chareyre B, Dorofeenko S, Duriez J, Gladky A, Kozicki J, Modenese C, Scholtès L, Sibille L, Stransky J, Thoeni K (2010) Yade reference documentation. In: Smilauer V (ed) The yade project. http://yade-dem.org/doc/

Yu HS, Yuan X (2006) On a class of non-coaxial plasticity models for granular soils. Proc R Soc A Math Phys Eng Sci 462(2067):725–748

A Framework to Assess the Controllability of Wetting on Unsaturated Collapsible Soils

C. Mihalache and G. Buscarnera

Abstract The paper focuses on the process of wetting-induced compaction in unsaturated soils. While such forms of irreversible deformation are referred to as *collapses*, a mechanical interpretation based on concepts of material stability has not yet been provided. Here we use a critical state plasticity model and an expression of second-order energy input accounting for changes in fluid pressures and/or volume fractions. Soaking-induced compaction is then simulated and interpreted in light of the controllability theory. Our results suggest that wetting-compaction is not necessarily associated with the loss of uniqueness of the incremental response. Compaction is indeed predicted not to be controllable only in soils that undergo considerable plastic strains upon wetting. In these cases, plastic models with enhanced hydraulic hardening predict the possibility of a loss of control of the water injection process, according to which the applied stress is no longer sustainable because of the degradation of the mechanical properties.

1 Introduction

The accumulation of volumetric strains due to changes in moisture plays a fundamental role in interpreting and modeling the mechanical response of unsaturated soils. Indeed, failure to predict the accumulation of compaction upon suction removal has been widely regarded as an evidence of the inapplicability of the effective stress principle to unsaturated soils (Jennings and Burland 1962), and has inspired the use of independent sets of stress variables and the adaptation of critical state models by expressing the yield domain as a function of suction and/or other hydrologic variables (Alonso et al. 1990; Gens et al. 2006).

C. Mihalache · G. Buscarnera (✉)
Northwestern University, Evanston, IL 60208, USA
e-mail: g-buscarnera@northwestern.edu

K.-T. Chau and J. Zhao (eds.), *Bifurcation and Degradation of Geomaterials in the New Millennium*, Springer Series in Geomechanics and Geoengineering,
DOI 10.1007/978-3-319-13506-9_45

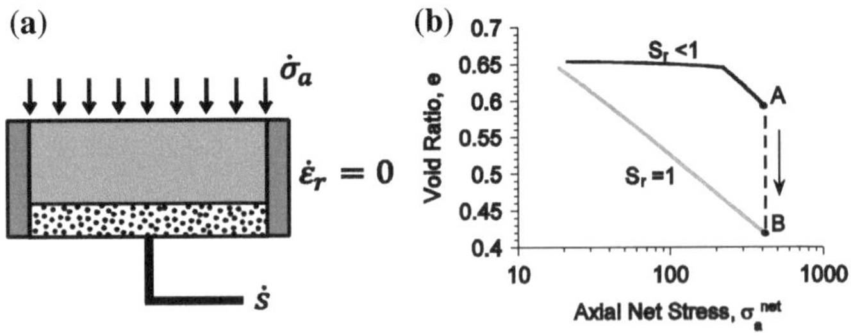

Fig. 1 Oedometer testing in unsaturated soils: **a** schematic of control parameters and **b** example of simulated wetting compaction

The phenomenology of wetting-compaction is typically explored via oedometric testing (Fig. 1a), from which matric suction has been observed to have two major effects on the compression behavior: (i) it increases both the stiffness and apparent pre-consolidation stress compared to saturated conditions, and (ii) its removal promotes inelastic compaction at constant state of stress. The latter stage of compaction is traditionally referred to as *wetting-collapse*. While this phenomenon has been widely studied in the literature and recent experimental findings have shown the possible formation of heterogeneous compaction zones upon saturation (Bruchon et al. 2013), its mechanical interpretation has received little attention. In particular, the literature does not address in detail whether such processes can be externally controlled by imposing prescribed combinations of stresses, strains and suction increments without causing a sharp increase in the rate of deformation (i.e., a passage to a non-quasistatic regime of deformation).

Here we use the second-order work input recently suggested by Buscarnera and di Prisco (2012) to inspect the predictions of wetting-compaction provided by a plasticity model for unsaturated soils (Buscarnera and Nova 2009). The goal is to provide a mathematical interpretation of the wetting-collapse process based on the concepts of controllability and loss of uniqueness/existence of the incremental response (Imposimato and Nova 1998).

2 Modeling Techniques

The simulation of wetting-compaction of unsaturated soils in the frame of elastoplastic models is typically simulated by using suction-dependent yield surfaces (Alonso et al. 1990). Here we use a particular model inspired by these concepts. Such model was developed by Buscarnera and Nova (2009) to reproduce the first-order features of unsaturated soil response. In particular, the suction-dependence of soil strength is incorporated by defining the constitutive functions in terms of the following average skeleton stress:

$$\sigma'_{ij} = \sigma_{ij} - S_r u_w \delta_{ij} - (1 - S_r) u_a \delta_{ij} \tag{1}$$

where σ_{ij} is the total stress, u_w and u_a are the water and air pressure, respectively, and S_r is the degree of saturation. The suction-dependence of the yield surface is incorporated in the hardening rule, as follows

$$\dot{p}'_s = \frac{1}{B_p} p'_s (\dot{\varepsilon}^p_v + \xi_s \dot{\varepsilon}^p_s) - r_{sw} p'_s \dot{S}_r \tag{2}$$

where p'_s is an internal variable describing the evolution of the size of the yield surface as a function of plastic strains and changes in degree of saturation. From (2) it is readily apparent that the variability in size of the elastic domain (and hence, the tendency to develop plastic strains upon saturation) is controlled by the parameter r_{sw}. Finally, the relation between suction and degree of saturation is simulated via a van Genuchten retention curve (van Genuchten 1980).

An example of simulated plastic compaction under oedometric conditions is provided in Fig. 1b, in which a set of constitutive parameters calibrated for a highly collapsible unsaturated soil has been used. In the following sections, we will mathematically inspect such model predictions by means of the concept of controllability (Imposimato and Nova 1998).

3 Theoretical Interpretation of Wetting-Collapse

Previous works by Nova and coworkers (Imposimato and Nova 1998) have elucidated the connection between second-order energy input and singularity of the principal minors of the constitutive matrix. Such a relation has been pivotal to correlate non-positive values of the second-order work to the potential loss of controllability of specific sets of loading variables, and hence with the potential failure of the specimen. In a hydro-mechanical context, the incremental response of a porous solid is controlled by both mechanical perturbations and hydrologic forcing (e.g., pore pressure fluctuations or fluid volume injection/extraction). As a consequence, additional control parameters must be included in the constitutive description, thus requiring an adaptation of the second-order energy input and control theory. Such an extension of the controllability theory for unsaturated soils has been recently discussed by Buscarnera and di Prisco (2012), leading to a second-order work expression that can be written as,

$$d^2W = \frac{1}{2}\dot{\sigma}^*_{ij}\dot{\varepsilon}_{ij} - \frac{1}{2}n\dot{s}\dot{S}_r = d^2W_M + d^2W_H \tag{3}$$

where $\dot{\sigma}^*_{ij} = \dot{\sigma}_{ij} - S_r\dot{u}_w\delta_{ij} - (1 - S_r)\dot{u}_a\delta_{ij}$ is a measure of the stress increments on the solid skeleton and $\dot{s} = \dot{u}_a - \dot{u}_w$ is the increment of suction. The first and second terms of Eq. (3) correspond to the mechanical (d^2W_M) and hydraulic (d^2W_H) components of second-order work, respectively.

It is possible to show that, by linking Eq. (3) to a hydro-mechanical constitutive law, the second-order work can be rewritten as follows:

$$d^2W = \frac{1}{2}\dot{\Sigma}\mathbf{D_{HM}}\dot{\mathbf{E}} = \frac{1}{2}\dot{\Sigma}\mathbf{D^S_{HM}}\dot{\mathbf{E}} \tag{4}$$

where $\dot{\mathbf{\Sigma}}$ and $\dot{\mathbf{E}}$ are two vectors collecting the hydro-mechanical variables, $\mathbf{D_{HM}}$ is the constitutive matrix and $\mathbf{D^S_{HM}}$ is its symmetric part. As discussed by Buscarnera and di Prisco (2012), a loss of positive definiteness of $\mathbf{D^S_{HM}}$ implies the possibility to have singularities of the principal minors of $\mathbf{D_{HM}}$. As a result, negative values of the energy input (3) can be explained as a reflection of a potential hydro-mechanical loss of controllability. For a specific set of control parameters, this condition can be expressed by means of the linear relation linking control parameters, $\dot{\boldsymbol{\phi}}$, and response variables, $\dot{\boldsymbol{\psi}}$:

$$\dot{\boldsymbol{\phi}} = \mathbf{X}\dot{\boldsymbol{\psi}} \tag{5}$$

where $\mathbf{X}$ is the constitutive matrix. A loss of control coincides with the singularity of $\mathbf{X}$ (i.e., with $\det \mathbf{X} = 0$).

In the following, the control conditions represented in Fig. 1a will be used to define two control matrices: one associated with the control of the vertical net stress and suction ($\det \mathbf{X}^S$, replicating usual laboratory conditions in experiments) and another associated with water-content control ($\det \mathbf{X}^W$, replicating abrupt soaking imposed via water volume injection). In both cases, oedometric conditions are considered. In other words, while the vertical net stress is used as a control parameter, the radial stresses are not controlled, but are indirectly recovered by constraining radial deformation (i.e., a mixed stress-strain control configuration).

Figure 2 illustrates the controllability analysis for the wetting-compaction process simulated in Fig. 1b. For the selected set of parameters, the second-order work in (3) has positive values. This results from a combination of negative values due to compaction during suction removal and positive values of the hydraulic contributions associated with the increase in degree of saturation. In addition, Fig. 2

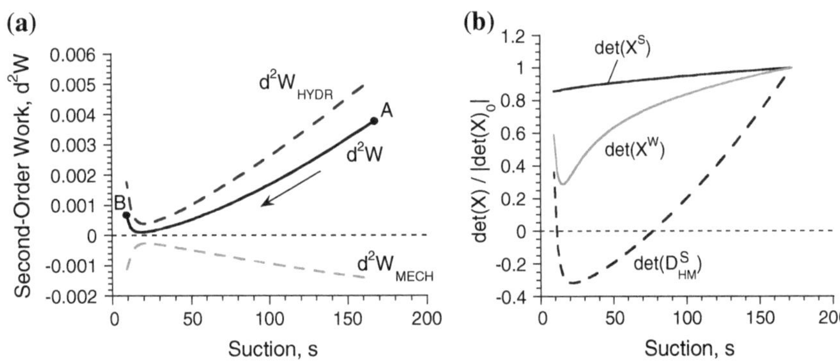

Fig. 2 Evolution of **a** second order work and **b** normalized controllability indices during a simulated wetting process imposed at constant net stress in an oedometer ($r_{sw} = 5.5$)

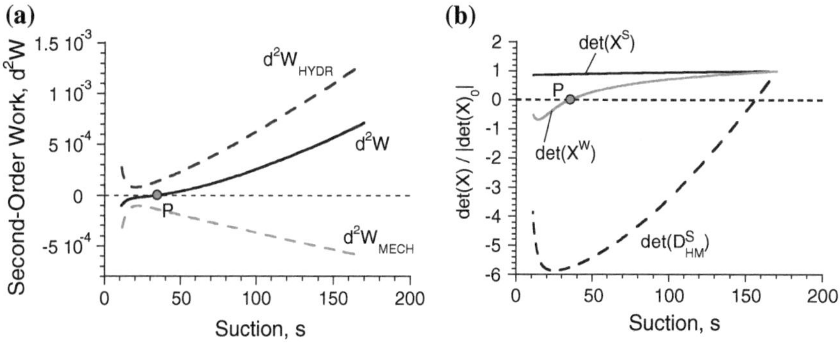

Fig. 3 Effect of an increase in r_{sw} on the controllability indices ($r_{sw} = 8.5$). P marks the state at which control of the axial net stress is lost under water injection

shows that positive values of d^2W during wetting correspond to positive values of the determinants associated with the two control scenarios. As a result, the analysis suggests that the simulation considered in the example is associated with a controllable compaction process

Nevertheless, the analysis indicates the loss of positive definiteness of $\mathbf{D}^{\mathbf{S}}_{\mathbf{HM}}$, and hence an increase in the instability potential during saturation. As discussed by Buscarnera and di Prisco (2012) this possibility is exacerbated by the tendency of the material to compact during wetting, and hence by hydro-mechanical coupling terms. As an example, Fig. 3 illustrates the effect of increasing the intensity of soil deterioration during wetting, here reflected by the constitutive parameter r_{sw}, in (2). The larger tendency of the soil to compact exacerbates the negative values of the mechanical contributions to d^2W and promotes negative values for both det $\mathbf{D}^{\mathbf{S}}_{\mathbf{HM}}$ and det $\mathbf{X}^W$. In other words, while wetting-compaction is predicted to be controllable via suction control (i.e., det $\mathbf{X}^S \neq 0$), a sudden soaking of the specimen (here replicated by a water-volume injection) is predicted to be capable of activating a loss of control, and hence the inability of the material to sustain the applied stresses.

4 Conclusions

A plasticity model for unsaturated soils and the controllability theory have been used to discuss the mechanical nature of wetting-compaction events. It has been shown that compactive strains generated during wetting are not necessarily associated with a collapse (i.e. with a loss of controllability of the external stresses). In other words, the tangent constitutive operator associated with one-dimensional compaction is not singular throughout the wetting stage. In order to interpret these processes as unstable modes of deformation, it is necessary to introduce a significant source of material deterioration, leading to a dominant role of the negative contributions of the second-order energy input. In these circumstances, negative values of second-order work are associated with a singularity of the control matrix for water-content control. In other words, the simulations suggest that the generation of a sharp loss of control of the loading parameters tends to be favored by the soaking of the specimen (i.e., by the sharp injection of water volume at a constant state of stress), while instabilities are predicted to be hindered by suction control.

Acknowledgments This work has been supported by the U.S. National Science Foundation, Geomechanics and Geomaterials Program, under Grant Nos. CMMI-1234031 and CMMI-1351534.

References

Alonso EE, Gens A, Josa A (1990) A constitutive model for partially saturated soils. Géotechnique 40(3):405–430

Bruchon JF, Pereira JM, Vandamme M, Lenoir N, Delage P, Bornert M (2013) Full 3D investigation and characterisation of capillary collapse of a loose unsaturated sand using X-ray CT. Granul Matter 15(6):783–800

Buscarnera G, di Prisco C (2012) Discussing the definition of the second-order work for unsaturated soils. Int J Numer Anal Meth Geomech 36(1):36–49

Buscarnera G, Nova R (2009) An elastoplastic strainhardening model for soil allowing for hydraulic bonding–debonding effects. Int J Numer Anal Meth Geomech 33(8):1055–1086

Gens A, Sánchez M, Sheng D (2006) On constitutive modelling of unsaturated soils. Acta Geotech 1(3):137–147

Imposimato S, Nova R (1998) An investigation on the uniqueness of the incremental response of elastoplastic models for virgin sand. Mech Cohesive-Frictional Mater 3(1):65–87

Jennings JEB, Burland JB (1962) Limitations to the use of effective stresses in partly saturated soils. Géotechnique 12(2):125–144

van Genuchten MT (1980) A closed-form equation for predicting the hydraulic conductivity of unsaturated soils. Soil Sci Soc Amer J 44:892–898

Model Tensile Cracking in Soil with Coupled Meshless-FEM Method

Chong Peng and Wei Wu

Abstract An adaptive, coupled method based on radial point interpolation meshless method (RPIM) and finite element method (FEM) is proposed for the simulation of tensile cracking in soil structure. Potential crack area in the initial FEM mesh is converted to RPIM nodes and refined for crack analysis. A crack tracking approach is adapted to model complex crack surface and eliminate node distribution bias. Numerical example is presented to verify the proposed method.

1 Introduction

Tensile cracking is a common problem in geotechnical structure, which may leads to catastrophic failure. Smeared crack model, which is based on continuum mechanics, models cracking by a band in which the material is weakened according to specific constitutive model (Rots and Blaauwendraad 1989). Although widely used, some problems still remain unsolved in smeared crack model. The crack is element-wise, thus the influence of element size on the cracking path is inevitable. As a consequence, accurate crack simulation requires very fine mesh. However the local mesh refinement remains difficult in the FEM, especially in three dimensions. The smeared crack model also suffers the shortcoming of mesh bias.

In meshless methods, only particle information is needed. Hence, crack simulation and adaptive analysis is relatively simple (Belytschko et al. 1996). Because of the low efficiency of meshless methods, its coupling with FEM is a natural way to make use of the advantages of these two methods. Among many meshless

C. Peng (✉) · W. Wu
Institut Für Geotechnik, Universität Für Bodenkultur, Vienna, Austria
e-mail: pengchong@boku.ac.at

K.-T. Chau and J. Zhao (eds.), *Bifurcation and Degradation of Geomaterials in the New Millennium*, Springer Series in Geomechanics and Geoengineering,
DOI 10.1007/978-3-319-13506-9_46

methods, the radial point interpolation method (RPIM) can be coupled with FEM easily, because its shape function is constructed by point interpolation (Wang and Liu 2002).

In this paper, an adaptive coupled RPIM-FEM method is proposed to simulate tensile cracking. The post cracking behavior of geo-material is modeled by an isotropic damage model. To overcome mesh bias, a crack tracking algorithm is introduced, making it possible to apply the proposed method to complicated problems.

2 Smeared Crack with Damage Model

An isotropic damage model is used to model the degradation of cracked material, which has the following constitutive equation

$$\boldsymbol{\sigma} = (1 - d)\bar{\boldsymbol{\sigma}} = (1 - d)\boldsymbol{D} : \boldsymbol{\varepsilon} \tag{1}$$

where $\bar{\boldsymbol{\sigma}}$ is the effective stress tensor, $\boldsymbol{D}$ the isotropic linear elastic constitutive tensor, $\boldsymbol{\varepsilon}$ the total strain and d the damage index. To split the tensile contribution from total effective stress, an equivalent stress is defined by Cervera and Chiumenti (2006)

$$\tau = \langle \bar{\sigma}_1 \rangle \tag{2}$$

where $\bar{\sigma}_1$ is the first principal effective stress, and the symbol $\langle \cdot \rangle$ is Macaulay brackets. The damage criterion is defined as

$$F = F(\tau, r) = \tau - r \leq 0 \tag{3}$$

where r is an internal stress-like variable. The evolution of r is given as

$$r = \max\{\sigma_t,\ \max(\tau)\} \tag{4}$$

where σ_t is the uniaxial tensile strength. The damage index d is defined as a function of the internal variable r

$$d(r) = 1 - \frac{\sigma_t}{r} \exp(-2H_s \frac{r - \sigma_t}{r}) \tag{5}$$

where H_s is a constant dependent on the mesh size used in the simulation. By the concept of fracture energy, H_s can be computed by the following equation

$$H_s = \frac{\sigma_t^2 l_{ch}}{2EG_f - \sigma_t^2 l_{ch}} \tag{6}$$

where E is the Young's modulus, G_f is the fracture energy. l_{ch} is the characteristic length of the mesh used in the simulation.

3 The Coupled RPIM and FEM Method

In RPIM, a field function $u(\boldsymbol{x})$ is approximated by radial basis functions $R_i(\boldsymbol{x})$ and a set of polynomial basis functions $P_j(\boldsymbol{x})$ (Liu et al. 2005)

$$u(\boldsymbol{x}) = \sum_{i=1}^{n} R_i(\boldsymbol{x})a_i + \sum_{j=1}^{m} P_j(\boldsymbol{x})b_j = \boldsymbol{R}^{\mathrm{T}}(\boldsymbol{x})\boldsymbol{a} + \boldsymbol{P}^{\mathrm{T}}(\boldsymbol{x})\boldsymbol{b} \tag{7}$$

where n is the number of nodes in the influence domain of the sample point $\boldsymbol{x}$, m is the number of polynomial basis function, a_i and b_j are the corresponding coefficients. By interpolating all the nodal values in the influence domain, the field function can be expressed as

$$u(\boldsymbol{x}) = [\boldsymbol{R}^{\mathrm{T}}(\boldsymbol{x}) \quad \boldsymbol{P}^{\mathrm{T}}(\boldsymbol{x})]\boldsymbol{G}^{-1}\begin{pmatrix} \boldsymbol{u}_e \\ \boldsymbol{0} \end{pmatrix} = \boldsymbol{N}(\boldsymbol{x})\boldsymbol{u}_e = \sum_{i=1}^{n} N_i u_i \tag{8}$$

where $\boldsymbol{G}$ is the interpolating matrix, $\boldsymbol{u}_e$ is the nodal value vector, $\boldsymbol{N}(\boldsymbol{x})$ is the resulting shape function vector composed of $N_i(\boldsymbol{x})$, $i = 1, 2, \ldots, n$.

Constructed by point interpolation, the shape functions of RPIM have the same Kronecker delta property as the FEM shape function. With this property, the coupling of RPIM and FEM is direct and simple. The problem domain can be divided into two subdomains, Ω_{RPIM} and Ω_{FEM}, connected by a set of interface nodes belonging to both sides. Due to the Kronecker delta property of both RPIM and FEM, C^0 continuity is achieved naturally along the interface.

To distinguish the RPIM region and FEM region, automatic identification of the potential crack zone is needed. The criterion is as follows:

$$\bar{\sigma}_1 > 0.85\sigma_t \tag{9}$$

In the distinguished potential crack zone, the original finite elements are converted to RPIM region. For better simulation of crack propagation, the RPIM area is automatically refined. More details about the adaptive coupling method can be found in Yuan et al. (2014).

4 Crack Tracking

The smeared crack suffers the drawback of mesh bias, that is, its solution is affected by the mesh discretization (Rots and Blaauwendraad 1989). The similar pathological dependency of solution is also found in meshless methods when smeared crack is adapted. A lot of remeshing methods and complicated constitutive models have been proposed to solve this problem. Recently Oliver et al. (2004) proposed a crack tracking method, which is based on the global stress

Fig. 1 Global crack tracking

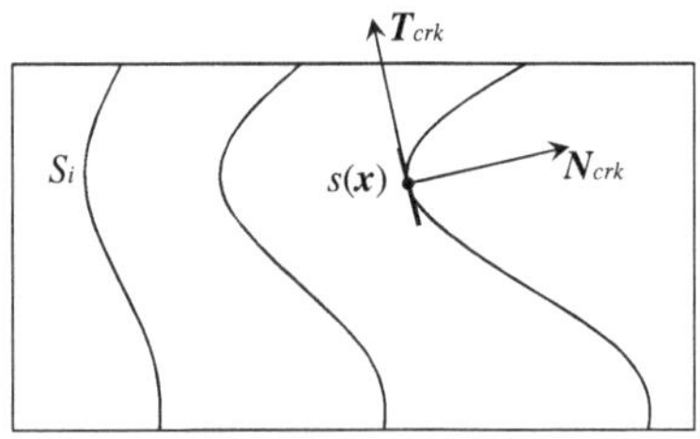

status, to handle the mesh bias problem. In our coupled meshless-FEM method, this global tracking method is implemented to simulate complicated curved cracking.

The global crack tracking method assumes that at arbitrary location $\boldsymbol{x}$, there is a potential crack surface depicted by a small crack plane $s(\boldsymbol{x})$. The potential crack propagation paths can be viewed as a collection of surfaces, which are the envelopes of the patches of the assumed crack planes. The collection of crack surfaces can also, be considered as the level contours of a scalar function $\phi(\boldsymbol{x})$. By this assumption any level contour $S_i = \{\boldsymbol{x} \in \Omega | \phi(\boldsymbol{x}) = \phi_i\}$ represents a potential crack path, as shown in Fig. 1. If we can find an appropriate field function, the crack propagation can be tracked easily.

According to the assumption, the level contours are the crack envelopes, comprised of patches of crack plane. As a result, at any location, the gradient of $\phi(\boldsymbol{x})$ is perpendicular to the crack plane, i.e., $\boldsymbol{N}_{crk} = \nabla\phi / \|\nabla\phi\|$. If the crack surfaces are considered as equal-potential surfaces, then the function is very similar to a potential function, like temperature or water pressure. By this concept, the function can be obtained by the following approach.

In the local coordinate defined on crack plane, we have $\boldsymbol{T}_{crk} \cdot \boldsymbol{N}_{crk} = 0$. Substituting $\boldsymbol{N}_{crk} = \nabla\phi / \|\nabla\phi\|$ into previous equation and pre-multiplying $\boldsymbol{T}_{crk}$, it yields

$$\boldsymbol{T}_{crk} \otimes \boldsymbol{T}_{crk} \cdot \nabla\phi = 0 \tag{10}$$

Define a flow $\boldsymbol{q} = -\boldsymbol{D}_{con} \cdot \nabla\phi$, where $\boldsymbol{D}_{con} = \boldsymbol{T}_{crk} \otimes \boldsymbol{T}_{crk}$ can be considered as a conductivity matrix. Therefore, the field function can be obtained by solving the following thermal-conduct like problem

$$\begin{aligned} \nabla \cdot \boldsymbol{q} = \nabla \cdot (-\boldsymbol{D}_{con} \cdot \nabla\phi) = 0 \quad \text{in } \Omega \\ q_v = \boldsymbol{q} \cdot \boldsymbol{v} = 0 \quad \text{in } \partial\Omega \end{aligned} \tag{11}$$

where $\boldsymbol{v}$ is the outward unit vector normal to the boundary $\partial\Omega$.

The above thermal-conduct like problem is linear and well posed, and can be solved with the same coupled FEM-RPIM method used to obtain displacement. In crack analysis, the crack tracking can be carried out once per step or every several steps. Once the value of $\phi(\boldsymbol{x})$ is obtained, the position of a crack can be identified. If an influence domain of an integral point is cracked, then the further propagation path of the crack is known to us: it lays on the iso-line of $\phi(\boldsymbol{x})$ which crosses this very influence of domain.

5 Numerical Example

A well-studied shear beam test is chosen to verify the proposed method. The geometry and configuration is shown in Fig. 2. The following material properties are used: Young's module $E = 40$ MPa, Poisson's ratio $\nu = 0.3$, tensile strength $\sigma_t = 6.0 \times 10^5$ Pa and mode I fracture energy $G_f = 25$ J/m^2.

Two simulations, with crack tracking and without crack tracking, are performed. In the analysis with crack tracking, the additional thermal-conduct like problem is solved in each step. Initially, a coarse FEM mesh is used. Figure 3 gives the result of crack tracking and damage index at the last step. Clearly, the crack tracking method captures the potential crack path well. With crack tracking, the crack starts from the notch and propagates towards the upper edge of the beam in a curved path.

The adaptive coupling process is clearly shown in the deformed geometries given in Fig. 4. The refined RPIM region starts from the notch and extends upwards, followed by the cracking. The node distribution in the RPIM region is very dense; therefore an accurate representation of the cracking can be obtained. With tracking the crack propagates along a curved path, while without tracking it goes along the alignment of the node distribution, which is obviously unphysical. The load displacement curve is given in Fig. 5.

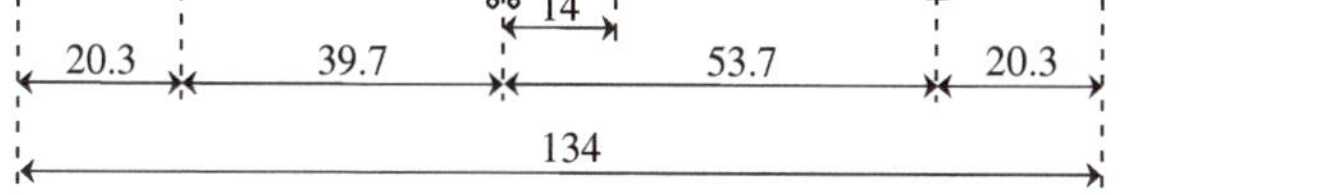

Fig. 2 Geometry of the shear beam (unit: mm)

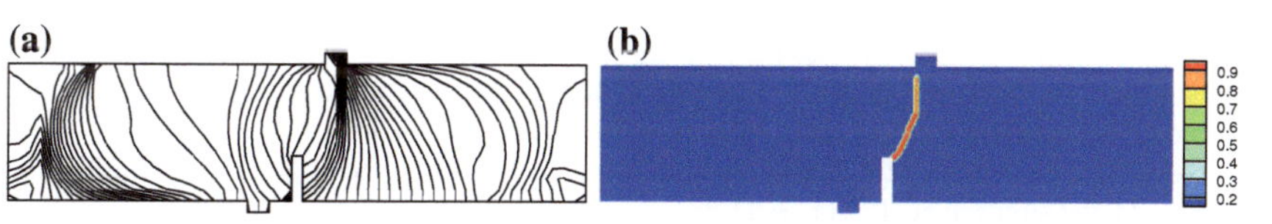

Fig. 3 Crack simulation with tracking: **a** crack tracking; **b** damage index

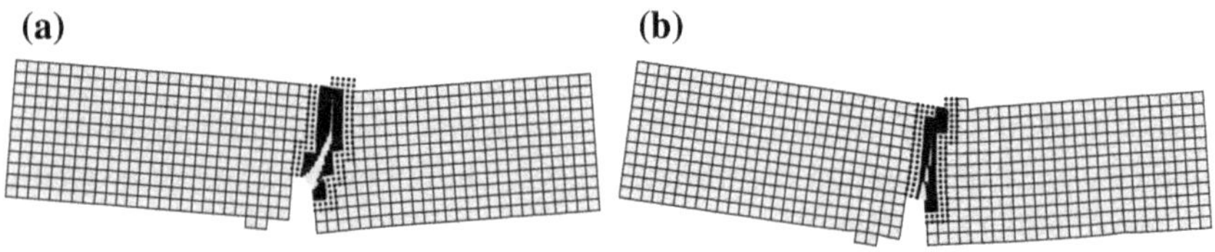

Fig. 4 Deformed mesh: **a** with crack tracking; **b** without crack tracking

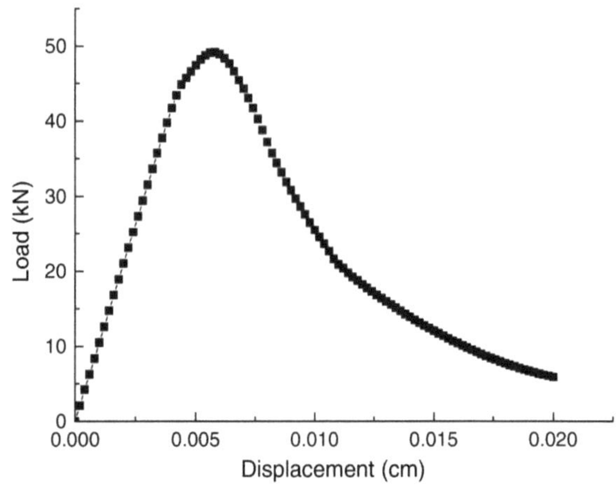

Fig. 5 Load versus displacement at the loading point

6 Conclusion

A coupled RPIM-FEM method is proposed for the simulation of tensile crack. With adaptive refinement and crack tracking it can simulate cracking problem with complicated propagation path. The computational cost of this method is relatively low because meshless method is only used in the crack region. The implementation of the method is quite simple and its extension to three dimensions is straightforward. Numerical example shows that the proposed method can be used to analysis the tensile cracking in geo-material.

References

Belytschko T, Krongauz Y, Organ D, Fleming M, Krysl P (1996) Meshless methods: an overview and recent developments. Comput Meth Appl Mech Eng 139(1):3–47

Cervera M, Chiumenti M (2006) Smeared crack approach: back to the original track. Int J Numer Anal Meth Geomech 30(12):1173–1199

Liu GR, Zhang GY, Gu YT, Wang YY (2005) A meshfree radial point interpolation method (RPIM) for three-dimensional solids. Comput Mech 36(6):421–430

Oliver J, Huespe AE, Samaniego E, Chaves EWV (2004) Continuum approach to the numerical simulation of material failure in concrete. Int J Numer Anal Meth Geomech 28(7–8):609–632
Rots J, Blaauwendraad J (1989) Crack models for concrete, discrete or smeared? fixed, multi-directional or rotating? Heron 34:1–59
Wang JG, Liu GR (2002) A point interpolation meshless method based on radial basis functions. Int J Numer Meth Eng 54(11):1623–1648
Yuan HN, Peng C, Lin Q, Zhang BY (2014) Simulation of tensile cracking in earth structures with an adaptive RPIM-FEM coupled method. KSCE J Civil Eng 18(7):2007–2018

Quantitative Observation of Strain Localisation in a Partially Saturated Triaxial Specimen Using Microfocus X-ray CT with Image Analysis

Y. Higo, F. Oka, R. Morishita and Y. Matsushima

Abstract Strain localisation behaviour in a partially saturated dense sand specimen during triaxial compression has been observed by microfocus X-ray CT and image analyses of CT images. Digital image correlation (DIC) has provided displacement and strain field in the specimen in a macroscopic point of view. In the meantime, trinarization by a region growing segmentation technique has given local void ratio and degree of saturation microscopically. Through these image analyses of CT images, the progressive strain localisation behaviour and microstructural changes in the shear band have been quantitatively studied.

1 Introduction

It is well known that loss of suction in partially saturated soils associated with the infiltration of water and/or shearing, induces a reduction in strength involving clearer shear banding with stronger strain softening than fully saturated or dry sand

Y. Higo (✉)
Department of Urban Management, Kyoto University, C1 Bld. C Cluster, Kyotodaigaku-Katsura 4, Nishikyo-ku, Kyoto 615-8540, Japan
e-mail: higo.yohsuke.5z@kyoto-u.ac.jp

F. Oka
Professor Emeritus, Kyoto University, Lab. 3, Association for Disaster Prevention Research 138-1 Tanaka-Asukai-Cho, Sakyo-ku, Kyoto 606-8226, Japan
e-mail: oka.fusao.38x@st.kyoto-u.ac.jp

R. Morishita
Former Student, Kyoto University (Presently, Japan Oil, Gas and Metal National Corporation), 2-10-1 Toranomon, Minato-ku, Tokyo 105-0001, Japan
e-mail: morishita-ryoichi@jogmec.go.jp

Y. Matsushima
Former Student, Kyoto University (Presently, Central Japan Railway Company), JR Central Towers, 1-1-4, Meieki, Nakamura-ku, Nagoya, Aichi 450-6101, Japan
e-mail: yoshiki_m0601@yahoo.co.jp

K.-T. Chau and J. Zhao (eds.), *Bifurcation and Degradation of Geomaterials in the New Millennium*, Springer Series in Geomechanics and Geoengineering,
DOI 10.1007/978-3-319-13506-9_47

(e.g., Cui and Delage 1996; Cunningham et al. 2003; Higo et al. 2011). From the geotechnical point of view, one of the most significant reasons for the collapse of earth structures such as river dikes and road embankments is the decrease in suction. It is, therefore, important to clarify the mechanism of the brittle failure of partially saturated soil, which is strongly related to the strain localisation phenomenon. In the present paper, the localised deformation in partially saturated sand has been investigated quantitatively using microfocus X-ray computed tomography (CT) and an image analysis of the CT images. Triaxial compression test on a partially saturated dense Toyoura sand specimen has been carried out under a low confining pressure. The macroscopic observation of the development of localised deformation, using microfocus X-ray CT, has been presented. And, the displacement field over the entire specimen has been quantified by an image analysis of the CT images with the digital image correlation (DIC) technique (Lenoir et al. 2007; Hall et al. 2010), through which the progressive development of shear bands has been discussed. In addition, the region of the localisation has been observed microscopically by the partial CT scanning with a micron scale of high spatial resolution. Changes in the microstructures have been discussed through trinarization of the CT images using a region growing segmentation technique (Higo et al. 2013, 2014), by which the soil particles, the pore water, and the pore air have been distinguished from each other.

2 Materials and Methods

The test sample used in this study is Toyoura sand. Toyoura sand is semi-angular in shape with an average diameter D_{50} of 0.185 mm. The μX-ray CT system used in this study is KYOTO-GEOμXCT (Higo et al. 2011). Since the triaxial cell can be mounted on a rotation table, specimens can be scanned during the triaxial test.

We used the conventional triaxial test procedure in which the moist-tamping method is applied to prepare the partially saturated specimens. The height of the specimen was 70 mm and the diameter was 35 mm. The initial void ratio was 0.668, corresponding to the relative density of 85 %, and the initial degree of saturation was 60 %. A confining pressure of 50 kPa was isotropically applied by air pressure without the use of any back pressure. The axial load was applied through a displacement control system with a constant axial strain rate of 0.5 %/min under drained conditions for air and undrained conditions for water. During the triaxial shearing, μX-ray CT scanning has been performed to view strain localisation behavior in non-destructive and three-dimensional manner. Total volume of pore water is constant throughout the test because of the undrained conditions, while the local degree of saturation varies due to deformation and induced pore water flow.

In order to quantify the displacement field of the specimen during triaxial compression, a digital image correlation (DIC) analysis has been applied (Lenoir et al. 2007; Hall et al. 2010; Higo et al. 2013). In addition, a region growing segmentation technique has been used to trinarize original CT images into solid phase, water phase and air phase. We have taken into account the uniform distribution of mixel (e.g., Kitamoto and Takagi 1995) when determining the tolerance value of the region growing method for partially saturated sand developed by Higo et al. (2014).

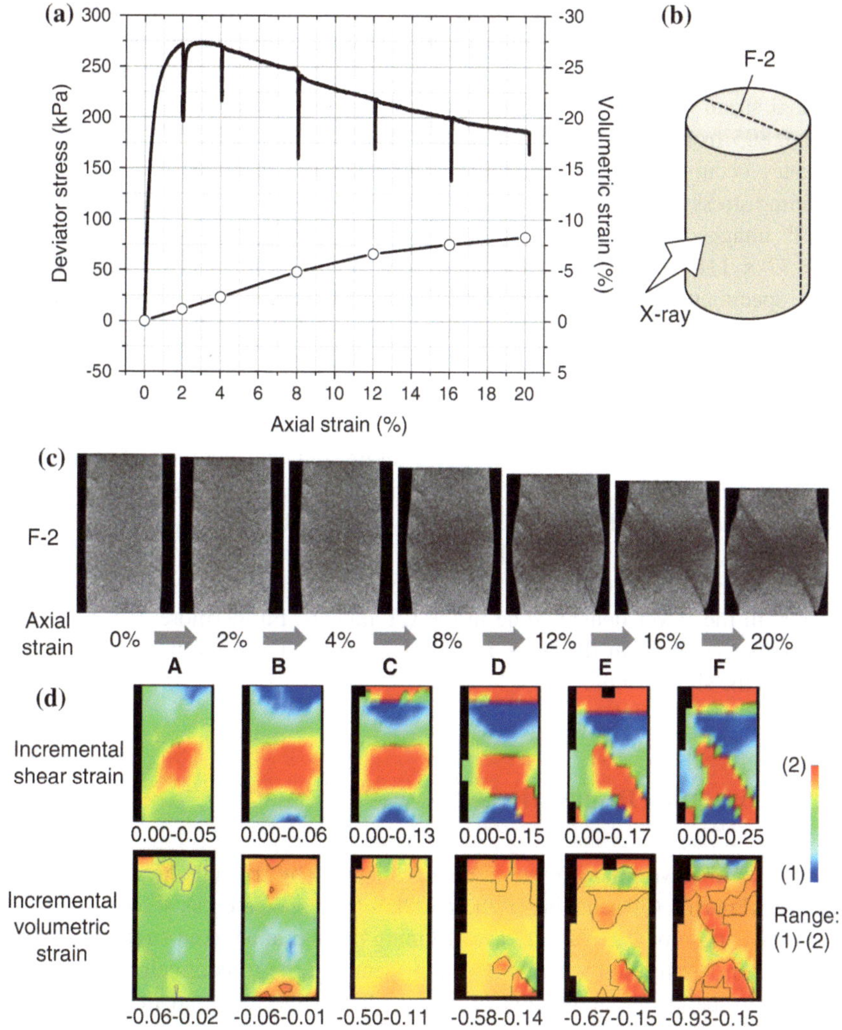

Fig. 1 Macroscopic response and strain localisation: **a** deviator stress-axial strain relation and volumetric strain-axial strain relation (*solid lines* deviator stress, *open marks with lines* volumetric strain, compression is positive), **b** location of the cross section, **c** CT images and **d** distribution of incremental shear strain and incremental volumetric strain obtained by DIC

3 Macroscopic Observations

Overall relation between the deviator stress and the axial strain and that between the volumetric strain and the axial strain are shown in Fig. 1a. Volumetric strain is given by the volumes of specimen measured using CT images, in which the volume is calculated by counting the numbers of voxels denoting the specimen multiplied by unit volume of the voxel (see Higo et al. 2011).

The peak deviator stress can be observed at an axial strain of about 3 %, and then the specimen exhibits strain-softening behavior until an axial strain of 20 %. Volume expansion occurs from the beginning to the end. The volumetric strain at an axial strain of 20 % is 8.24 %. Consequently, the overall void ratio increased up to 0.805 from 0.668 at the initial state. These results suggest that the positive dilatancy occurs due to the triaxial shearing since the specimen is dense and the confining stress is rather small.

CT images are demonstrated in Fig. 1c. Voxel size of the images is $77 \times 77 \times 118$ μm. At an axial strain of 4 %, just after the peak stress, the center of the specimen becomes slightly darker in color which denotes the lower density. The lower density region becomes wider at an axial strain of 8 %, and then we can observe the localisation of the lower density region with band-like shape at an axial strain of 12 %. Finally, the color at the center of the specimen and the band-like localisation zones goes darker further.

Distributions of the incremental shear strain and the incremental volumetric strain obtained by the DIC analysis are shown in Fig. 1d. It has been confirmed that shear bands develop in the specimen because the regions with high shear strain levels, indicated by red color, are in between the two rather discontinuous surface of displacement gradient. In addition, the location of the shear bands coincides with the lower density zone in the CT images. Furthermore, the incremental volumetric strain in the shear band is expansive. Thus, the positive dilatancy occurs in the shear bands.

4 Microscopic Observations

In order to observe microstructural changes caused by triaxial shearing, we have performed partial CT scanning at the middle of the specimen as shown in Fig. 2a at an axial strain of 0 and 20 %. The scanning portion at an axial strain of 20 % is in the strain localisation zone. During the scanning, the triaxial cell was removed in order to avoid the attenuation of X-rays due to the cell, and then placed the specimen as close to the X-ray source as possible in order to achieve high spatial resolution. The confining pressure of 50 kPa was released, but was replaced by the vacuum of 20 kPa applied to the inside of the specimen.

Figure 2b, c show the partial CT images and their trinarized images, respectively, at axial strains of 0 and 20 %. Voxel size of the images is $4.2 \times 4.2 \times 7.0$ μm. The middle of the specimen was chosen as the location of the partial CT scan where it is seen at an axial strain of 20 % that the volume expansion due to the shear deformation occurs.

Trinarized images provide the volume of solid phase, water phase and air phase, by which the local void ratio and the local degree of saturation in the area of the partial CT scan have been calculated. In Table 1, those local values obtained by the calculation and the measured global values for the whole specimen are listed. Before the test, we have obtained the similar local void ratio to the global one, while the local degree of saturation is larger than the global one. This is partly because of the

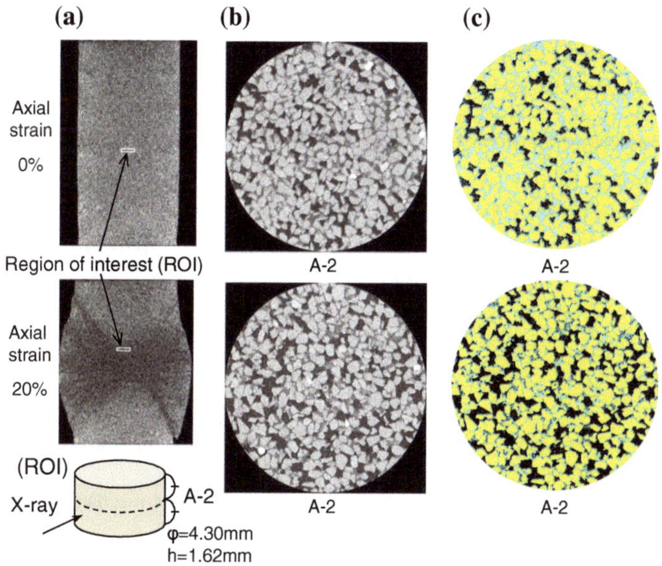

Fig. 2 Partial CT images and their trinarized images at initial state and at an axial strain of 20 %

Table 1 Local void ratio and the local degree of saturation in the area of the partial CT scan

	Before the test (axial strain: 0 %)		After the test (axial strain: 20 %)	
	Local (calculated)	Global (measured)	Local (calculated)	Global (measured)†
Void ratio	0.653	0.668	0.874	0.805
Degree of saturation (%)	68.9	60.0	43.2	49.7

†Global values after the test were obtained by the volumetric strain and their initial values

nonhomogeneous distribution of pore water when preparing the specimen. The local void ratio after the test is larger than the global one. This means that volume expansion due to the positive dilatancy is more significant than that for whole specimen. On the other hand, the local degree of saturation is smaller than the global ones. In the images the blue portion denoting pore water is seen to be narrow compare to that before the test. Namely, the pore water in the strain localisation zone significantly decreases from the initial state, in the case of dilative deformation.

5 Conclusions

Strain localisation in partially saturated dense sand specimen during triaxial compression has been observed using μX-ray CT with image analysis. It is confirmed that the shear bands developed in the specimen by the DIC analysis.

Trinarization by a region growing segmentation technique has provided the local void ratio and the degree of saturation. In the shear bands with positive dilatancy, it is found that the pore water significantly reduces from the initial value. It is known that suction level of very loose sand is almost null even if the degrees of saturation is small (e.g., Nishimura et al. 2011). This is probably because water retention ability of the loose particulate structure is low. It is therefore possible that the suction level in the shear bands is not high enough to strengthen the overall response, which may cause the stronger softening of partially saturated sand than fully saturated or air-dried sand.

References

Cui YJ, Delage P (1996) Yielding and plastic behaviour of an unsaturated compacted silt. Géotechnique 46(2):291–311

Cunningham MR, Ridley AM, Dineen K, Burland JB (2003) The mechanical behaviour of a reconstituted unsaturated silty clay. Géotechnique 53(2):183–194

Hall SA, Bornert M, Desrues J, Pannier Y, Lenoir N, Viggiani G, Bésuelle P (2010) Discrete and continuum analysis of localised deformation in sand using X-ray μCT and volumetric digital image correlation. Géotechnique 60(5):315–322

Higo Y, Oka F, Kimoto S, Sanagawa T, Matsushima Y (2011) Study of strain localization and microstructural changes in partially saturated sand during triaxial tests using microfocus X-ray CT. Soils Found 51(1):95–111

Higo Y, Oka F, Sato T, Matsushima Y, Kimoto S (2013) Investigation of localized deformation in partially saturated sand under triaxial compression by microfocus X-ray CT with digital image correlation. Soils Found 53(2):181–198

Higo Y, Oka F, Morishita R, Matsushima Y, Yoshida Y (2014) Trinarization of μX-ray CT images of partially saturated sand at different water-retention states using a region growing method. Nucl Instrum Meth Phys Res B 324(1):63–69

Kitamoto A, Takagi M (1995) Mixture density estimation under the existence of mixels. In: Geoscience and remote sensing symposium, IGARSS '95, Quantitative remote sensing for science and applications, International, vol 1, pp 121–123

Lenoir N, Bornert J, Desrues J, Bésuelle P, Viggiani G (2007) Volumetric digital image correlation applied to X-ray microtomography images from triaxial compression tests on argillaceous rock. Strain 43:193–205

Nishimura T, Koseki J, Rahardjo H (2011) Effect of density for sand on hysteresis in soil-water characteristic curve. In: Jotisankasa, Sawangsuriya, Soralump and Mairaing (eds.), Unsaturated soils: theory and practice, Kasetsart University, Thailand, pp 391–396

Simple Modeling of Time-Dependent Behavior for Structured Soils

Teruo Nakai, Hossain Md. Shahin and Hiroyuki Kyokawa

Abstract A simple model to describe time-dependent behavior of various soils in 1D stress conditions is presented in this paper. The model is formulated not using the usual viscoplastic theories such as over-stress type and non-stationary flow surface type but utilizing the subloading surface concept by Hashiguchi (1980), and paying attention to the experimental results that the normally consolidation line (NCL) on the e–ln σ plane shift depending on the strain rate. The present model can describe various time-dependent behaviors not only of normally consolidated soil but also of over consolidated and naturally deposited soils in the same manner without violating the objectivities. The 1D model can easily be extended to the 3D using the t_{ij} concept (Nakai and Mihara 1984).

1 Introduction

Most of the previous viscoplastic models have been formulated based on the over-stress theory or the non-stationary flow surface theory. However, these models have some problems in their formulation such as including non-objective time variable and lacking of loading condition. In the present study, unified 1D formulation of time-dependent behavior for normally consolidated soil to structured soil is presented. To describe the behavior of structured soil, a state variable ρ which represents the influence of density and another state variable ω which represents the bonding effect as well as the state variable ρ are introduced.

T. Nakai (✉)
Nagoya Office, Geo-Research Institute, 1-1-8-706 Fikiage, Chikusa-ku, Nagoya, Japan
e-mail: nakai.teruo@nitech.ac.jp

H.M. Shahin
Nagoya Institute of Technology, Gokiso-cho, Showa-ku, Nagoya, Japan

H. Kyokawa
IIS, University of Tokyo, 4-6-1 Komaba, Meguro-ku, Tokyo, Japan

K.-T. Chau and J. Zhao (eds.), *Bifurcation and Degradation of Geomaterials in the New Millennium*, Springer Series in Geomechanics and Geoengineering,
DOI 10.1007/978-3-319-13506-9_48

2 One-dimensional Modeling for Structured Soil

Figure 1 shows a typical e–ln σ relation of natural clay schematically. Asaoka et al. (2002) developed a model to describe such structured soils, introducing not only subloading surface concept but also superloading surface concept to the Cam-clay model. In their modeling, a factor related to the over consolidation ratio (corresponding to non-real density but imaginary density) has been introduced to increase the stiffness, and a factor related to the soil skeleton structure has been introduced to decrease the stiffness.

Figure 2 shows the change of void ratio when the stress condition moves from the initial state I ($\sigma = \sigma_0$) to the current state P ($\sigma = \sigma$). Here, e_0 and e are the initial and current void ratios of structured soil, and e_{N0} and e_N are the corresponding void ratios on the normally consolidation line (NCL). It can be understood that the structured soil is stiffer than non-structured over consolidated soil, when the initial void ratio e_0 are the same. Then, the change in void ratio for structured soil indicated by the arrow with solid line is smaller than that for non-structured over consolidated soil (arrow with broken line). Now, the state variable ρ (= $e_N - e$) which

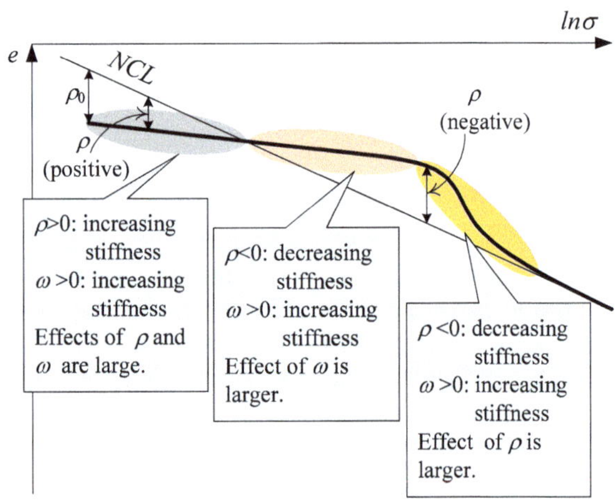

Fig. 1 Void ratio (e): ln σ relation in structured

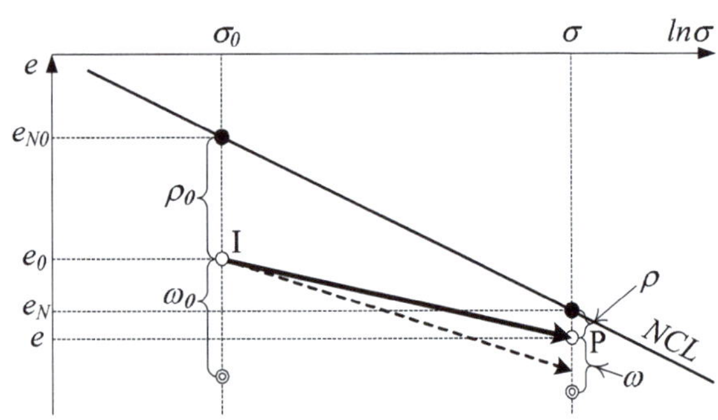

Fig. 2 Change of void ratio in structured soil

represents the effect of the density is defined by the difference between the void ratio on NCL and the current void ratio of the soil at the same stress. The initial value of ρ is expressed as $\rho_0 = e_{N0} - e_0$. To describe such increase in stiffness of structured soil, we introduce an imaginary increase of density ω (initial value ω_0) which represents the effect of the bonding, in addition to the real density factor ρ.

When stress condition moves from σ_0 to σ, the plastic change of void ratio $(-\Delta e)^p$ for the clay is obtained on referring to Fig. 2.

$$\begin{aligned}(-\Delta e)^p &= (-\Delta e) - (-\Delta e)^e = \{(e_{N0} - e_N) - (\rho_0 - \rho)\} - (-\Delta e)^e \\ &= \lambda \ln \frac{\sigma}{\sigma_0} - (\rho_0 - \rho) - \kappa \ln \frac{\sigma}{\sigma_0}\end{aligned} \tag{1}$$

Here, $(-\Delta e)^e$ is the elastic change of void ratio, and λ and κ denote compression and swelling indices. Therefore, yield function for the soil is expressed as:

$$F + \rho = H + \rho_0 \quad \text{or} \quad f = F - \{H + (\rho_0 - \rho)\} = 0 \tag{2}$$

in which, F and H denote the terms of stress and plastic change in void ratio.

$$F = (\lambda - \kappa) \ln \frac{\sigma}{\sigma_0}, \quad H = (-\Delta e)^p \tag{3}$$

From the consistency condition ($df = 0$) at the occurrence of plastic deformation with satisfying Eq. (2), the following equation is obtained:

$$df = dF - \{dH - d\rho\} = (\lambda - \kappa)\frac{d\sigma}{\sigma} - \left\{d(-\Delta e)^p - d\rho\right\} = 0 \tag{4}$$

Now, it can be considered that the evolution rule of ρ with the development of plastic deformation for structured soil is determined using both ρ and ω. Then, the evolution rules of ρ and ω can be given in the following form, using increasing functions $G(\rho)$ and $Q(\omega)$ which satisfy $G(0) = 0$ and $Q(0) = 0$, respectively:

$$d\rho = -\{G(\rho) + Q(\omega)\} \cdot d(-e)^p \tag{5}$$

The evolution rule of ω is also given as follows using the same function $Q(\omega)$:

$$d\omega = -Q(\omega) \cdot d(-e)^p \tag{6}$$

Equations (4) and (5) give the increment of the plastic change in void ratio as

$$d(-e)^p = \frac{\lambda - \kappa}{1 + G(\rho) + Q(\omega)} \cdot \frac{d\sigma}{\sigma} \tag{7}$$

The increment of total change in void ratio is expressed as the summation of the plastic components $(-\Delta e)^p$ and the elastic component $(-\Delta e)^e$.

$$d(-e) = d(-e)^p + d(-e)^e = \left\{\frac{\lambda - \kappa}{1 + G(\rho) + Q(\omega)} + \kappa\right\}\frac{d\sigma}{\sigma} \tag{8}$$

Positive ρ and ω have effects to increase stiffness of soil, because $G(\rho)$ and $Q(\omega)$ are positive when both ρ and ω are positive. At the first stage ($\rho > 0$ and $\omega > 0$), the stiffness is much larger than that of remolded NC soil because of the positive values of $G(\rho)$ and $Q(\omega)$. When the current void ratio becomes the same as that on NCL ($\rho = 0$), the stiffness of the structured soil is still greater than that of remolded NC soil (because of $\omega > 0$). In this stage ($\rho < 0$ and $\omega > 0$), the effect to increase the stiffness by positive value of ω is larger than the effect to decrease the stiffness by negative value of ρ. After this stage, the effect of ω becomes small with the development of plastic deformation. The effect of ρ to reduce the stiffness becomes prominent. Finally, the void ratio approaches to that on NCL, because ρ and ω converge to zero. Such feature of structured soil can be described by extending the defined domain of the increasing function $G(\rho)$ to the negative value of ρ—e.g., $G(\rho)$ and $Q(\omega)$ are given by increasing functions. Although $Q(\omega)$ is monotonically approaching to zero with the development of plastic deformation, $G(\rho)$ has an effect for the current void ratio to approach to that on NCL, regardless the sign of ρ.

3 One-dimensional Modeling of Time-Dependent Behavior

Time-dependent behavior of soils has been formulated employing elasto/viscoplastic theories such as non-stationary flow surface model (e.g., Sekiguchi 1977) and overstress model (e.g., Adachi and Oka 1982). We will formulate the time-dependent behavior of soils not using the past elasto/viscoplastic theories but by an extension of the above-mentioned formulation for the structured soils. Figure 3 shows a time-dependent behavior of normally consolidated soil in 1D condition schematically. It is well known that the normally consolidation line (NCL) shifts due to strain rate (rate of plastic change in void ratio $(-\dot{e})^p$), and the void ratio (e) changes linearly against time on a log scale ($\ln t$) under creep condition. In order to model these features, introducing a state valuable ψ which is given by a function of strain rate as shown in Fig. 4. Here, ψ_0 is the initial value of ψ, and point I and point P indicate the initial state ($\sigma = \sigma_0$, $e = e_0$ and $\psi = \psi_0$) and the current state($\sigma = \sigma$, $e = e$ and $\psi = \psi$), respectively. By referring to this figure, the plastic change of void ratio $(-\Delta e)^p$ in which the above features should be considered is expressed as.

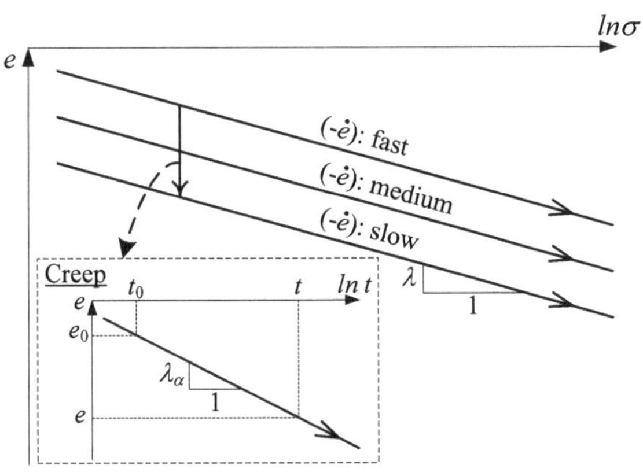

Fig. 3 Time dependent behavior of normally consolidated clay

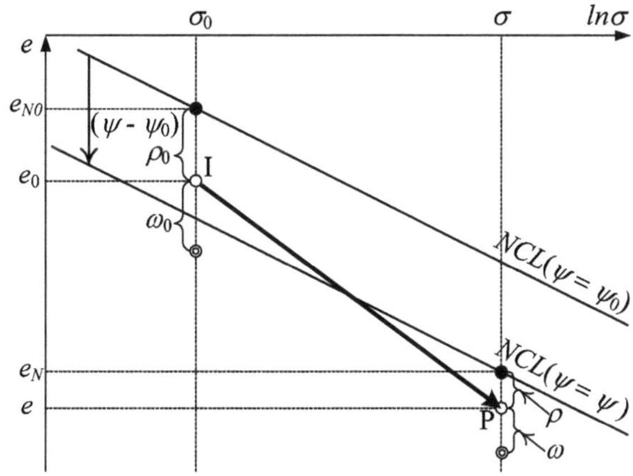

Fig. 4 Change of void ratio in structured clay with time dependent behavior

$$
\begin{aligned}
(-\Delta e)^p &= (-\Delta e) - (-\Delta e)^e = \{(e_{N0} - e_N) - (\rho_0 - \rho)\} - (-\Delta e)^e \\
&= \left\{\lambda \ln \frac{\sigma}{\sigma_0} + (\psi - \psi_0) - (\rho_0 - \rho)\right\} - \kappa \ln \frac{\sigma}{\sigma_0} \\
&= \lambda \ln \frac{\sigma}{\sigma_0} - (\rho_0 - \rho) - (\psi_0 - \psi) - \kappa \ln \frac{\sigma}{\sigma_0}
\end{aligned} \tag{9}
$$

Therefore, the following equation holds between F and H:

$$
F + \rho + \psi = H + \rho_0 + \psi_0 \quad \text{or} \quad f = F - \{H + (\rho_0 - \rho) + (\psi_0 - \psi)\} = 0 \tag{10}
$$

Consistency condition ($df = 0$) gives

$$
df = dF - \{dH - d\rho - d\psi\} = (\lambda - \kappa)\frac{d\sigma}{\sigma} - \left\{d(-e)^p - d\rho - d\psi\right\} = 0 \tag{11}
$$

Now, referring to Fig. 3, we can give the distance ($\psi - \psi_0$) in Fig. 4 as a function of the elapsed time t or the rate of plastic change in void ratio $(-\dot{e})^p$:

$$
\left\{
\begin{array}{l}
\psi - \psi_0 = \lambda_\alpha \ln \frac{t}{t_0} = \lambda_\alpha \ln t - \lambda_\alpha \ln t_0, \quad \text{or,} \\
\psi - \psi_0 = \lambda_\alpha \ln \frac{(-\dot{e})_0^p}{(-\dot{e})^p} = \{-\lambda_\alpha \ln(-\dot{e})^p\} - \left\{-\lambda_\alpha \ln(-\dot{e})_0^p\right\}
\end{array}
\right. \tag{12}
$$

Here, λ_k is the secondary compression index, and subscript 0 means the value at initial state. The increment $d\psi$ is expressed as

$$
d\psi = \frac{\partial \psi}{\partial t} dt = \lambda_\alpha \frac{1}{t} dt = (-\dot{e})^p dt \tag{13}
$$

It is assumed that Eq. (12) holds for the normally consolidated soil and the over consolidated soil and naturally deposited soil. Substituting Eqs. (5) and (13) into Eq. (11), we can obtain the increment of the plastic change of void ratio as:

$$d(-e)^p = \frac{(\lambda - \kappa)\frac{1}{\sigma}d\sigma + (-\dot{e})^p \cdot dt}{1 + G(\rho) + Q(\omega)} \cong \frac{(\lambda - \kappa)\frac{1}{\sigma}d\sigma + (-\dot{e})^{p*} \cdot dt}{1 + G(\rho) + Q(\omega)} \quad (14)$$

Here, $(-\dot{e})^{p*}$ denotes the rate of the plastic change of void ratio in the previous calculation step. Finally, the total increment of void ratio is given as:

$$d(-e) = d(-e)^p + d(-e)^e = \left(\frac{\lambda - \kappa}{1 + G(\rho) + Q(\omega)} + \kappa\right)\frac{d\sigma}{\sigma} + \frac{(-\dot{e})^{p*}}{1 + G(\rho) + Q(\omega)}dt \quad (15)$$

For easy and simple numerical calculations, the known rate $(-\dot{e})^{p*}$ in the previous step is used instead of the current rate, as described in Eqs. (14) and (15). The error caused using the previous rate is negligible in the calculations, because an incremental method with small step is used, and some error in the present step is automatically corrected in the next step. The validity of the proposed time-dependent model is checked by some simulations of one-dimensional constant strain rate consolidation tests and conventional oedometer tests. Assuming Fujinomori clay which was used in the previous experimental verification of constitutive models, following material parameters are employed in the numerical simulations—compression index $\lambda = 0.104$, swelling index $\kappa = 0.010$ and void ratio on NCL at $\sigma = \text{Pa} = 98$ kPa (atmospheric pressure) $N = 0.83$. The parameter for density $a = 100$ and the degradation parameter of bonding $b = 40$ are used.

Figure 5a shows e-logσ relations for different strain rates in normally consolidated clay under drained condition. Here, the coefficient of secondary compression $\lambda_\alpha = 0.0030$. It is seen that with the increase of strain rates the resistance to compression increases, and the lines of constant strain rate are parallel to each other which are commonly seen in laboratory tests. The solid line designated as "no creep" is the results without time effect. When the strain rate changes at a

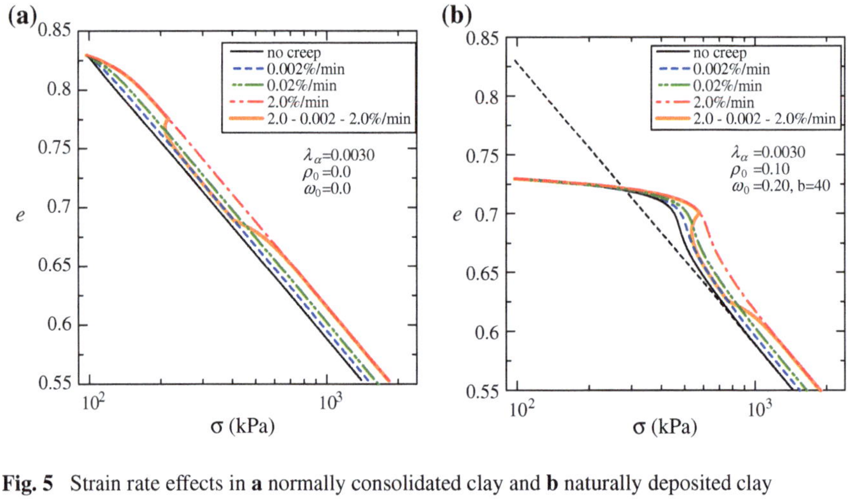

Fig. 5 Strain rate effects in **a** normally consolidated clay and **b** naturally deposited clay

certain point, the curve follows exactly the same path the new rate is supposed to follow. This is valid for both increasing and decreasing the strain rates, and in the case where the strain rate is increased the simulation gradually reaches the target curve following the phenomenon of 'isotache'. It can be said that the proposed model can well produce the strain rate dependency. Figure 5b illustrates the results of structured clay ($\omega_0 = 0.20$). The initial void ratios are 0.73. It is found that the apparent pre-consolidation stress P_c increases with the increase of strain rate.

4 Conclusions

As one-dimensional model, a simple method to describe the behavior of structured soil is presented by using the state variables of density (ρ) and bonding (ω), and their monotonous evolution rules. Next, introducing the state variable (ψ) which fixes the position of NCL depending on the strain rate, the model is extended to one considering time-dependent behavior for not only normally consolidated soil but also for over consolidated soil and structured soil. These models can easily be extended to 3D ones by introducing the t_{ij} concept. The detailed explanation of the present 1D and 3D modeling is presented in Nakai et al. (2011) and Nakai (2012).

References

Adachi T, Oka F (1982) Constitutive equation for normally consolidated clays based on elasto/viscoplasticity. Soils Found 22(4):57–70

Asaoka A, Noda T, Yamada E, Kaneda K, Nakano M (2002) An elasto-plastic description of two distinct volume change mechanisms of soils. Soils Found 42(5):47–57

Hashiguchi K (1980) Constitutive equation of elastoplastic materials with elasto-plastic transition. J Appl Mech ASME 102(2):266–272

Nakai T (2012) Constitutive modeling of geomaterials: principles and applications. CRC Press, Boca Raton, p 376

Nakai T, Mihara Y (1984) A new mechanical quantity for soils and its application to elastoplastic constitutive models. Soils Found 24(2):82–94

Nakai T, Shahin HM, Kikumoto M, Kyokawa H, Zhang F, Farias MM (2011) A simple and unified one-dimensional model to describe various characteristics of soils. Soils Found 51(6):1129–1148

Sekiguchi H (1977) Rheological characteristics of clays. In: Proceedings of 9th ICSMFE, vol 1, Tokyo, pp 289–292

The Role of Microstructure in the Liquefaction Mechanism

G. Veylon and F. Nicot

Abstract Microdirectional models involving a mesoscopic scale constitute a good alternative to phenomenological models. In the H-microdirectional model, the granular assembly is modeled by a distribution of hexagonal patterns of grains in contact oriented in space. In this paper, we investigate the micromechanical origin of the liquefaction observed from the H-microdirectional model in the light of an analysis of individual behavior of each hexagon. It is shown that the principal direction of anisotropy determines the liquefaction susceptibility of a soil, whereas the degree of anisotropy affects the amplitude of the post-peak loss of shear strength.

1 Introduction

Predicting soil instability remains a major challenge today, from both academic and engineering purposes. Liquefaction was broadly observed and described on loose materials loaded in isochoric conditions. For example, over undrained triaxial loading paths, the deviatoric stress reaches a peak, and then decreases sometimes to zero (liquefaction), or not (partial liquefaction). The microstructural reasons behind are not fully understood. In addition to the influence of the initial void ratio, other microstructural features are likely to govern the macroscopic behavior, and the ability of the material to liquefy. In this perspective,

G. Veylon (✉)
Irstea, Civil Engineering Group, OHAX, 3275 route de Cézanne, CS 40061,
13182 Aix-en-Provence Cedex 5, France
e-mail: guillaume.veylon@irstea.fr

F. Nicot
Irstea, Geomechanics Group, ETNA, 2 rue de la Papeterie, BP 76,
38402 Saint-Martin d'Hères, France
e-mail: francois.nicot@irstea.fr

K.-T. Chau and J. Zhao (eds.), *Bifurcation and Degradation of Geomaterials in the New Millennium*, Springer Series in Geomechanics and Geoengineering,
DOI 10.1007/978-3-319-13506-9_49

developing a micromechanically-based model is appealing since an effort is made to relate the macroscopic behavior to microscopic features. This novel road enables interpreting the main constitutive features from microstructural, clear arguments. The H-microdirectional model (Nicot and Darve 2011) was developed along this line of reasoning. Three basic scales are considered: the granular assembly scale (referred to as the macroscopic scale), an intermediate scale made up of hexagonal patterns of contacting grains (mesoscopic scale), and the elementary contact scale between adjoining grains (microscopic scale). This manuscript investigates how the topologic properties on the intermediate scale can determine the macroscopic susceptibility of the specimen to static liquefaction.

2 The H-microdirectional Model

The H-microdirectional model (Nicot and Darve 2011) was developed in a two-dimensional reference frame $(\vec{x}_1, \vec{x}_2)$. The granular assembly is described as a distribution of regular hexagonal patterns of circular grains with a radius r_g, oriented along the symmetry direction $\vec{n}$. In order to take into account the geometric effects of the angularity of real grains, no rolling is assumed to affect the grains (Oda et al. 1982).

The incremental deformation of each hexagon is assumed to derive from the macroscopic strain increment applied at the specimen scale. If we choose the principal directions as the reference frame, the incremental deformation of each hexagon is determined as:

$$\delta l_n = -l_n \delta\varepsilon_{ij} n_i n_j = -l_n \left(\delta\varepsilon_1 \cos^2\theta + \delta\varepsilon_2 \sin^2\theta \right) \quad (1)$$

$$\delta l_t = -l_t \delta\varepsilon_{ij} t_i t_j = -l_t \left(\delta\varepsilon_1 \sin^2\theta + \delta\varepsilon_2 \cos^2\theta \right) \quad (2)$$

where $\delta\varepsilon_1$ and $\delta\varepsilon_2$ are the principal components of the incremental strain tensor and θ is the orientation angle between $\vec{x}_1$ and $\vec{n}$ (see Fig. 1). Given the symmetries of the hexagonal pattern, its geometric description is defined by three geometrical variables: d_1 the branch vector length between grains 1 and 2, d_2 the branch vector length between grains 2 and 3 and α the opening angle of the hexagon (see Fig. 2).

It can be shown (Nicot and Darve 2011) that the principal components S_1 and S_2 of the Cauchy stress acting on the intermediate scale are:

$$v(\vec{n})S_1 = 4N_1 d_1 \cos^2\alpha + 4T_1 d_1 \cos\alpha \sin\alpha + 2N_2 d_2 \quad (3)$$

$$v(\vec{n})S_2 = 4N_1 d_1 \sin^2\alpha - 4T_1 d_1 \cos\alpha \sin\alpha + 2N_3 l_t \quad (4)$$

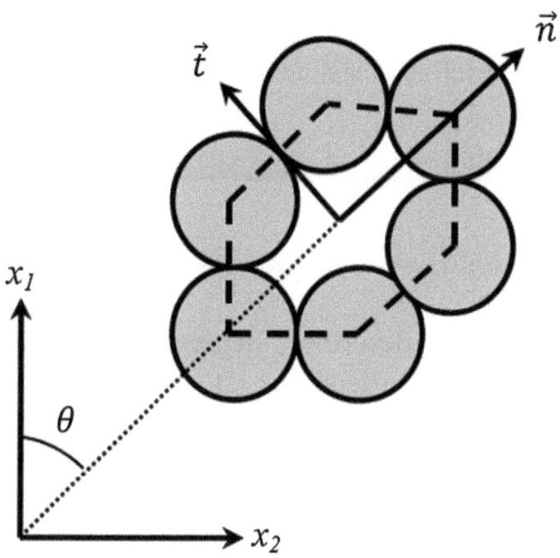

Fig. 1 Hexagonal pattern of contacting particles

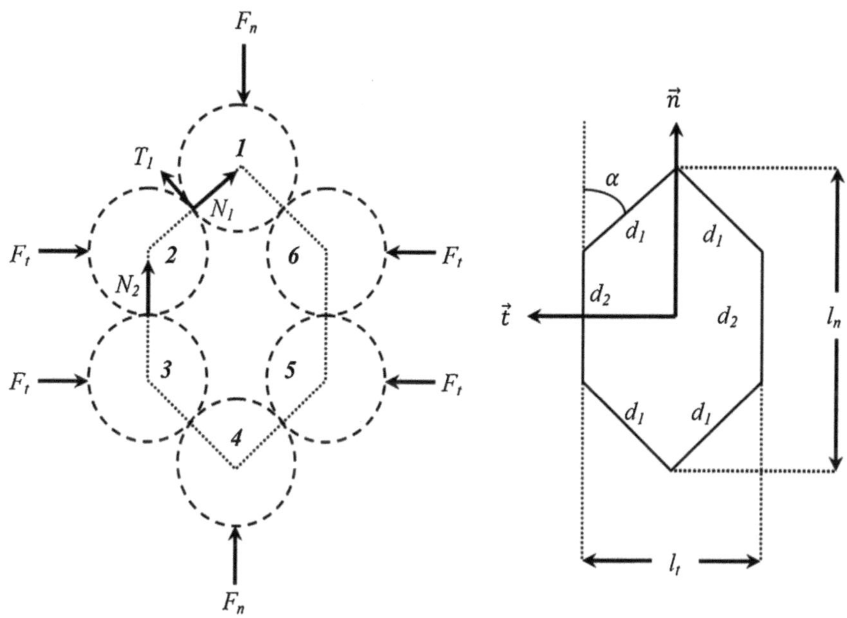

Fig. 2 Contact and external forces and geometrical description of hexagonal pattern

where $v(\vec{n})$ stands as the surface of the hexagon oriented along the direction $\vec{n}$ and N_3 the normal contact force between grains 2 and 6 ($\alpha \leq \pi/6$). In addition to the parameters describing the local behavior at the contact scale, the H-microdirectional model requires other parameters to describe the microstructure of the granular assembly: the opening angle distribution $\alpha(\theta)$ and the distribution of hexagons over the two-dimensional plane $\omega(\theta)$. Thus, anisotropy can be introduced by a second order tensorial development of functions $\alpha(\theta)$ and $\omega(\theta)$:

$$\omega(\theta) = \frac{N_H}{2\pi}(1 + a_\omega \cos 2(\theta - \beta_\omega)) \tag{5}$$

$$\alpha_0(\theta) = \overline{\alpha_0}(1 + a_\alpha \cos 2(\theta - \beta_\alpha)) \tag{6}$$

where a_ω (<1) is a parameter describing degree of anisotropy of the distribution and β_ω is the major principal orientation of the distribution. The parameter a_α describes the degree of anisotropy of the initial opening angle distribution and β_α is the direction of maximum opening angle. The parameter $\overline{\alpha_0}$ represents the mean value of initial opening angle. The opening angle is bounded by a minimum value of $\pi/6$ and a maximum value of $\pi/2$.

3 Micromechanical Analysis

In the present study, the mechanical response of the granular assembly will be assessed through two loading phases: (i) isotropic consolidation to a mean pressure equal to 100 kPa, (ii) biaxial test under constant volume conditions ($\delta\varepsilon_2 = -\delta\varepsilon_1$). The response of the granular assembly will be represented in the (ε_1, t) and (s', t) planes where $s' = (\sigma_1 + \sigma_2)/2$ and $t = \sigma_1 - \sigma_2$.

3.1 Mono-crystalline Models

This paragraph aims to analyze the influence of the microstructural parameters at the hexagon scale considering mono-crystalline models. Let a granular assembly be composed of N_H hexagons which are all oriented along the same direction defined of the angle θ and have the same initial opening angle α_0. In the present study, the influence of initial opening angle and orientation of the hexagons are analyzed. As the number of hexagons only influences the quantitative response of the sample, its influence will not be discussed. In the following, the parameters for model simulations are summarized in Table 1.

The opening angle of hexagons is of crucial interest regarding the volumetric behavior of the granular assembly. Depending on the initial configuration of the hexagons, the material can exhibit a contractive behavior or a completely dilative

Table 1 Model parameters for model simulations

Test	r_g (mm)	k_n (kN/m)	k_t (kN/m)	φ_g (°)	N_H
R0	0.175	124.6	52.3	30	2×10^6

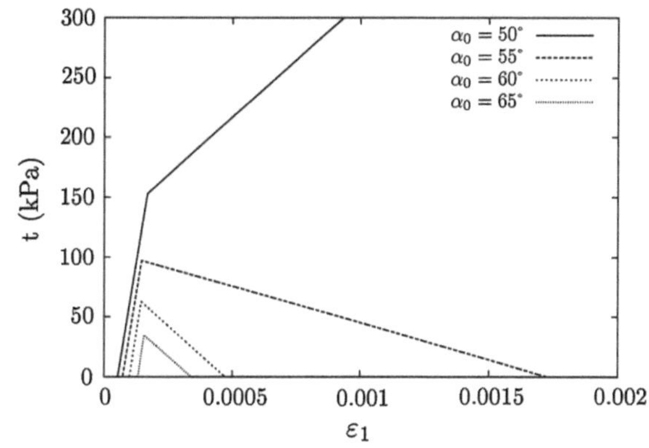

Fig. 3 Influence of initial opening angle α_0 of hexagons on stress deformation curves for $\theta = 0°$

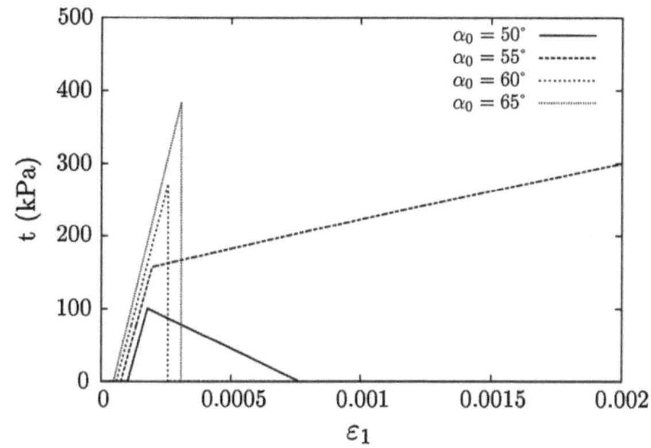

Fig. 4 Influence of initial opening angle α_0 of hexagons on stress deformation curves for $\theta = 90°$

response under constant volume biaxial test. When the hexagons are oriented in the principal loading direction, they are subjected to compression (Fig. 3). Whatever the initial opening angle, the elastic moduli are substantially the same but the peak strength decreases as the initial opening angle of the hexagons increases. Similarly, from a threshold value of the opening angle (around 53.5°), the bifurcation point is followed by a descending branch (softening regime).

When the hexagons are oriented in the lateral direction (minor principal direction), they undergo a tensile loading (Fig. 4). The elastic moduli are also poorly influenced by the initial opening angle of the hexagons. The maximum shear strength in the elastic regime increases with the initial opening angle of the hexagons. Once reached the limit of the elastic domain, it can be observed: (i) a gradual softening behavior for opening angles lower than the threshold defined above (around 53.5°), (ii) a sudden collapse of the material shear strength for opening angles exceeding 60°, (iii) a hardening behavior for intermediate values of the opening angle (between the threshold value and 60°).

3.2 Poly-crystalline Models

3.2.1 Distribution of Initial Opening Angles

According to Eq. (6), the anisotropy can be introduced by varying the initial opening angle (see Fig. 2) with the orientation of the hexagons defined by vector $\vec{n}$ (see Fig. 1). It is worth noting that the distribution of hexagons is isotropic. The average initial opening angle is taken equal to $\overline{\alpha_0} = 50°$. Two principal direction of anisotropy are examined: (i) $\beta_\alpha = 0°$ where hexagons oriented along the principal loading direction are in compression regime, while those that are oriented in the perpendicular direction are elongated; (ii) $\beta_\alpha = 90°$ which is the opposite configuration.

First of all, it is shown (Fig. 5) that no liquefaction was triggered for $\beta_\alpha = 90°$, whatever the degree of anisotropy (only one curve with $a_\alpha = 0.1$ is reported in Fig. 6). In contrast, liquefaction was triggered for $\beta_\alpha = 0°$ for some values of a_α. Thus, the orientation of anisotropy strongly governs the onset of liquefaction. The influence of the degree of anisotropy is examined by assuming $\beta_\alpha = 0°$ and varying the a_α parameter between 0 and 0.2 (Fig. 5).

When the distribution of opening angles is isotropic, the stress-strain curve shows a plateau followed by a hardening phase. As the initial opening angle increases, a peak shear strength is observed followed by a softening phase. For $a_\alpha = 0.05$, there is an increase of shear resistance after reaching a minimum shear strength. It is worth noting that increasing the degree of anisotropy of the material reduces the peak shear strength and extends the softening phase towards larger strains.

3.2.2 Distribution of Orientations

As seen in Eq. (5), anisotropy can also be introduced by varying the distribution of hexagons with their orientation (initial opening angle is assumed constant). As for the opening angle anisotropy model, the principal direction of anisotropy

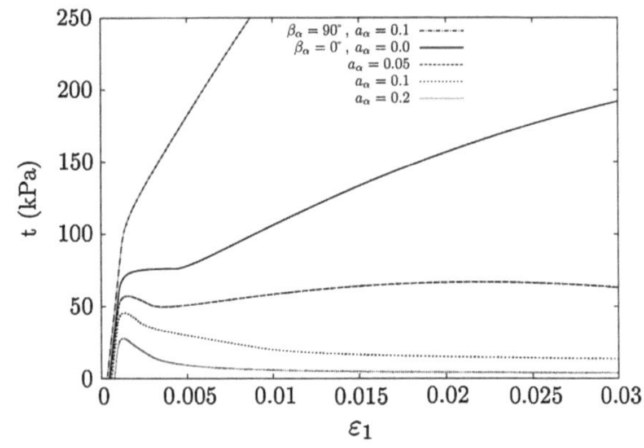

Fig. 5 Stress-deformation curves for different distributions of opening angle

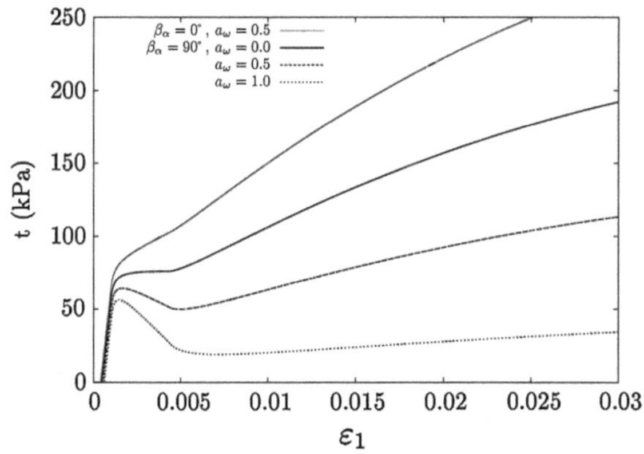

Fig. 6 Stress-deformation curves for different distributions of hexagons

determines the onset of liquefaction: no liquefaction was triggered for $\beta_\omega = 0^\circ$ whereas liquefaction was triggered for $\beta_\alpha = 0^\circ$.

Moreover, the anisotropy degree influences the peak shear strength and the extension of the softening phase. However, from a quantitative point of view, these effects are more limited than those observed by considering the opening angle anisotropy. As the softening phase is immediately followed by a hardening phase, it was not possible to reproduce a complete liquefaction of the material just by introducing anisotropy in the hexagon distribution.

4 Conclusions

In this paper, we studied the influence of microstructure anisotropy on the material response by using the H-microdirectional model (Nicot and Darve 2011) and varying the orientation distribution of hexagons and their initial opening angle. We showed that by just changing the microstructural parameters, it is possible to model a wide range of soil behavior types observed experimentally (Castro 1969; Konrad 1990; Kramer and Seed 1988; Ishihara 1993; Benahmed et al. 2004). The importance of the principal direction and degree of microstructural anisotropy on the onset of liquefaction was highlighted. In particular, it is shown that the principal direction of anisotropy determines the liquefaction susceptibility of a soil, whereas the degree of anisotropy affects the amplitude of the post-peak loss of shear strength.

References

Benahmed N, Canou J, Dupla JC (2004) Structure initiale et propriétés de liquéfaction statique d'un sable. Comptes Rendus de Mécanique 332(11):887–894

Castro G (1969) Liquefaction of sands. In: Press HU (ed) Harvard soil mechanics series. Massachusset, Cambridge

Ishihara K (1993) Liquefaction and flow failure during earthquakes. Géotechnique 43(3):351–415

Konrad JM (1990) Minimum undrained strength versus steady-state strength of sands. J Geotech Eng 116(6):948–963

Kramer SL, Seed HB (1988) Initiation of soil liquefaction under static loading conditions. J Geotech Eng 114(4):412–430

Nicot F, Darve F (2011) The H-microdirectional model: accounting for a mesoscopic scale. Mech Mater 43:918–929

Oda M, Konishi J, Nemat-Nasser S (1982) Experimental micromechanical evaluation of strength of granular materials: effects of particle rolling. Mech Mater 1(4):269–283

Dense Granular Flows: Structure and Microscopic Origins of Internal Friction

E. Azéma and F. Radjai

Abstract By means of three-dimensional Contact Dynamics simulations, we investigate the microstructural properties of sheared granular packings by varying systematically the inertial number I from the quasi-static to the collisional regime. A detailed analysis of the connectivity of the particles reveals that the transition between the regimes is correlated to profound changes of the nature of the contact network. In particular, we show that the increase of internal friction angle with inertial number is mainly a consequence of contact anisotropy. As the inertial number increases, the role of force chains decreases and that of friction mobilization increases as long as percolating force chains are present in the system. A detailed analysis of the spatial correlations of floating particles, i.e., particles without bearing contacts, reveals the occurrence of "fluidized" volumes in the dense flow regime, which increase in size and coalesce at the approach of the collisional regime.

1 Introduction

Over the past, important advances have been made, both experimentally and numerically, in the physical understanding of granular flows from a continuum point of view (Azema et al. 2012; Baran et al. 2006; Pouliquen 1999; Savage 1984; Silbert et al. 2001). In particular, it is now well accepted that the local rheological law, i.e. the relations between normal stress σ_n, shear stress σ_t, shear rate $\dot{\gamma}$ and solid fraction can be formulated in terms of the dimensionless inertial number

E. Azéma (✉) · F. Radjai
LMGC, Université Montpellier 2, CNRS, Place Eugene Bataillon,
34095 Montpellier Cedex 05, France
e-mail: emilien.azema@univ-montp2.fr

F. Radjai
e-mail: franck.radjai@univ-montp2.fr

K.-T. Chau and J. Zhao (eds.), *Bifurcation and Degradation of Geomaterials in the New Millennium*, Springer Series in Geomechanics and Geoengineering,
DOI 10.1007/978-3-319-13506-9_50

$I = \dot{\gamma} d \sqrt{\rho/\sigma_n}$, where d is the mean diameter of the particles (da Cruz et al. 2005; GDR-MiDi 2004). According to this description, quasi-static critical state regime corresponds to very low values of I. The dense flow regime is identified with a steady flow characterized by increasing friction μ and decreasing packing fraction ν as I increases. The fully collisional flow regime occurs at larger values of I, and it corresponds to a saturation of μ (da Cruz et al. 2005; GDR-MiDi 2004; Pouliquen 1999).

This empirical model of steady granular flows, in combination with continuum conservation equations, correctly predicts the velocity and stress fields in various flow geometries (GDR-MiDi 2004; Forterre and Pouliquen 2008). However, beyond those "broadly defined" rheological features, the grain-scale mechanisms leading to such variations of the effective friction are still poorly understood, and furthermore difficult to analyze experimentally. It is therefore necessary to consider primary mechanisms at the natural scale of grains and their interactions in order to connect particle-scale mechanisms with the rheology. In this paper, we use contact dynamics simulation to analyze the grain-scale origins of the rheology of granular flows. In the following, we first introduce the numerical approach, then in Sect. 3, we focus on shear strength and packing fraction for a wide range of I. The microstructure is analyzed in Sect. 4 in terms of connectivity, fabric and force anisotropies. Stress partition has been successfully applied to quasistatic deformations (Azéma et al. 2013, 2009; Ouadfel and Rothenburg 2001) but never to inertial flows. As we shall see below, the relative importance of local mechanisms evolves with I and therefore the flow structure undergoes qualitative transitions that underlie the evolution of the effective friction.

2 Numerical Procedures

Contact dynamics simulations were carried out with $N_p = 24{,}000$ spheres. The packing is sheared between two parallel rough walls by imposing a constant horizontal velocity V_y on the top wall and periodic boundary conditions along the flow in the y direction and along the transversal x direction. The packing is confined by a constant compressive stress σ_{zz} applied on the top wall along the z-direction; see Fig. 3b. A small polydispersity $(1 \pm 0.1)d$ in particle diameters is introduced to prevent crystallization at the walls. The gravity is set to zero in order to avoid strain localization at the boundaries and to ensure uniform stress field in the bulk. In the contact dynamics method, the particles are treated as perfectly rigid so that I is the only relevant dimensionless parameter of flow (Radjaï and Richefeu 2009). Contact dissipation is modeled in terms of normal and tangential restitution coefficients e_n and e_t as well as a friction coefficient μ_s between particles. We set $\mu_s = 0.4$ and $e_n = e_t = 0$. This choice corresponds to a highly dissipative packing (da Cruz et al. 2005; GDR-MiDi 2004).

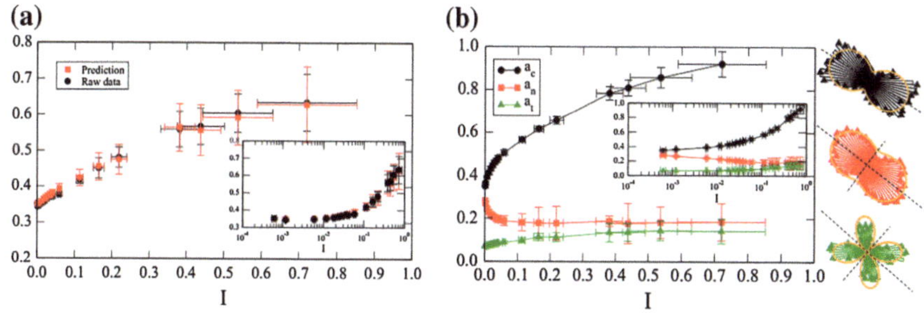

Fig. 2 **a** Effective friction coefficient μ (*black circles*) together with its harmonic approximation (Eq. 2) (*red squares*) as a function of inertial number I. **b** Evolution of anisotropy parameters as a function of I in log-linear and linear representation. The polar diagrams of $P_n(\theta)$ (*black circle*), $\langle f_n \rangle(\theta)$ (*red square*) are shown for $I \sim 0.1$ together with their fits (*plain orange line*) (Eq. 1)

In general, a_c varies oppositely with the coordination number z, which declines in our simulations from 4.5 to 1.1 as I varies from 10^{-4} to 0.7. This is because the contact anisotropy is mainly a consequence of the loss of contacts in the extension direction (Azéma et al. 2013). The normal force anisotropy a_n reflects the force chains, which are increasingly destabilized by particle inertia causing a_n to decrease. Hence, the friction force anisotropy a_t, which reflects friction mobilization $(\langle |f_t| \rangle / \langle f_n \rangle \propto a_t)$, grows and re-stabilizes the force chains and thus takes more actively part in force transmission (Azéma et al. 2013).

Since the contact anisotropy seems to be sensibly important with respect to the effective friction, we now turn to the connectivity of the contact network in order to obtain morphological clues to the evolution of the force network. At lowest order, the connectivity of the particles is characterized by the proportion P_0 of floating particles (i.e. particle with no bearing forces). Figure 3a displays P_0 as a function of I. It can be seen that, $P_0 \simeq 0.07$ for $I < 0.01$ and increases slightly to 0.2 at $I = 0.1$. But remarkably, beyond $I = 0.1$, P_0 undergoes a sharp increase until $P_0 \simeq 0.9$ for our highest values of I. In other words, the flow is dominated by collisions. At $I \simeq 0.1$ a fraction $P_0^* = 0.2$ of particles is floating and this fraction of particles disconnected from the force network is consistent with the site percolation threshold of a hexagonal compact packing. This means that, beyond $I \simeq 0.1$, the effect of inertia may be analyzed more sensitively in terms of the evolution of floating particles.

Figure 3b shows four snapshots of floating particles for different values of I. At low I the floating particles are mostly isolated, but they tend to cluster into fluidized zones as I increases. For each floating particle i, let $P_0^i(I, R)$ be the proportion of floating particles for a spherical volume of radius R centered on i. We determine the size R^i of the fluidized zone by requiring $P_0^i(I, R^i) = P_0^*$. In words, this is the size of the spherical volume in which the density of floating particles is equal to the percolation threshold P_0^*. Figure 3 displays the clustering length R_{mean} as a function of I. We see that R_{mean} is nearly constant and equal to $2d$ for $I < 0.1$, as expected for isolated floating particles. Thereafter, R_{mean} grows rapidly with I and

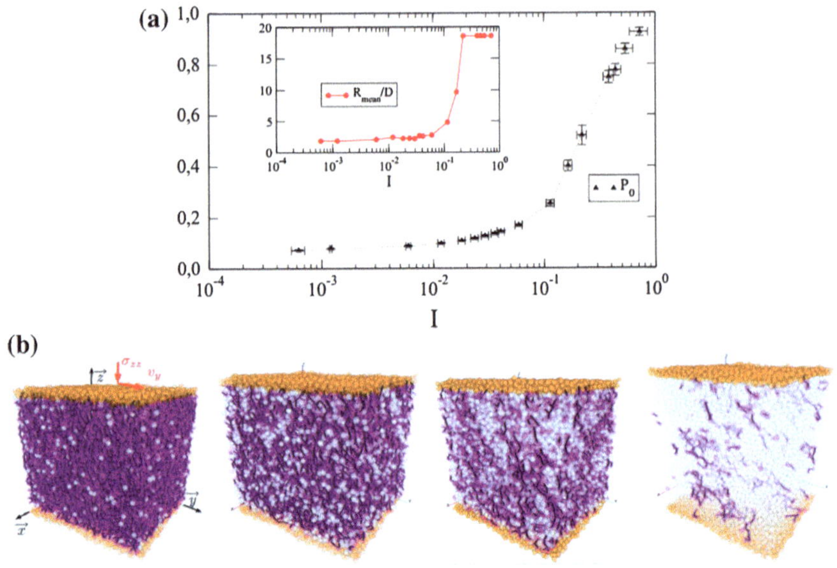

Fig. 3 **a** Proportion of floating particles as a function of *I*. The *inset* shows the mean size of critical floating clusters as a function of *I*. **b** Floating particles are in *grey*, non-floating in *violet* for $I \sim 10^{-3}$; 0.1; 0.21; 0.6

reaches the system size for $I \simeq 0.3$. This evolution reflects the coalescence process of fluidized zones and a transition to the collisional regime at $I \simeq 0.3$ where $P_0 > 0.6$.

5 Conclusions

To summarize, by means of additive decomposition of the shear stress, we have numerically shown that the contact anisotropy provides the main contribution to the effective shear friction of the flow. The contact anisotropy grows due to enduring force chains and by loss of contacts for $I < 0.1$ and due to both impulsive force chains and shear-induced collisions of floating particles beyond $I = 0.1$. Hence, in the whole range $I > 0.1$, the microstructure may be described as composed of short-lived and impulsive force chains embedded in a "soup" of floating particles. Thus sheared granular systems appears as two phase materials involving fluidized zones made of floating particles and solid skeleton made of non-floating particles. The influence of material parameters such as particle shape and the coefficient of friction will be investigated in a forthcoming work.

References

Azema E, Descantes Y, Roquet N, Roux JN, Chevoir F (2012) Discrete simulation of dense flows of polyhedral grains down a rough inclined plane. Phys Rev E 86:031303

Azéma E, Radjai F, Dubois F (2013) Packings of irregular polyhedral particles: strength, structure, and effects of angularity. Phys Rev E 87:062203

Azéma E, Radjai F, Peyroux R, Saussine G (2009) Quasistatic rheology, force transmission and fabric properties of a packing of irregular polyhedral particles. Mech Mater 41(6):729–741

Baran O, Ertas D, Halsey T, Grest G, Lechman J (2006) Velocity correlations in dense gravity-driven granular chute flow. Phys Rev E 74:05130200

da Cruz F, Emam S, Prochnow M, Roux JN, Chevoir F (2005) Rheophysics of dense granular materials: discrete simulation of plane shear flows. Phys Rev E 72:021309

Forterre Y, Pouliquen O (2008) Flows of dense granular media. Annu Rev Fluid Mech 40:1–24

GDR-MiDi (2004) On dense granular flows. Euro Phys J E 14(4): 341–365

Ouadfel H, Rothenburg L (2001) 'Stress-force-fabric' relationship for assemblies of ellipsoids. Mech Mater 33(4):201–221

Pouliquen O (1999) Scaling laws in granular flows down rough inclined planes. Phys Fluids 11(3):542–548

Radjaï F, Richefeu V (2009) Contact dynamics as a nonsmooth discrete element method. Mech Mater 41(6):715–728

Savage S (1984) The mechanics of rapid granular flows. Adv Appl Mech 24:289–366

Silbert L, Ertas D, Grest G, Halsey TC, Levine D, Plimpton SJ (2001) Granular flow down an inclined plane: bagnold scaling and rheology. Phys Rev E 64:051302

Dynamic Analysis of Strain Localization in Water-Saturated Elasto-Viscoplastic Material

F. Oka, B. Shahbodagh Khan, M. Mirjalili and S. Kimoto

Abstract Although strain localization has been studied for the last four decades, there are still many problems, in particular for the dynamic localization problem. In the present study, we have numerically analyzed a dynamic strain localization problem for the water-saturated elasto-viscoplastic constitutive material. For the material model, the nonlinear kinematic hardening rule and softening due to the structural degradation of soil skeleton are considered. In order to numerically simulate the large deformation phenomenon in strain localization analysis, the dynamic finite element formulation for a two-phase mixture using the updated Lagrangian method is adopted. The shear band development is discussed through the distributions of the viscoplastic shear strain in a material.

1 Introduction

Strain localization or shear banding is a phenomenon in materials, in which plastic deformations localize into finite narrow bands of intense straining. The phenomenon is described as a precursor to fracture and, in general, as a characteristic feature of inelastic deformations. Compared with the study under quasi-static

F. Oka (✉) · S. Kimoto
Kyoto University, Kyoto 615-8540, Japan
e-mail: oka.fusao.38x@st.kyoto-u.ac.jp

B. Shahbodagh Khan
School of Civil and Environmental Engineering, University of New South Wales, Sydney, NSW 2052, Australia

M. Mirjalili
ABRAND Civil Design Inc., Tehran, Iran

S. Kimoto
Department of Civil and Earth Resources Engineering, Kyoto University, Kyoto 615-8540, Japan

K.-T. Chau and J. Zhao (eds.), *Bifurcation and Degradation of Geomaterials in the New Millennium*, Springer Series in Geomechanics and Geoengineering,
DOI 10.1007/978-3-319-13506-9_51

conditions, the strain localization under dynamic loading conditions (Loret and Prevost 1991; Bigoni 2012) are not well studied. The aim of the present study is to extend the finite element analysis of shear banding in water-saturated elasto-viscoplastic clays to the analysis under dynamic transient loading conditions. The development of shear bands is examined numerically without resorting to the usual bifurcation or stability analysis. A cyclic elasto-viscoplastic constitutive model is adopted to simulate the behavior of clay under dynamic loading conditions. The structural degradation of the soil skeleton is considered in the constitutive model as the strain softening with respect to the viscoplastic strain. The finite element equations based on the *u-p* formulation and the Biot's type two-phase mixture theory are derived in the framework of finite deformation theory with updated Lagrangian description.

2 Cyclic Elasto-Viscoplastic Constitutive Model

The cyclic elasto-viscoplastic constitutive model presented in this paper follows the concept of structural degradation as described by Kimoto and Oka (2005), with the inclusion of the nonlinear kinematic hardening rule (Armstrong and Frederick 1966; Oka and Kimoto 2012). The model considers the structural degradation of the soil skeleton by shrinkage of both the static yield surface and the overconsolidation boundary surface with respect to the accumulation of viscoplastic strain. For the kinematical hardening, we adopted two hardening variables for both the changes of stress ratio and the mean effective stress (Kimoto et al. 2013).

3 Finite Element Formulation

For the dynamic analysis of strain localization, the *u-p* formulation based on the updated Lagrangian method is employed with the Jaumann rate of Cauchy stress tensor. The grain particles in the soil skeleton are assumed to be incompressible. The distribution of porosity in the soil and the distribution of body force in space are assumed to be smooth. In addition, the density levels of the soil skeleton and the pore fluid are assumed to be spatially homogeneous. An isoparametric 20-node hexahedron element with a reduced Gaussian integration is adopted for the displacement, the velocity, and the acceleration of the solid skeleton in three-dimensional analysis. For coupled formulation, the restriction imposed by Babuska-Brezzi condition does not allow the use of equal-order interpolation for all field variables. Therefore, the 8-node hexahedron element is used for the pore pressure.

4 Equation of Motion for Two-Phase Mixture

In nonlinear dynamic finite element analysis involving large deformations with incremental law, it is necessary to linearize the equation of motion. But, solving the total form of equation of motion is more fundamental than solving the incremental form. It is possible to use an incremental formulation if the calculation error is completely zero. But if not, the error will accumulate. In the present formulation, no rate-form equation of motion is adopted in the updated Lagrangian scheme.

In the Lagrangian formulation, as the reference state is referring to the configuration at time t, the equation of motion at time $t + \Delta t$ becomes

$$[\rho(a_i - b_i) - \Pi_{ji,j}]_{t+\Delta t} = 0 \quad (1)$$

in which ρ is the mass density of the fluid saturated soil, a_i is the acceleration vector of the soil skeleton, b_i is the body force, and Π_{ij} is the nominal stress tensor. For simplicity, we neglect a notation of time $t + \Delta t$ in the following. The boundary conditions for the mixture are considered as

$$\Pi_{ji}N_j = \bar{t}_i \quad \text{on } \Gamma_t \quad (2)$$

$$v_i = \bar{v}_i \quad \text{on } \Gamma_u \quad (3)$$

where Γ_t is the traction boundary surface, Γ_u is the displacement boundary surface, N_j is an outward unit normal vector of the surface Γ_t, $\bar{t}_i$ is the traction vector at time $t + \Delta t$, v_i is the velocity vector of the solid phase, and superimposed bar indicates the prescribed values.

The nominal stress Π_{ij} at time $t + \Delta t$ can be approximated by

$$\Pi_{ji|t+\Delta t} \cong \Pi_{ji|t} + (\Delta t)\dot{\Pi}_{ji} \quad (4)$$

where superimposed dots indicate time differentiations.

In the updated Lagrangian method the configuration at time t, i.e. the latest known configuration, is considered as a reference configuration, and the configuration at time $t + \Delta t$ is unknown. A weak form of Eq. (1) becomes

$$\int_V \rho a_i \delta v_i dV + \int_V T_{ij|t} \delta L_{ij} dV + (\Delta t) \int_V \dot{\hat{S}}_{ji} \delta L_{ij} dV = \int_{\Gamma_t} \bar{t}_i \delta v_i d\Gamma + \int_V \rho b_i \delta v_i dV \quad (5)$$

where $\dot{\hat{S}}_{ij}$ is the nominal stress rate tensor with respect to the configuration at time t, $\dot{\hat{S}}'_{ij}$ is the effective nominal stress rate tensor, T_{ij} is the Cauchy stress tensor, L_{ij} is the velocity gradient tensor of the solid phase and the changes in area can be estimated by Nanson's formula in continuum mechanics.

5 Numerical Analysis

Weak forms of the continuity equations for water and air and the rate type of conservation of momentum are discretized in space and solved by the finite element method. In the finite element formulation, the independent variables are the pore water pressure, the pore air pressure, and the nodal velocity. A 20-node hexahedron element with a reduced Gaussian integration is used for the displacement, and eight nodes are used for the pore water pressure and the pore air pressure, shown in Fig. 1. The backward finite difference method is used for the time

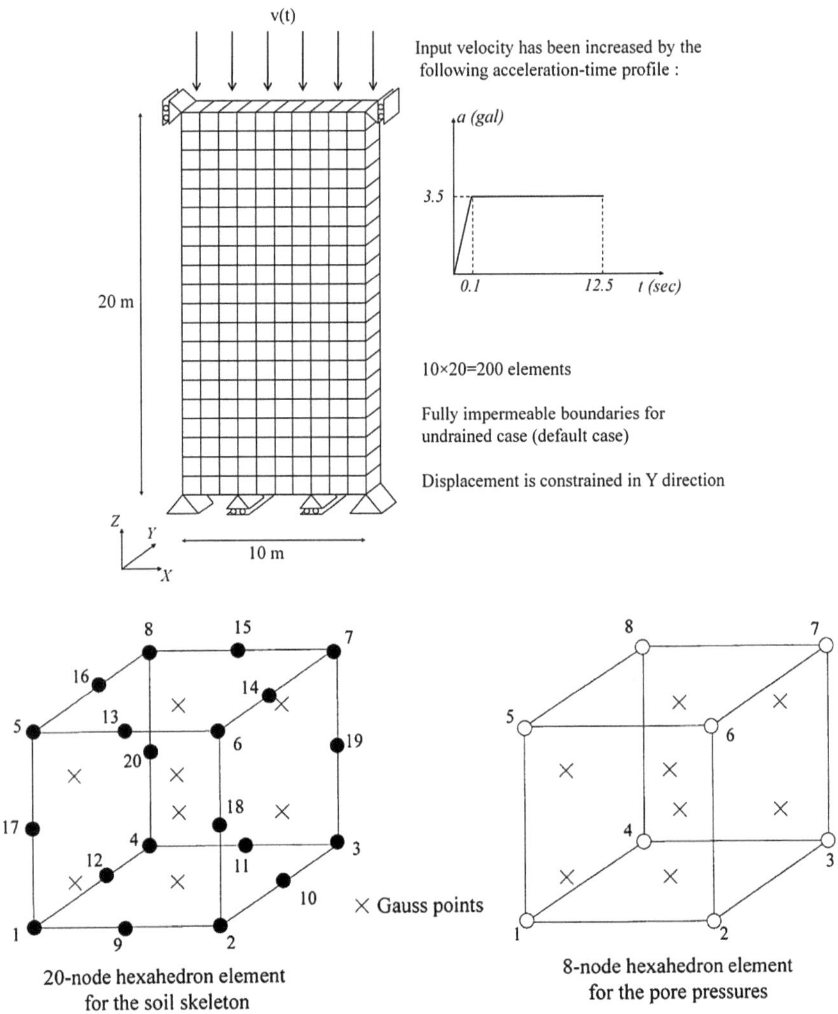

Fig. 1 Isoparametric elements and finite element mesh with boundary conditions

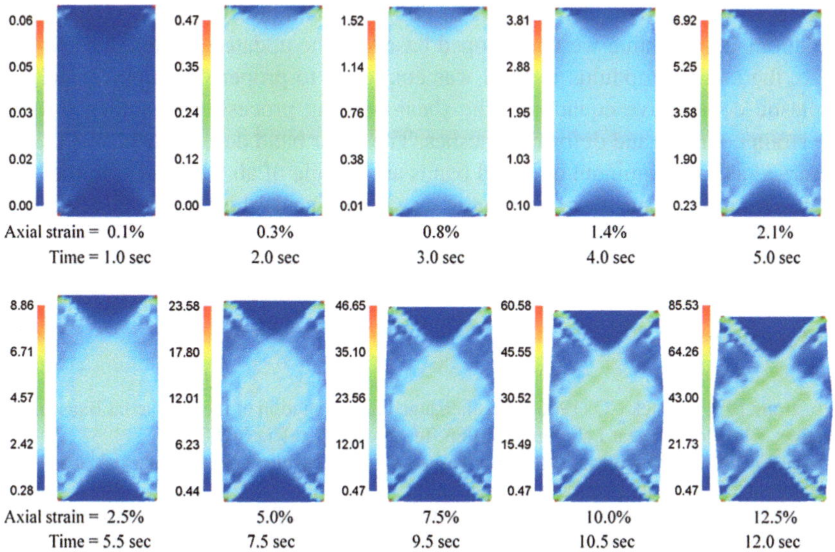

Fig. 2 Distribution of accumulated viscoplastic shear strain

discretization. The finite element mesh, the boundary conditions and the acceleration-time profile for the simulations are shown in Fig. 1.

The shear band development is studied through the evaluation of the results for the accumulated viscoplastic shear strain, the axial strain and the pore water pressure. Figure 2 demonstrates the distribution of accumulated viscoplastic shear strain in the specimen through the loading. The strain localization emerges from the trigger points at the corners of the specimen, which eventually narrows to four distinct diagonal bands by increasing the overall axial displacement. The distinct shear banding initially appears at time 5 s under the overall axial strain of about 2.1 %.

For saturated clay, Oka et al. (1995) demonstrated that by using the viscoplastic model, the preferred orientation of shear bands is 45° under plane strain locally undrained condition, i.e., $k = 0$. In the present analysis, since the coefficient of permeability is very small, and the load is applied in very short time duration, the condition is very similar to locally undrained conditions, which makes the shear bands occur at an inclination angle of about 45°.

6 Conclusions

A computational framework was presented for dynamic strain localization and deformation analyses of water-saturated clay. The dynamic behavior of clay was represented by a cyclic elasto-viscoplastic constitutive model which incorporates the nonlinear kinematic hardening rule and the structural degradation. For the large deformation dynamic analysis of two-phase mixture, the finite element formulations in *u-p* frame were developed based on the updated-Lagrangian scheme. Also, the tangent modulus method was employed to properly implement the viscoplastic constitutive equations. The shear banding process was studied through the strain contours and deformed meshes. The shear band development was clearly observed as four dominant diagonal bands at an angle of about 45°. From the local stress-strain relations, more brittleness was found inside the shear bands than outside of them.

References

Armstrong PJ, Frederick CO (1966) A mathematical representation of the multiaxial Bauschinger effect. Berkeley Nuclear Laboratories 1966. Technical report RD/B/N 731

Bigoni D (2012) Nonlinear solid mechanics: bifurcation theory and material instability. Cambridge University Press, Cambridge

Kimoto S, Oka F (2005) An elasto-viscoplastic model for clay considering destructuralization and consolidation analysis of unstable Behavior. Soils Found 45(2):29–42

Kimoto S, Khan BS, Mirjalili MM, Oka F (2013) A cyclic elasto-viscoplastic constitutive model for clay considering the nonlinear kinematic hardening rules and the structural degradation. Int J Geomech (ASCE). doi:10.1061/(ASCE)GM.1943-5622.0000327

Loret B, Prevost JH (1991) Dynamic strain localization in fluid-saturated porous media. J Eng Mech 117(4):907–922 (ASCE)

Oka F, Adachi T, Yashima A (1995) A strain localization analysis of clay using a strain softening viscoplastic model. Int J Plast 11(5):523–545

Oka F, Kimoto S (2012) Computational modeling of multiphase geomaterials. CRC Press, Taylor & Francis Group, Boca Raton, London and New York

Micromechanical Modelling of Granular Materials Through Fabric and Coordination Number Anisotropies

Mehdi Pouragha and Richard Wan

Abstract Granular geomaterials involve microphysics at the various scales that give rise to distinct constitutive characteristics such as steady states, plastic limit states and unstable material behaviour. This paper is concerned with the statistics of coordination number and fabric anisotropy at the local level to impart microstructural information to the macroscopic description of the state of the material. As such the evolutions of coordination number (connectivity) distribution and anisotropy (contact normal) distribution curves with deformation history and their correlations are determined through DEM (Discrete Element Modelling). These results are useful as they enter as joint probability distribution functions into homogenization expressions during upscaling to a continuum constitutive model enriched with microstructural information.

1 Introduction

Statistical approaches are readily amenable to the analysis of a wide range of microstatics and micro-kinematics phenomena in granular media. For instance, stresses in a granular assembly have been successfully described through statistical representations of coordination number and fabric anisotropy which define average contact density and contact orientation distribution respectively (Bathurst and Rothenburg 1990). As efforts along these lines, Pouragha and Wan (2013) and Wan and Pouragha (2014) have recently shown that similar relations can be

M. Pouragha · R. Wan (✉)
University of Calgary, Calgary, Canada
e-mail: wan@ucalgary.ca

M. Pouragha
e-mail: mehdi.pouragha@gmail.com

K.-T. Chau and J. Zhao (eds.), *Bifurcation and Degradation of Geomaterials in the New Millennium*, Springer Series in Geomechanics and Geoengineering,
DOI 10.1007/978-3-319-13506-9_52

Fig. 2 DEM simulation of biaxial test for dense and loose packings. **a** Stress-strain-volumetric responses. **b** Deviatoric strain and second-order worked fields

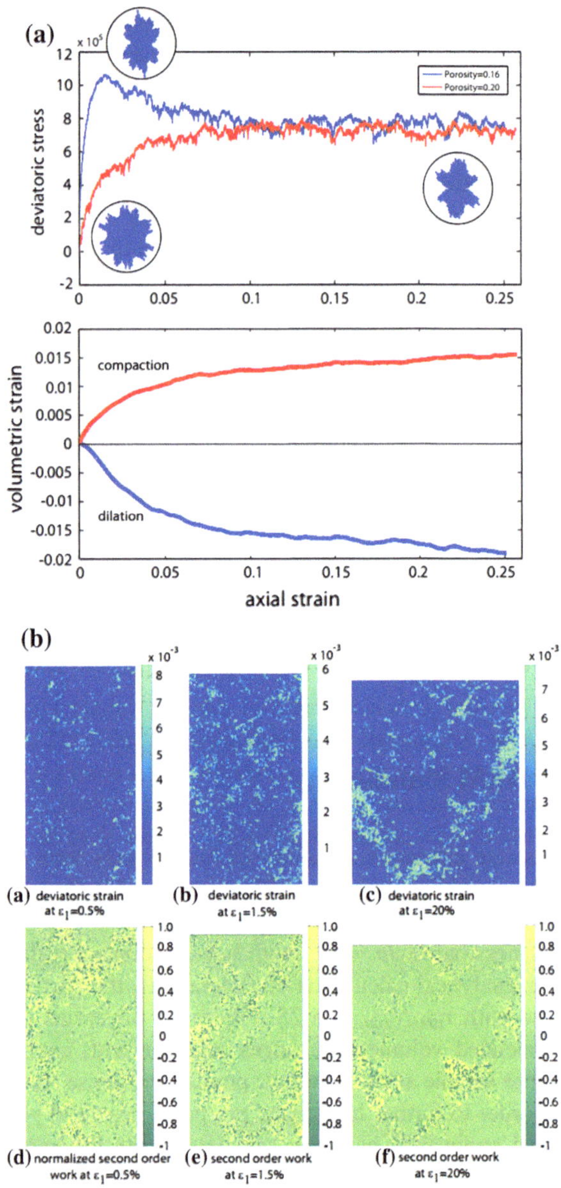

4 Cell Micro-variables Distribution

A circular REV containing around 6,500 particles has been extracted from the center of the sample for which the tessellation network has been constructed and subsequently cell coordination number and anisotropy have been calculated for all the cells. The cell coordination number defined in Eq. (1) can be shown to

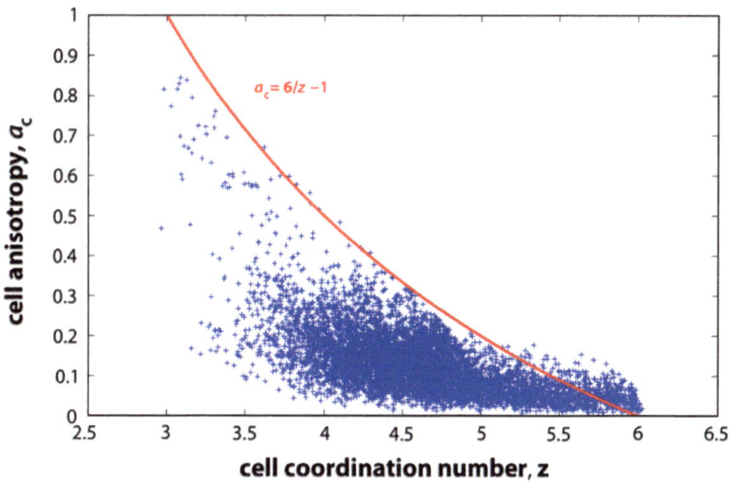

Fig. 3 Population of local anisotropy and local coordination number within the assembly at peak condition

follow a Gaussian distribution curve which is consistent with the already known distribution of the coordination number in granular material (Rosato and Yacoub 2000). On the other hand, the asymptotic behaviour of the stress-strain curve in granular materials is known to originate from fabric anisotropy being limited due to geometric exclusions. The limit is shown to be a function of packing density, and hence coordination number at the so called steady state (Radjai 2008). The relation can be simply explained considering the maximum contact density that is plausible in a neighbourhood. Due to geometric exclusions, for a mono-dispersed assembly of particles, the maximum density of contact cannot exceed $3/\pi$ contacts per radian. Considering the contact normal probability being approximated by first order harmonic series as $P(\theta) = (z/2\pi)(1 + a_c \cos 2\theta)$, with a_c being the second order fabric anisotropy and z the coordination number, the maximum contact density can be calculated as $(z/2\pi)(1 + a_c)$ which together with the geometric exclusion condition determines an upper limit for fabric anisotropy as:

$$a_c^{\max} = \frac{6 - z}{z} \qquad (3)$$

The current simulations confirm that the above relation is valid for cell parameters defined for Dirichlet cells in Eqs. (1) and (2) as shown in Fig. 3.

If we superimpose the coordination number distribution curve onto the anisotropy limit, we can identify different deformation regimes: one in which anisotropy evolves due to strength mobilization and another one where anisotropy reaches its limit to cause asymptotic behaviour. This local interdependency between coordination number and anisotropy has been investigated in Wan and Pouragha (2014) which provides a better understanding of the various deformation regimes within the assembly.

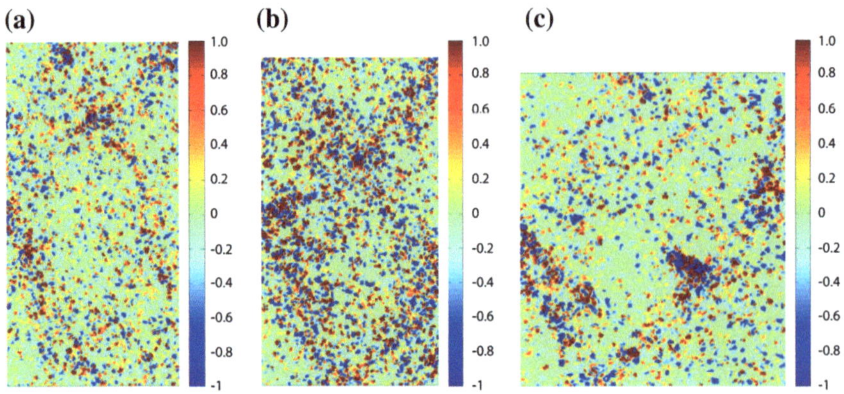

Fig. 4 Local anisotropy a_c field at various strain levels for the dense sample. **a** Local anisotropy work at $\varepsilon_1 = 0.5\ \%$. **b** Local anisotropy at $\varepsilon_1 = 1.5\ \%$. **c** Local anisotropy at $\varepsilon_1 = 20\ \%$

Furthermore, being defined locally for each particle, the cell coordination number and anisotropy provide the opportunity to investigate the correlation between these parameters and instability patterns within the assembly. Figure 4 shows the spatial change in local cell anisotropy which clearly shows the same pattern of localization as the strain and second order work. It should be noticed that strain fields (Fig. 2b) are calculated from the particles displacement, and the second order work field is found using the information about interparticle forces and relative movements at the contact, while the local anisotropy field carries only the information about the cell network and their evolution and as simple as it is, yet it can capture the instability pattern.

5 Conclusions

New field micro-variables such as local coordination number and fabric (contact) anisotropy defined over a (Dirichlet) tessellation of a particulate system have been proposed to describe the deformation of granular media. The statistics and correlation between the two proposed fields during loading history have been analysed through DEM computations of simple biaxial tests on both dense and loose packings. A limiting anisotropy value of granular fabric is confirmed which leads to the explanation of two regimes of deformation involving first mobilization of strength and then asymptotic behaviour in the limit. The evolution of the two field variables also reveals complex patterned failure structures pertaining to material instability in the form of strain localization. This suggests that these two local fields are rich in that they encapsulate both statistical and kinematical information. They are also plausible starting point for developing purely micromechanical models of granular media with only a very few intrinsic parameters.

Acknowledgments This work is jointly funded by the Computer Modelling Group (CMG) and the Natural Science and Engineering Council of Canada through a CRD grant.

References

Bathurst RJ, Rothenburg L (1990) Observations on stress-force-fabric relationships in idealized granular materials. Mech Mater 9(1):65–80

Nicot F, Hadda N, Bourrier F, Sibille L, Wan R, Darve F (2012) Inertia effects as a possible missing link between micro and macro second-order work in granular media. Int J Solids Struct 49(10):1252–1258

Pouragha M, Wan R (2013) Interrelationship between coordination number and fabric anisotropy at the particle scale as a constitutive requirement for modelling granular materials. Karakov, Poland

Radjai F (2008) Particle-scale origins of shear strength in granular media. arXiv:0801.4722v1 [cond-mat.soft]

Rosato AD, Yacoub D (2000) Microstructure evolution in compacted granular beds. Powder Technol 109(1–3):255–261

Wan R, Pouragha M (2014) Continuum representation of granular fabric and connectivity. Contin Mech Therm. doi:10.1007/s00161-014-0370-9

Repetitive Compression Tests on a Carbonate Sand

Guien Miao and David W. Airey

Abstract Existing studies on breakage suggest that the ultimate state is dependent on the particular particle arrangement which results from a given test methodology. Further breakage of the particles is possible when the particle arrangement is disrupted and then reloaded. Thus, there is uncertainty about the concept and utility of an ultimate grading. This paper investigates this issue by presenting data from one-dimensional compression experiments on an initially uniform grading of a carbonate sand. The samples were reconstituted after each cycle of loading, allowing for rearrangement of the particles in order to promote breakage, and tested until there was no change in the grading and compression behaviour (i.e. at the ultimate state for repetitive compression). For samples which were repetitively compressed under high stresses, the grading was similar and the compression curves converged. This represents a limit to the amount of breakage which is possible under repetitive compression and is not necessarily associated with a fractal dimension of 3 nor is it necessarily mono-fractal.

1 Introduction

Empirical studies (McDowell and Daniell 2001; Altuhafi and Coop 2011) have shown that the breakage of sands produces self-similar particles. The largest particles are protected from breakage by the cushioning effect, whereby the presence of smaller particles increases the coordination number and reduces the stress experienced by larger particles. In an initially uniform sand, this promotes the breakage of smaller particles over larger particles despite particle strength

G. Miao (✉) · D.W. Airey
School of Civil Engineering, University of Sydney, Sydney, NSW, Australia
e-mail: guien.miao@sydney.edu.au

D.W. Airey
e-mail: david.airey@sydney.edu.au

K.-T. Chau and J. Zhao (eds.), *Bifurcation and Degradation of Geomaterials in the New Millennium*, Springer Series in Geomechanics and Geoengineering,
DOI 10.1007/978-3-319-13506-9_53

increasing with decreasing size. Although a fractal is, in theory, associated with particles of all sizes, in empirical studies, self-similarity can only be described over a particular range of sizes. This can be captured by using Eq. 1 (Einav 2007), which describes the particle size distribution with a fractal dimension of α as

$$F(d) = \frac{d^{3-\alpha} - d_m^{3-\alpha}}{d_M^{3-\alpha} - d_m^{3-\alpha}} \tag{1}$$

where d is the reference particle size, d_M is the maximum particle size and d_m is the minimum particle size. Moreover, a single curvature (mono-) fractal grading results when the probability of fracture of all particles is equal (Palmer and Sanderson 1991). Both theoretical (Tyler and Wheatcraft 1992) and empirical studies (Lőrincz et al. 2005) have shown that the maximum breakage occurs with the maximum possible fractal dimension of 3, which corresponds to a straight-line particle size distribution on a semi-logarithmic plot.

Monotonic compression of any given sand results in a particular arrangement of particles in space. Due to limited particle rotation, this does not achieve the same level of breakage as monotonic shearing (Altuhafi and Coop 2011; Miao and Airey 2013). Reconstituting a compressed sample allows for the re-arrangement and rotation of particles and thus allows more breakage than a single loading in monotonic compression. Repetitive compression tests (Lőrincz et al. 2005; Uygar and Doven 2006) involve material being remixed and reloaded several times to the same maximum load. The tests in Uygar and Doven (2006) consisted of 5 cycles to a maximum stress of 14 MPa on a carbonate sand and, as they did not continue to the ultimate state, the compression curves continued to change. The compression tests in Lőrincz et al. (2005) were on performed Holocene fluvial sand to a maximum stress of approximately 12.5 MPa and showed that the grading evolved towards a fractal dimension of 3. The final state in these tests represents the ultimate state for repetitive compression; however, it is unclear whether the fluvial sand would have broken down further under higher stresses. Furthermore, there was a significant loss of fines as a result of sieve analyses between cycles. This would bias the final grading towards the breakage limit as the reduction in the cushioning of the larger particles would cause additional breakage. Further study of the ultimate state of repetitive compression for higher stresses and when the fines are not lost is required.

2 Methodology

The sand used in this study was Barry's Beach carbonate sand sized between 1.18 and 2.36 mm. The brittleness and the irregular shape of the particles allowed for a higher amount of breakage for a given load and thus the final state to be reached more quickly. Samples weighing 30 g were placed in a one-dimensional compression cylinder with an internal diameter of 35 mm. Initial void ratio was not controlled as the compression behaviour at high stresses (described by a limiting compression curve, LCC) is unaffected by initial void ratio (Pestana and Whittle 1995). Each sample was compressed to a pre-determined stress of 1, 10, 50 or 95 MPa, at a constant displacement rate of 1 mm/min. The sample was then unloaded and reconstituted. Segregation was minimised by mixing samples within the compression cylinder for tests at low stresses and, for tests at high stresses, the amount of breakage in the first cycle creates a well-graded material, so the likelihood of segregation is reduced. The compression process was repeated until no change in the compression behaviour (final void ratio and compression/decompression curves) was noted. The grading of the sample was then determined with a sieve-analysis (for particles sized greater than 75 μm) and with the use of a Morphologi G3 (for particles sized smaller than 75 μm). Minimum and maximum densities were obtained after cycles 1, 3 and 6 for samples which had been repetitively compressed to a normal stress 95 MPa. Due to the relatively small sample size, modified tests were used: for the minimum density test, the sample was placed loosely in a container with a volume of 20.5 cm^3 and the mass was recorded; and, for the maximum density test, the sample was placed in the one-dimensional compression cylinder, vibrated and the final height of the sample was measured.

3 Breakage Behaviour with Increasing Cycles

Figure 1 shows typical compression behaviour with increasing cycles of compression. The compression curves are shown on semi-log axes in order to highlight changes in the LCC. In Fig. 1, it can be seen that, with increasing cycles of compression, the void ratio at 95 MPa decreases to an asymptotic value and the slope of the LCC becomes less steep, eventually collapsing onto a single line. While the change in the LCC at early cycles is associated with the occurrence of breakage, the similarity of the final LCCs is associated with the cessation of breakage. As reconstitution of the sample and reloading to the specified maximum stress results in no change to the compression behaviour, this therefore corresponds to an ultimate state for repetitive compression. At the ultimate state of repetitive compression, void space can be recovered by reconstitution. Furthermore, it

Fig. 1 The compression curves for repetitive compression under a maximum stress of 95 MPa

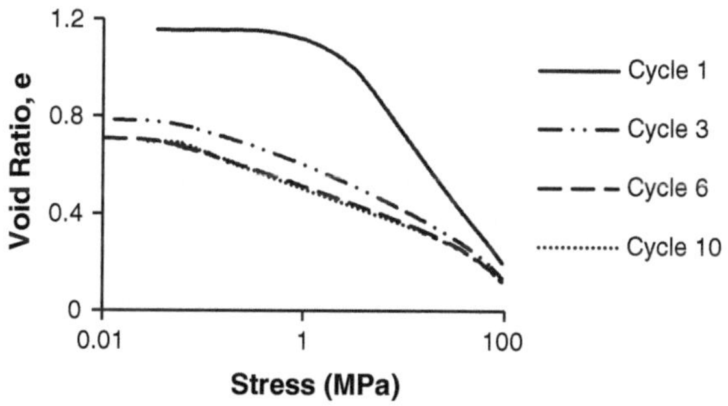

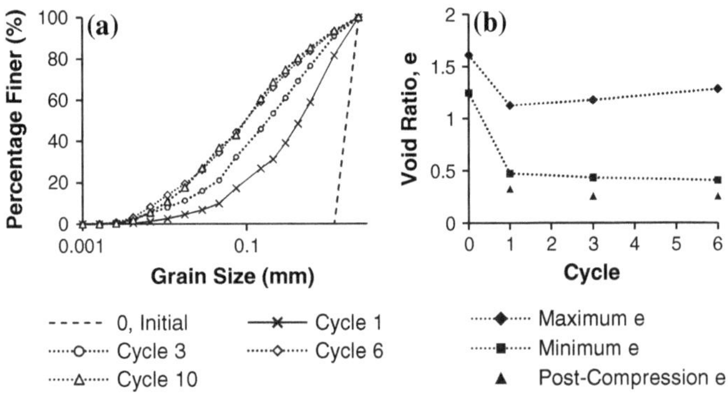

Fig. 2 For maximum normal stress of 95 MPa, **a** the evolution of the grading with increasing cycles and **b** the change in minimum and maximum void ratio with increasing cycles

can be seen that the stiffness of the material increases with each cycle. Breakage results in a reduction in particle irregularity and a more well-graded particle size distribution which, in turn, produces a better-packed material. Studies have shown that decreasing particle irregularity increases the soil stiffness (Cho et al. 2006) and that a well-graded sand will have a lower LCC gradient (compression index, C_c) than a uniform grading of the same material (Grine and Glendinning 2007).

Figure 2a shows the evolving grading with increasing cycles of breakage. It can be seen that the greatest amount of breakage occurs within the first cycle and that the difference in grading between cycles 6 and 10 is minimal. The behaviour of the compression curves shown in Fig. 1 reflects the evolution of the grading; the greatest change in void ratio occurs in the first cycle and the difference in the compression behaviour between cycles 6 and 10 is minimal. As a mono-fractal distribution occurs when each grain has an equal probability of fracture (Palmer and Sanderson 1991), the transition from single-curvature at low cycles to double-curvature for higher cycles shows that there is still preferential breakage of the larger particles, despite the formation of smaller particles which provide cushioning.

Figure 2b shows the change in void ratios with increasing cycles of compression. It can be seen that, due to breakage, the minimum and maximum void ratios after loading are lower than the original material. With increasing cycles, the minimum void ratios decrease; however, the maximum void ratios increase after the initial decrease. The initial increase in fines content causes a decrease in the maximum void ratio as the fines lie within a matrix of coarse particles (i.e. filling the inter-granular voids) while a further increase will result in the fines becoming part of the matrix (Kuerbis et al. 1988). Furthermore, each post-compression void ratio is lower than the corresponding minimum void ratio of the assembly. The application of normal stress causes voids to be filled by either the generation of smaller particles (breakage) or by forcing existing small particles into these voids (rearrangement). Without this normal stress, as in the maximum density test, the potential for breakage or rearrangement is much lower and thus the void ratio is greater.

4 Effect of Normal Stress

Figure 3 shows that the ultimate grading and compression behaviour under repetitive compression is dependent on normal stress. In Fig. 3a, it can be seen that, for low normal stresses, the final grading is further from the theoretical limit while, for high normal stresses, the grading is closer to the limit. As can be seen in Fig. 3b, this trend corresponds to a smaller decrease in the void ratio for the lower normal stresses and a greater decrease in the void ratio for the higher normal stresses. It can also be seen that the compression curves of the initial cycle of loading converge onto the LCC. Furthermore, the similarity of the final gradings and compression curves for stresses of 50 and 95 MPa suggests that the limit of breakage has been reached. Although the compression curve for the 10 MPa test shows similarity to the compression curves for higher stresses, the void ratio at 10 MPa is slightly higher. This similarity, in spite of the significant difference in the grading, nevertheless suggests that the compression behaviour of a well-graded soil is not very sensitive to the grading.

As repetitive compression at higher stresses would not result in further breakage, repetitive compression at 95 MPa corresponds to a limit to breakage for one-dimensional compression. As can be seen in Fig. 3a, this state does not correlate well with the theoretical limit. The empirical limit is clearly not associated with a fractal dimension of 3 and the grading of the soil is no longer a single-curvature (mono-fractal) grading. This contrasts with Lőrincz et al. (2005) in which a mono-fractal grading with a fractal dimension of 3 was obtained. The single-curvature grading in their study could reflect the relatively high individual particle strengths of calcite and quartz particles which promote the breakage of smaller particles, while the double-curvature grading in this study reflects the relatively weak individual particle strengths of bioclastic carbonate particles which promote the breakage of larger particles. Furthermore, the difference in the fractal dimensions can be attributed to the retention of the fines throughout the compression cycles in this

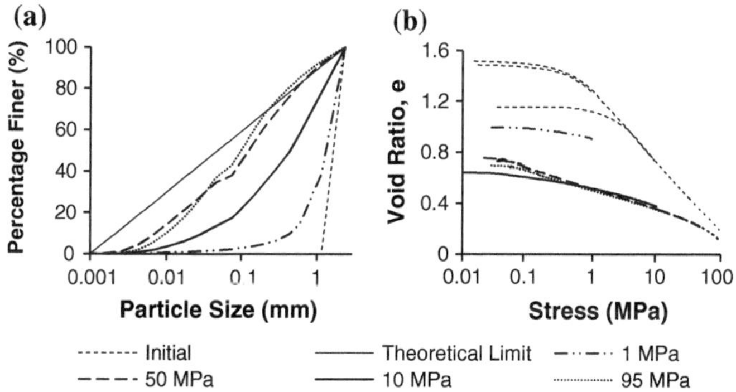

Fig. 3 For repetitive compression tests, the effect of normal stress on **a** the final grading and **b** the first and final compression curves

study, in contrast to the loss of fines in Lőrincz et al. (2005). The limiting grading in this study generally has a lower fractal dimension than 3, which can be attributed to a reduction in breakage due to the presence of the fines.

5 Conclusion

The ultimate states for repetitive compression are dependent on the applied normal stress for low stresses; however, little difference was noted between the gradings and compression behaviour when the soil was compressed at high stresses. The limiting state is not necessarily associated with a fractal dimension of 3 nor is it necessarily mono-fractal. A fractal dimension of 3 is associated with a loss of fines during the breakage process and a mono-fractal may be associated with higher individual particle strengths.

References

Altuhafi FN, Coop MR (2011) The effect of mode of loading on particle-scale damage. Soils Found 51:849–856

Cho GC, Dodds J, Santamarina JC (2006) Particle shape effects on packing density, stiffness, and strength: natural and crushed sands. J Geotech Geoenvironment 132:591–602

Einav I (2007) Breakage mechanics—part I: theory. J Mech Phys Solids 55:1274–1297

Grine K, Glendinning S (2007) Creation of an artificial carbonate sand. Geotech Geol Eng 25:441–448

Kuerbis R, Negussey D, Vaid YP (1988) Effect of gradation and fines content on the undrained response of sand. Paper presented at conference on hydraulic fill structures, Colorado State University, Fort Collins, pp 15–18

Lőrincz J, Imre E, Gálos M et al (2005) Grading entropy variation due to soil crushing. Int J Geomech 5:311–319

McDowell GR, Daniell CM (2001) Fractal compression of soil. Géotechnique 51:387

Miao G, Airey DW (2013) Breakage and ultimate states for a carbonate sand. Géotechnique 63:1221–1229

Palmer AC, Sanderson TJO (1991) Fractal crushing of ice and brittle solids. Proc R Soc London A: Math Phys Sci 433:469–477

Pestana JM, Whittle AJ (1995) Compression model for cohesionless soils. Géotechnique 45:611–631

Tyler SW, Wheatcraft SW (1992) Fractal scaling of soil particle-size distributions: analysis and limitations. Soil Sci Soc Am J 56:362–369

Uygar E, Doven AG (2006) Monotonic and cyclic oedometer tests on sand at high stress levels. Granular Matt 8:19–26

The Evolution of Shear Bands in Sand: Numerical Investigations Based on an Elasto-Plastic Cosserat Continuum Approach

B. Ebrahimian and M.I. Alsaleh

Abstract This research is focused on the numerical investigations of the evolution of shear bands and polar effects within a planar layer of cohessionless and dry sand material under shearing. In this regard, micro-polar (Cosserat) continuum is used to account for micro-rotations, couple stresses and size effect in sand. In particular, extending the non-polar version of the employed elasto-plastic soil model is presented within the framework of Cosserat continuum. Non-linearity is considered in constitutive relations and geometry for the finite element implementation. It is demonstrated that strain localization with a finite thickness occurs under large quasi-static shearing. Location, thickness and evolution of strain localization are strongly affected by the rotating resistance of boundary grains of sand layer and the boundary conditions of entire system. The localization patterns are different within finite and infinite shear layers, depending on the prescribed lateral boundary conditions.

1 Introduction

So far, limited quantitative information has been provided on the kinematics of soil grains within the induced shear bands formed along the bounding interfaces. Herein, a narrow layer of sand material, located between two parallel rigid boundaries of different surface roughness, is simulated under the conditions of plane

B. Ebrahimian (✉)
School of Civil Engineering, Faculty of Engineering, University of Tehran, Tehran, Iran
e-mail: bebrahimian@ut.ac.ir; ebrahimian.babak@gmail.com

B. Ebrahimian
The Highest Prestigious Scientific and Professional National Foundation, Iran's National Elites Foundation (INEF), Tehran, Iran

M.I. Alsaleh
R&D, Virtual Product Development Technology, Caterpillar Inc., Mossville, IL 61552, USA

K.-T. Chau and J. Zhao (eds.), *Bifurcation and Degradation of Geomaterials in the New Millennium*, Springer Series in Geomechanics and Geoengineering,
DOI 10.1007/978-3-319-13506-9_54

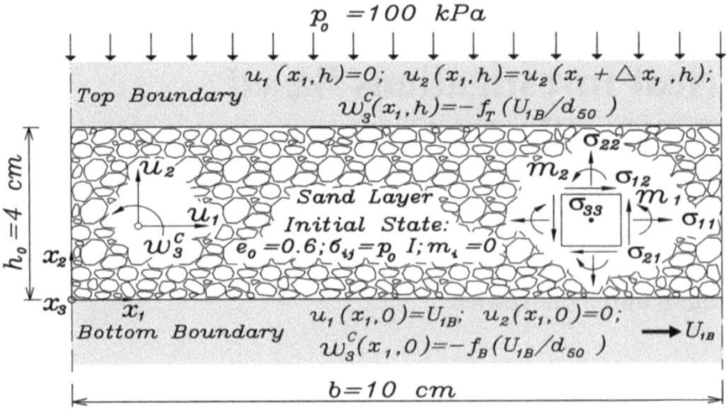

Fig. 1 Modeling an infinite layer of Cosserat sand material under plane shearing and constant vertical pressure located between two parallel rigid boundaries of different surface roughness

shearing, constant vertical pressure and free dilatancy, Fig. 1. The enhanced micro-polar elasto-plastic Lade's model is used to describe the mechanical behavior of sand (Ebrahimian et al. 2012b). The model takes into account sliding and rotating resistances of soil grains, non-symmetric shear stresses, curvatures, couple stresses, pressure level, void ratio and mean grain diameter (Ebrahimian et al. 2012b). Mean grain diameter is considered as an internal length. Therefore, the results of finite element calculations will be mesh independent provided the element size is small enough (Ebrahimian and Noorzad 2013). The rotating resistance of soil grains along the boundaries of sand layer can be naturally modeled with the additional non-standard Cosserat boundary conditions (Ebrahimian et al. 2012b). In particular, it is investigated how boundary conditions and rotating resistance of boundary grains of a micro-polar sand layer influence the evolution of shear localization and polar effects. The distribution of state variables and polar quantities across the height of an infinite sand layer under large shearing are presented based on the initiated shear displacement at the bottom boundary.

2 Constitutive Relations and Finite Element Formulation

In 2D Cosserat continuum, each material point has three degrees of freedom (Fig. 1): u_i $(i = 1, 2)$ and ω_3^c are displacement components and Cosserat rotation, respectively. The former describes macro-motion of material point and the latter its micro-motion. Accordingly, the objective strain rate tensor is defined as

$$^{n+1}_{\ \ n}\dot{\gamma}_{ij} = {}^{n+1}_{\ \ n}\dot{E}_{ij} + \left({}^{n}\dot{\Omega}_{ij} - {}^{n}\dot{\Omega}^{c}_{ij}\right) \tag{1}$$

where, $\dot{\gamma}$ is spatial objective strain rate tensor; $\dot{E}$ is classical strain rate tensor; $\dot{\Omega}_{ij} = 1/2\left(v_{i,j} - v_{j,i}\right)$ and $\dot{\Omega}^c_{ij} = -e_{ijk}\omega^c_k$ are classical spin and Cosserat spin tensors, respectively; and e_{ijk} is Ricci permutation tensor. The curvature vector of deformation is given by ${}^{n}\kappa_{ij} = {}^{n}\omega^{c}_{j,i}$. Micro-polar Lade's model, employed in this study, is enhanced through the second stress and deviatoric stress invariants in order to incorporate the internal length (Ebrahimian et al. 2012b). The model has a non-linear elasticity function and assumes non-associative flow rule and non-linear plastic work-based hardening function (Ebrahimian et al. 2012b). The second invariant of deviatoric stress tensor is enhanced to account for the effect of couple stresses (Ebrahimian et al. 2012b) and defined as :

$$J_2' = \left\{ \left[(\sigma_{11} - \sigma_{22})^2 + (\sigma_{33} - \sigma_{22})^2 + (\sigma_{11} - \sigma_{33})^2 \right] + \left((\sigma_{12} + \sigma_{21})/2 \right)^2 \right\} + \left(m_1^2 + m_2^2 \right) \Big/ l^2 \tag{2}$$

where, σ_{ij} is stress; m_i is couple stress; and l is internal length equal to d_{50} (d_{50} is the mean grain diameter). Correspondingly, the second stress invariant is also enhanced to incorporate the effect of couple stresses (Ebrahimian et al. 2012b) and given by :

$$I_{II} = 1/2(\sigma_{12}\sigma_{21} - \sigma_{11}\sigma_{22} - \sigma_{11}\sigma_{33} - \sigma_{22}\sigma_{33}) - (m_1 m_2) \Big/ l^2 \tag{3}$$

Incorporating Cosserat couple stresses and rotations, the virtual work equation is obtained through the updated Lagrangian reference (Alsaleh et al. 2006) as follows:

$$\int\limits_{{}^{n}V} \left({}^{n+1}S_{ij}\delta\left({}^{n+1}_{n}\dot{\gamma}_{ij} \right) + {}^{n+1}m_i \delta^{n+1}{}_{n}\dot{\kappa}_i \right) dV = {}^{n+1}R \tag{4}$$

$$R = \int\limits_{{}^{n+1}S} \left({}^{n+1}T_i \delta v_i + {}^{n+1}m \delta\dot{\omega}^c \right) d^{n+1}S + \int\limits_{{}^{n+1}V} \left(\rho \dot{v}_i \delta v_i + I \ddot{\omega}^c \delta\dot{\omega}^c \right) d^{n+1}V \tag{5}$$

where, S_{ij} is second Piola-Kirchhoff stress; R is external virtual work; $T_i = \sigma_{ij}n_j$ is traction force; σ_{ij} is Cauchy stress tensor; $M = m_i n_i$ is traction couple stress; $\rho\dot{v}_i$ is inertial forces; $I\ddot{\omega}^c$ is inertial couples; ρ is micro-medium density; and I is first moment of inertial for micro-medium. Regarding numerical simulations, the proposed elasto-plastic Cosserat model is implemented in a non-linear finite element code. This model can capture shear localization with finite thickness and follow the approaching of stationary states in the localized shear band (Ebrahimian et al. 2012a, b; Ebrahimian and Noorzad 2013).

3 Modeling Large Shearing of a Planar Sand Layer

The non-standard micro-polar boundary conditions i.e. couple stress and Cosserat rotation should be introduced along the boundaries of a Cosserat sand layer in addition to the stress and displacement boundary conditions defined in non-polar

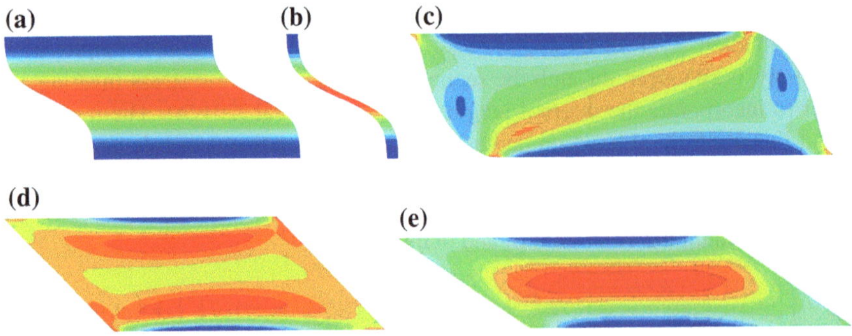

Fig. 2 The effects of prescribed lateral boundary conditions on the deformation pattern and evolution of void ratio; **a, b** lateral symmetric conditions for an infinite shear layer; **c** lateral constant pressures; **d, e** lateral linear displacement fields

continuum. The relevant kinematic and static quantities of 2D Cosserat continuum are represented in Fig. 1 with respect to a Cartesian coordinate system. The Cosserat boundary conditions allow simulating different boundary roughness with considering grain rotations. Very rough boundaries can capture the adjoining small grains so that neither sliding nor rotating may occur (Ebrahimian et al. 2012b; Tejchman 1997). An empirical relation is assumed between the boundary displacement and its corresponding Cosserat rotation for rough and medium rough boundaries and quasi-static processes (Tejchman 1997) as : $\omega_3^c = -f\left(u_{1B}/d_{50}\right)$. In this relation, the ratio of surface roughness to mean grain diameter ($f = r_w/d_{50}$) is considered as a suitable measure for evaluating Cosserat rotation in comparison to the relative shear displacement of soil grains parallel to the boundary. A relative smoother bounding surface is corresponding to a greater value of f. The boundary conditions assumed along the top and bottom surfaces of an infinite Cosserat sand layer are shown in Fig. 1. The possibility of relative displacement, resulted from lower skin frictions, is not considered along the bottom boundary ($x_2 = 0$) (Ebrahimian and Bauer 2012). Regarding the lateral infinite extended narrow sand layer, the field quantities are independent of the coordinate in the shearing direction (Huang and Bauer 2003). The obtained results are related to the simulations of: 1) a shear specimen with different lateral boundary conditions; 2) an infinite shear layer with three different boundary roughness: *Case I*: $f_T = 0$ and $f_B = 0.0001$ (top surface is very rough and bottom surface is rough); *Case II*: $f_T = 0$ and $f_B = 0.1$ (top surface is very rough and bottom surface is medium rough); and *Case III*: $f_T = f_B = 0.1$ (top and bottom surfaces are medium rough). Therefore, the micro-polar boundary conditions of entire layer are non-symmetric in *Cases I* and *II*.

Figure 2 shows the contour plot of void ratio within a shear specimen for three different prescribed lateral boundary conditions. In all calculations, the same boundary conditions are applied along the top and bottom surfaces. For an infinite shear layer, a single column of elements is sufficient to consider the layer behavior (Fig. 2a, b). If shearing occurs under the constant lateral pressure, deformations

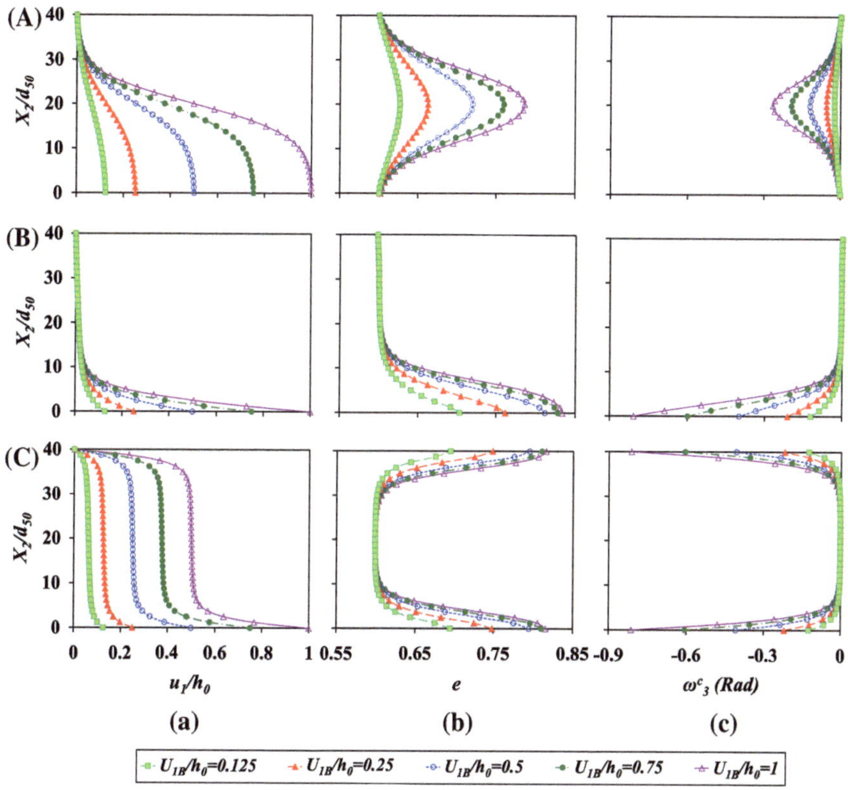

Fig. 3 Distribution of field quantities, **a** u_1/h_0, **b** e and **c** ω_3^c across X_2/d_{50} for: **A** *Case I* $f_T = 0$ and $f_B = 0.0001$, **B** *Case II* $f_T = 0$ and $f_B = 0.1$ and **C** *Case III* $f_T = f_B = 0.1$

are localized in a diagonally inclined zone. The specimen cannot be maintained in equilibrium under larger shearing, Fig. 2c. The zone of higher void ratios (looser state) is eccentrically located and slightly inclined in the linear displacement field of lateral boundaries (Fig. 2d). The results of shear specimen with finite width are strongly affected by the assumed lateral boundary conditions and the initial specimen size, (Fig. 2d, e).

Figure 3 shows the development of normalized horizontal displacement (u_1/h_0), void ratio (e) and Cosserat rotation (ω_3^c) across the normalized height (X_2/d_{50}) of an infinite sand layer. Higher values of proportionality coefficient (*Cases II* and *III*: $f = 0.1$) are corresponding to the medium rough surface conditions. In such cases, shear deformations are localized into a zone adjacent to the boundaries, forming shear bands, Fig. 3B, C. Shear band is nearly located at middle of the layer in the rough bottom surface (*Case I*: $f = 0.0001$), (Fig. 3A). The similar pattern is seen for very rough top and bottom surfaces of the layer as well (Ebrahimian et al. 2012a, b; Ebrahimian and Bauer 2012). In contrast to non-polar continuum, u_1/h_0 across X_2/d_{50} is no longer linear from the beginning of shearing

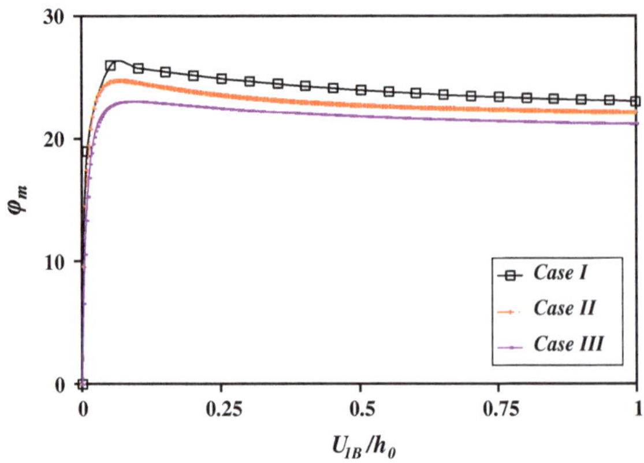

Fig. 4 Evolution of mobilized friction angle (φ_m) of sand material along the bottom surface$_0$ for: *Case I* $f_T = 0$ and $f_B = 0.0001$, *Case II* $f_T = 0$ and $f_B = 0.1$ and *Case III* $f_T = f_B = 0.1$

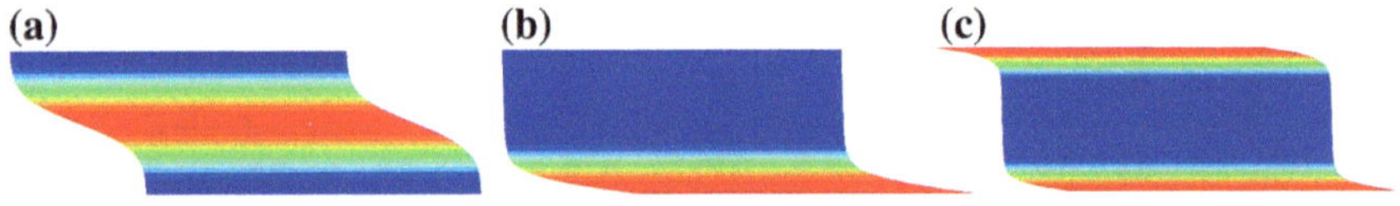

Fig. 5 Deformed shape of an infinite sand layer at residual state for: **a** *Case I* $f_T = 0$ and $f_B = 0.0001$, **b** *Case II* $f_T = 0$ and $f_B = 0.1$ and **c** *Case III* $f_T = f_B = 0.1$

in the proposed micro-polar continuum approach, Fig. 3a. According to Fig. 3b, c, significant grain rotations, accompanied with dilatancy, are observed within the shear band. The obtained results are in accordance with the experimental observations (Hall et al. 2010). Two shear bands are developed near the top and bottom surfaces of the layer for symmetric boundary constraints (*Case III*: $f_T = f_B = 0.1$). This result indicates that deformations are concentrated in the area where large grain rotations occur. The evolution of mobilized friction angle of sand material (φ_m) along the bottom surface of layer versus the normalized applied horizontal displacement is shown in Fig. 4. It is indicated that the wall friction angle (φ_m) increases as the proportionality coefficient decreases. This implies that the rougher surfaces create the higher values of wall friction angle. In all *Cases*, the wall friction angle firstly increases up to a peak state and then it decreases asymptotically to a critical state (Fig. 4). More pronounced peak state is observed in the evolution curves for surfaces with higher roughness (comparing *Case I* with *Cases II* and *III*). The peak and residual wall friction angles are 26 and 23, 24.7 and 22.1, 23 and 21.2, respectively, for *Case I* ($f_T = 0$ and $f_B = 0.0001$), *Case II* ($f_T = 0$ and $f_B = 0.1$) and *Case III* ($f_T = f_B = 0.1$), respectively. The deformed configuration of sand layer at the residual state is given for different values of proportionality coefficient, Fig. 5. The brighter zones, in the plot, are of higher void ratios or where failure may start. The predicted thicknesses of shear band, derived from the distributions of void ratio and Cosserat rotation, are about $23d_{50}$, $11d_{50}$ and $7d_{50}$ for *Cases I*, *II* and *III*, respectively. As a result, the shear band is narrower in the lower rotating resistance of boundary grains. Comparing the deformed configurations, presented in Fig. 5, the location and thickness of shear band depend significantly on the prescribed boundary conditions.

4 Conclusions

When relative large grain rotations are allowed along a surface, shear band is formed near the surface; otherwise, it is located apart from the surface. Shear strain localization occurs as a result of prior inhomogeneous deformation, accompanied with large Cosserat rotations and volume dilation. The rougher the surface of boundary is, the thicker the developed shear band. Due to localization phenomenon occurred within the layer, an initial homogeneous and non-polarized material becomes inhomogeneous and polarized from the beginning of shear deformations. The FE results show that the proposed elasto-plastic Cosserat model can appropriately capture the evolution of polar effects which are noticeable within the shear bands.

Acknowledgments The first author wants to express his sincere gratitude to the Iran's National Elites Foundation (INEF) for the moral support and encouragement.

References

Alsaleh MI, Voyiadjis G, Alshibli K (2006) Modeling strain localization in granular materials using micropolar theory: mathematical formulations. Int J Numer Anal Meth Geomech 30(15):1501–1524

Ebrahimian B, Bauer E (2012) Numerical simulation of the effect of interface friction of a bounding structure on shear deformation in a granular Soil. Int J Numer Anal Meth Geomech 36(2):1486–1506

Ebrahimian B, Noorzad A (2013) Numerical investigations of shear strain localization in an elasto-plastic Cosserat material. In: Proceedings of 18th international conference on soil mechanics and geotechnical engineering, paper no. TC103-P17, pp 703–706, Paris, France

Ebrahimian B, Noorzad A, Alsaleh MI (2012a) Effects of periodic fluctuations of micro-polar boundary conditions on shear localizations in granular soil–structure interaction. Int J Numer Anal Meth Geomech 36(7):855–880

Ebrahimian B, Noorzad A, Alsaleh MI (2012b) Modeling shear localization along granular soil–structure interfaces using elasto-plastic Cosserat continuum. Int J Solids Struct 49:257–278

Hall SA, Bornert M, Desrues J, Pannier Y, Lenoir N, Viggiani G, Bésuelle P (2010) Discrete and continuum analysis of localised deformation in sand using X-ray μCT and volumetric digital image correlation. Géotechnique 60(5):315–322

Huang W, Bauer E (2003) Numerical investigations of shear localization in a micro-polar hypoplastic material. Int J Numer Anal Meth Geomech 27:325–352

Tejchman J (1997) Modelling of shear localisation and autogeneous dynamic effects in granular bodies. Publication series of the institute of soil and rock mechanics. In: Gudehus G, Natau O (eds), Karlsruhe: University Karlsruhe 140:1–353

MIX
Papier aus verantwortungsvollen Quellen
Paper from responsible sources
FSC® C105338

If you have any concerns about our products,
you can contact us on
ProductSafety@springernature.com

In case Publisher is established outside the EU,
the EU authorized representative is:
Springer Nature Customer Service Center GmbH
Europaplatz 3, 69115 Heidelberg, Germany

Printed by Libri Plureos GmbH
in Hamburg, Germany